UNDERSTANDING INORGANIC CHEMISTRY
The Underlying Physical Principles

JACK BARRETT B.Sc., Ph.D., M.R.S.C., C.Chem.
Lecturer in Inorganic Chemistry
King's College London

ELLIS HORWOOD
NEW YORK LONDON TORONTO SYDNEY TOKYO SINGAPORE

First published in 1991 by
ELLIS HORWOOD LIMITED
Market Cross House, Cooper Street,
Chichester, West Sussex, PO19 1EB, England

A division of
Simon & Schuster International Group
A Paramount Communications Company

Typeset in Times by Ellis Horwood Limited
Printed and bound in Great Britain
by Hartnolls, Bodmin, Cornwall

British Library Cataloguing in Publication Data

Barrett, Jack
Understanding inorganic chemistry: The underlying physical principles. —
(Ellis Horwood series in inorganic chemistry)
I. Title. II. Series
546.01

ISBN 0–13–932682–0 (Library Edn.)
ISBN 0–13–932690–1 (Student Pbk. Edn.)

Library of Congress Cataloging-in-Publication Data

Barrett, Jack
Understanding inorganic chemistry: the underlying physical principles / Jack Barrett
p. cm. — (Ellis Horwood series in inorganic chemistry)
Includes bibliographical references and index.
ISBN 0–13–932682–0 (Library Edn.)
ISBN 0–13–932690–1 (Student Pbk. Edn.)
1. Chemistry, Inorganic. I. Series.
QD151.5.B37 1991
546–dc20 91–13350
CIP

Table of contents

Preface

The aim of this book is to describe the major theories that attempt to explain the observed properties of the elements and their compounds. The guiding philosophy is that summed up by Hinshelwood's statement: 'Science is not the mere collection of facts, which are infinitely numerous and mostly uninteresting, but the attempt by the human mind to order these facts into satisfying patterns'. In his book entitled *The Structure of Physical Chemistry*, published in 1951, he complained about the trend, in university teaching of the subject, of solving the problem of its diversity by judicious specialization. The difficulties of understanding the subject, he claimed, could be overcome by a suitable intensity of application. Since 1951 there has been an explosion of knowledge in all branches of chemistry and in particular in the field of inorganic chemistry, with the development of theories which were not current in 1951.

The modern ideas of inorganic chemistry cover very broad areas, ranging from the detailed understanding of chemical bonding to the kinetic and thermodynamic aspects of the function of metals in biological chemistry. The explosion in the factual content of the subject since 1951 has forced university courses to contain a smaller fraction of that content. That is to be expected and is not too serious. Courses that essentially deal with principles and understanding have a greater pool of examples to depend upon. The really serious trend in the present-day teaching of chemistry is the poorer treatment of the basic principles of the subject which is offered to students. There has been a trend towards making the subject more palatable and more 'relevant'. What has happened is that the 'hard bits' have been either toned down or even omitted altogether.

All that seems a very gloomy outlook upon current teaching of the subject. It is even worse in the so-called 'Life Sciences' where great areas of understanding are ignored so that more factual material can be accommodated.

The explosion in factual material is not going to stop. The situation is not going to improve. What must be saved is room for sufficient coverage of the hard physical principles upon which a proper understanding of any part of chemistry is based, whether it be in the domain of a chemistry degree or whether it be in the life sciences.

In chemistry the explanation of any observations must be tailored to the student's capability of appreciating it at their particular level of development. At the school level, some earlier theories are put forward as present-day explanations. There is a case for the further study of chemical theory at the postgraduate level.

The level of treatment in this book is by no means as 'deep' as it might be. It is deep enough for the student to realize that there are problems in understanding the elements and their properties and that there are routes to a deeper understanding should they find this necessary. The book is aimed at those majoring in chemistry and students of the subject whose main interests lie in some part of the life sciences.

The treatment of the subject in this book is designed to give as good a description of currently held theory as is possible without being too mathematical. Each chapter ends with a summary of the subject matter and there is a further reading guide at the end of the book. The book contains the minimum amount of chemical theory and understanding that should be possessed by anyone graduating as a chemist or as a life scientist. It is based upon lectures given by the author.

ACKNOWLEDGEMENTS

I thank John Burgess (Editor of the Ellis Horwood Series in Inorganic Chemistry), Martin Hughes (my professional colleague at King's) and Paul O'Brien (of Queen Mary and Westfield Colleges, University of London) very much for their many comments and suggestions that have contributed to the arrangement and accuracy of the subject matter of the book. Any remaining inaccuracies are entirely the author's responsibility.

I wish to express my thanks to Dr S. M. Colwell of the University of Cambridge, England, for supplying me with a copy of MICROMOL (as described by S. M. Colwell and N. C. Handy in the *Journal of Molecular Structure*, *Theochem.*, **170**, 197–203 (1988)), with which I carried out the *ab initio* calculations included in the text.

Jack Barrett
King's College London 1991

1

The electronic configurations and electronic states of atoms

1.1 INTRODUCTION

Atoms consist of a nucleus of Z protons together with $(A - Z)$ neutrons, and Z extra-nuclear electrons, Z being the **atomic number** and A the **mass number** (the whole number nearest to the exact **relative atomic mass**, RAM) of any given atom. The relative atomic mass of an atom is its mass with respect to that of the carbon-12 atom (taken to be exactly 12.0000). Naturally occurring atoms are usually a mixture of **isotopes** — atoms which have identical values of Z but have different A values because their nuclei possess different numbers of neutrons. The conventional manner of symbolizing an isotope is to indicate the Z value as a left-hand subscript, and the A value as a left-hand superscript to the element's chemical symbol, ${}^{A}_{Z}X$. This leaves the right-hand side for any charge (as a superscript) or for the atomicity of molecular species (as a subscript), e.g. ${}^{35}_{17}Cl_2^+$ represents fully the symbol for a singly ionized dichlorine molecule composed of the ${}^{35}Cl$ isotope.

The three isotopes of the hydrogen atom are thus symbolized as: ${}^{1}_{1}H$, ${}^{2}_{1}D$ (deuterium), and ${}^{2}_{1}T$ (tritium). These isotopes are given separate chemical symbols because of their special importance in chemistry. It is perfectly correct to write the symbol for deuterium as ${}^{2}_{1}H$; the isotopes of all other atoms are symbolized by the normal chemical symbol for the particular atom. The two main isotopes of the uranium atom are written as: ${}^{235}_{92}U$ and ${}^{238}_{92}U$.

In chemistry the importance of the nucleus of an atom is that it supplies the vast majority of the mass of the atom and that it determines the number of extra-nuclear electrons, Z, in the neutral atom.

The chemistry of an element is determined by the manner in which the extra-nuclear electrons are arranged, and this chapter deals with such arrangements and their chemical consequences, leading to a general understanding of the structure of the periodic classification of the elements and of the variations in some atomic properties.

1.2 THE HYDROGEN ATOM

The hydrogen atom is the simplest atom in that it possesses a single extra-nuclear electron. The basis of chemical theory is dependent upon a thorough understanding of this atom.

Important experimental evidence for the possible electronic arrangements in the hydrogen atom was provided by its emission spectrum. Essentially it consists of **lines** of a particular frequency rather than being the continuous emission of all possible frequencies. Such discrimination of emission frequencies leads to the concept of there being discrete **energy levels** within the atom which are permitted for electron occupation.

Detailed analysis of the frequencies of the lines in the emission spectrum of the hydrogen atom leads to the formulation of the Rydberg equation:

$$\nu_{ij} = cR\left(\frac{1}{n_i^2} - \frac{1}{n_j^2}\right) \tag{1.1}$$

where the frequency, ν_{ij}, relates to an electronic transition from the level j to the level i (j being larger than i), the terms n_i and n_j being particular values of the **principal quantum number**, n. The other terms in the equation are c, the velocity of light, and R, the Rydberg constant, which is a proportionality constant, and has a value of $1.097 \times 10^7\ \mathrm{m^{-1}}$. The terms in parenthesis in equation (1.1) are dimensionless, and if R is multiplied by c ($\mathrm{m\ s^{-1}}$) the result is a frequency ($\mathrm{s^{-1}}$ or hertz, (Hz)).

The line spectrum of the hydrogen atom supplies the basis of the concept of the **quantization of electron energies** — that the permitted energies for the electron in the atom of hydrogen are **quantized**. They have particular values so that it is not possible for the electron to possess any other values for its energy than those given by the Rydberg equation. As that equation implies, the electron energy is dependent upon the particular value of the quantum number, n. To convert the Rydberg equation into one which relates directly to electron energies it is essential to multiply both sides of equation (1.1) by Planck's constant, h ($6.626 \times 10^{-34}\ \mathrm{J\ s}$), and by the Avogadro number, N (to obtain the energy in units of J mol). Such a procedure makes use of the Einstein equation relating frequency of electromagnetic radiation, ν, to energy, E:

$$E = h\nu \tag{1.2}$$

The Rydberg equation in molar energy units is:

$$E_{ij} = Nh\nu = NhcR\left(\frac{1}{n_i^2} - \frac{1}{n_j^2}\right) \tag{1.3}$$

Bohr proposed the interpretation of the spectral lines arising from electronic transitions within the hydrogen atom. The Bohr equation expresses the idea that E_{ij}

represents the difference in energy between the two levels, ΔE, and may be written in the form:

$$E_{ij} = \Delta E = E_j - E_i = Nh\nu_{ij} \quad (1.4)$$

A combination of equations (1.3) and (1.4) gives:

$$E_j - E_i = NchR\left(\frac{1}{n_i^2} - \frac{1}{n_j^2}\right) \quad (1.5)$$

Equation (1.5) may be regarded as being the difference between the two equations:

$$E_j = -NchR/n_j^2 \quad (1.6)$$

and

$$E_i = -NchR/n_i^2 \quad (1.7)$$

so that, by an inductive process, the general relationship may be written as:

$$E = -NchR/n^2 \quad (1.8)$$

Equation (1.8) describes the permitted quantized energy values for the electron in the atom of hydrogen. Some of these values are shown in the diagram of Fig. 1.1, together with the possible electronic transitions which form part of the emission spectrum of the atom. The reference zero for the diagram is that corresponding to the complete removal (**ionization**) of the electron from the influence of the nucleus of the atom. The Lyman transitions are observed in the far ultraviolet region of the electromagnetic spectrum (wavelengths below 200 nm). The Balmer transitions are found mainly in the visible region, the Paschen transitions being in the infrared. There are other series of transitions of lower energies still which are all characterized by their different final values of n.

The ionization energy of the hydrogen atom from its **ground state** (the lowest energy electronic state) may be calculated from equation (1.5). The process of ionization is equivalent to the electronic transition between the energy levels corresponding to $n_i = 1$ and $n_j = \infty$ (since $E_j = 0$ when $n_j = \infty$). Using these values in equation (1.5) produces the equation:

$$I_H = E_\infty - E_1 = NchR = 1312 \text{ kJ mol}^{-1} \quad (1.9)$$

The Rydberg equation refers to emissive transitions in which electrons are transferred from higher to lower energy levels. Equation (1.9) is a calculation of the energy **released** when an electron with $n_j = \infty$ falls to the lowest energy level ($n_i = 1$). The ionization energy is normally defined as the energy **required** to effect the reverse process.

The values of the constants used in the above equations, N, c, h, and R, are known to a high degree of accuracy so the actual permitted energies for the electron

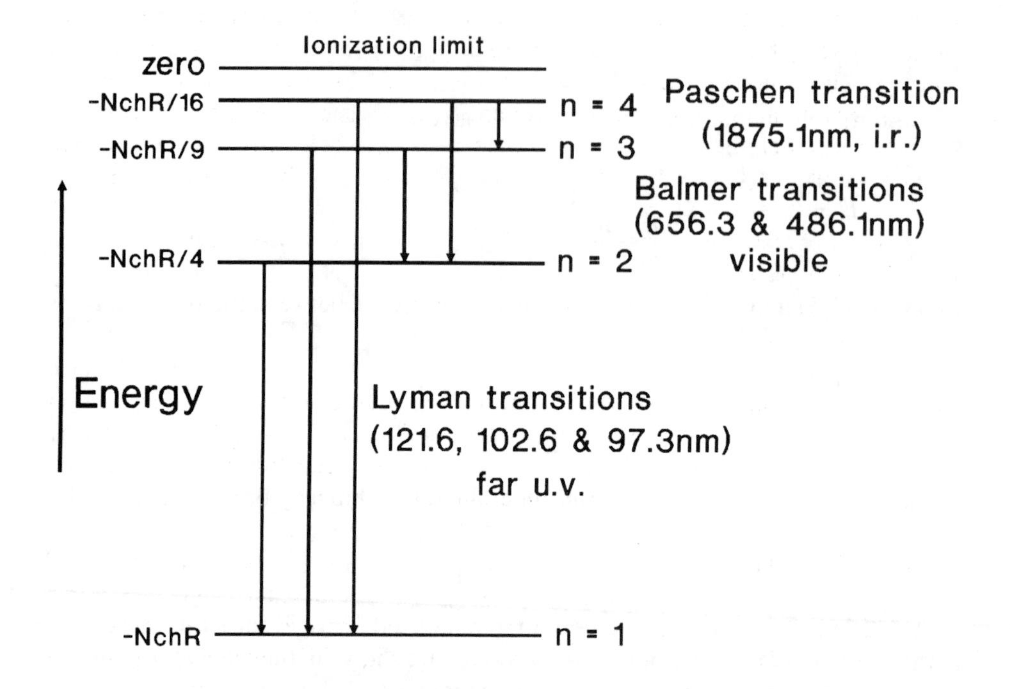

Fig. 1.1 — Some electronic energy levels of the hydrogen atom and some electronic transitions.

in a hydrogen atom are also known very accurately. The consequence of such accurate knowledge is that the **position** of the electron within the atom is very uncertain. This situation is an example of the application of the **Heisenberg uncertainty principle** (or Principle of Indeterminacy) which may be stated as: 'It is impossible to determine the position and momentum of an electron simultaneously'. Symbolically the principle may be written as:

$$\Delta p . \Delta q \sim h/2\pi \qquad (1.10)$$

In equation (1.10) Δp represents the uncertainty (error in determining) in the momentum, p ($p =$ mass $\times$ velocity), and Δq the uncertainty in position of the electron. The two uncertainties bear an inverse relationship to each other so that if the energy (related to the momentum) is known with a high degree of precision then the uncertainty in position will be correspondingly large. That is the situation with electrons in atoms (and molecules). The basis of the uncertainty principle is understandable if the problem of observing the electron is considered. Any process of observation involves the use of electromagnetic radiation.

Eyesight depends upon the process of detecting the results of the reflexion of **quanta** (or **photons**) of electromagnetic radiation (light) from the body to be observed. In the case of massive bodies (such as everyday objects) the act of observing them does not alter their energies appreciably nor does it affect their positions. This is not so with elementary particles such as electrons whose energies and positions **are** altered by the act of attempting to observe them. The consequence of being in considerable ignorance about the position of an electron in an atom is that calculations of the **probability** of finding an electron in a given position must be made.

The general mathematical expression of the problem may be written as one form of the **Schrödinger** equation:

$$H\psi = E\psi \tag{1.11}$$

The equation implies that if the operations represented by H are carried out on the function, ψ, the result will contain knowledge about ψ and its associated permitted energies. The term represented by ψ is the **wavefunction** which is such that its square, ψ^2, is the probability density. The value of $\psi^2\,\mathrm{d}\tau$ represents the probability of finding the electron in the volume element, $\mathrm{d}\tau$ (which may be visualized as the product of three elements of the Cartesian axes: $\mathrm{d}x\,.\,\mathrm{d}y\,.\,\mathrm{d}z$). Although the solution of the Schrödinger equation for any system containing more than one electron requires the iterative techniques available in computers, it may be solved for the hydrogen atom (and for hydrogen-like atoms such as He^+, Li^{2+}, ...) by analytical means, the energy solutions being represented by the equation:

$$E = -\frac{N\mu e^4}{8\varepsilon_0^2 h^2}\left(\frac{1}{n^2}\right) \tag{1.12}$$

where μ is the reduced mass of the system, defined by the equation:

$$1/\mu = 1m_e + 1/m_n \tag{1.13}$$

in which m_e is the mass of the electron and m_n is the mass of the nucleus; e is the electronic charge (1.602×10^{-19}C) and ε_0 is the permittivity of a vacuum (8.854×10^{-12} J^{-1} C^2 m^{-1}). By comparing equations (1.8) and (1.12) it may be concluded that the value of the Rydberg constant is given by:

$$R = \mu e^4/8\varepsilon_0^2 ch^3 \tag{1.14}$$

The value of the Rydberg constant was initially established by analysis of experimental data concerning the emission spectra of hydrogen and hydrogen-like atoms, the theoretical calculation of its value being carried out almost fifty years later.

Equation (1.12) expresses the permitted energies for the electron in the hydrogen atom. The second part of the information deriving from the solution of the

Schrödinger equation is that concerned with ψ and its distribution in the space around the nucleus of the hydrogen atom. The solutions for ψ are characterized by the values of three quantum numbers (three essentially because of the three spatial dimensions), and every allowed set of values for the quantum numbers describes what is termed an **atomic orbital**.

The quantum rules

The quantum rules are statements of the permitted values of the quantum numbers, n, l and m.

(i) The **principal quantum number**, n (the same n as in equation (1.12)), has values which are integral and non-zero:

$$n = 1, 2, 3, 4, \ldots$$

It defines groups of orbitals which are distinguished, within each group, by the values of l and m.

(ii) The **secondary** or **orbital angular momentum quantum number**, l, as its name implies, describes the orbital angular momentum of the electron, and has values which are integral and may be zero:

$$l = 0, 1, 2, 3, \ldots, \quad (n-1)$$

There is an important restriction upon the values of l which relates to the value of n for any particular case. This is that, for a given value of n, the maximum permitted value of l is $(n-1)$, so that for a value of $n=3$, l may have only the values 0, 1 or 2.

(iii) The **magnetic quantum number**, m, so called because it is related to the behaviour of electronic energy levels when subjected to an external magnetic field, has values which are dependent upon the value of l. The permitted values are:

$$m = l, l-1, l-2, \ldots 0, \ldots -(l-2), -(l-1), -l.$$

For instance a value of $l=2$ would yield five different values of m: 2, 1, 0, -1 and -2. In general there are $2l+1$ values of m for any given value of l. In the absence of an externally applied magnetic field, the orbitals possessing a given l value would have identical energies. Orbitals of identical energy are described as being **degenerate**. In the presence of a magnetic field the degeneracy of the orbitals breaks down in that they then have different energies. The breakdown of orbital degeneracy (for a given l value) in a magnetic field is the explanation of the **Zeeman effect**. This is the observation that in the presence of a magnetic field the atomic spectrum of an element has more lines than in the absence of the field. In discussing such a phenomenon it is normal to refer to the values of m with their appropriate l values as subscripts; m_l.

The values of l and m are dependent upon the value of n, and so it can be concluded that the value of n is concerned with a particular set of atomic orbitals all characterized by the given value of n. For any one value of n there is the possibility of more than one permitted value of l (except that when $n=1$, l can only be zero). There is an important notation which is used to distinguish orbitals with different l values, consisting of a code letter associated with each value. The code letters are shown in Table 1.1.

Table 1.1 — Code letters for l values

Value of l	Code letter
0	s
1	p
2	d
3	f
4	g

It should be realized that Table 1.1 is only a portion of an infinite set of values of l but that the ones contained in it are the only ones of any interest for the majority of applications to known atoms. The selection of the code letters seems, and is, illogical in that the first four are the initial letters of the words: sharp, principal, diffuse and fundamental — words used historically to describe aspects of line spectra. The fifth letter follows on alphabetically, the sixth being h.

The numerical value of n and the code letter for the value of l are sufficient for a general description of an atomic orbital. The main differences between atomic orbitals with different l values are concerned with their spatial orientations, and those with different n values have different sizes. The main importance of the m values is that they indicate the number, $(2l+1)$, of differently spatially oriented orbitals for the given l value. For instance, if $l=2$ there are five different values of m corresponding to five differently spatially oriented d orbitals. The number of differently spatially oriented orbitals for particular values of l are given in Table 1.2.

Table 1.2 — Number of atomic orbitals for a given l value

Value of l	No. of orbitals
0	1
1	3
2	5
3	7

Application of all the above rules allows the compilation of the types of atomic orbital and the number of each type:

For $n=1$ there is the single 1s orbital ($l=0$, $m=0$).
For $n=2$ there is one 2s orbital ($l=0$, $m=0$), plus

the three 2p orbitals ($l = 1$, $m = 1, 0$ or -1).
For $n = 3$ there is one 3s orbital, plus
three 3p orbitals, plus
the five 3d orbitals ($l = 2$, $m = 2, 1, 0, -1$ or -2).
For $n = 4$ there is one 4s orbital, plus
three 4p orbitals, plus
five 4d orbitals, plus
the seven 4f orbitals ($l = 3$, $m = 3, 2, 1, 0, -1, -2$ or -3) and so on ... .

For every increase of one in the value of n there is an extra type of orbital. The number of atomic orbitals associated with any value of n is given by n^2 so that for $n = 1$ there is one orbital (1s), for $n = 2$ there are four orbitals (2s plus three 2p), for $n = 3$ there are nine orbitals (3s plus three 3p plus five 3d) and for $n = 4$ there are sixteen orbitals (4s plus three 4p plus five 4d plus seven 4f). The sequence can be extended to infinity, but in practice it is only necessary to consider values of n up to seven, and not all the orbitals so described are needed for the electrons in known atoms in their ground electronic states. Higher values of the n quantum number are sometimes required to interpret the emission spectra of the heavier elements. As far as the hydrogen atom is concerned the orbitals having a particular value of n all have the same energy — they are **degenerate**. Level one ($n = 1$) is singly degenerate, level two ($n = 2$) has a **degeneracy** of four, and so on. The degeneracy of any level is given by the value of n^2. Such widespread degeneracy of electronic levels in the hydrogen atom is the basis of the simplicity of the diagram of those levels shown in Fig. 1.1. The orbital energies are determined solely by the value of n in equation (1.8).

The spatial orientations of the atomic orbitals of the hydrogen atom are very important in the consideration of the interaction of such orbitals in the production of molecular ones. There are various methods of representing the spatial orientations of orbitals. The formal method is to express the orbital in mathematical terms but, apart from knowledge of the sign of ψ, much understanding of orbital properties may be gained by the use of pictorial representations. Pictorial representations are based upon contours of absolute values of ψ (i.e. with the sign of ψ being ignored), but the sign of ψ is indicated in the various parts of the contour diagram. Those for s, p, and d orbitals are shown in Fig. 1.2. As may be seen from the diagram, the s orbital is spherically symmetrical and the sign of ψ is positive. The three p orbitals are directed along the x, y and z axes and are described respectively as p_x, p_y and p_z. They each consist of two lobes, one of which has positive ψ values, the other having negative ψ values. The subscript descriptions of the d orbitals are chosen because such expressions are an important part of their respective mathematical formulations. It is generally accepted that the orbitals of polyelectronic atoms have similar spatial distributions as those of the hydrogen atom. The spatial orientations and the signs of ψ are of vital importance in the understanding of chemical bonding.

1.3 POLYELECTRONIC ATOMS

Serious complications arise in atoms that contain more than one electron. Even an atom possessing two electrons is not treatable by the analytical mathematics that

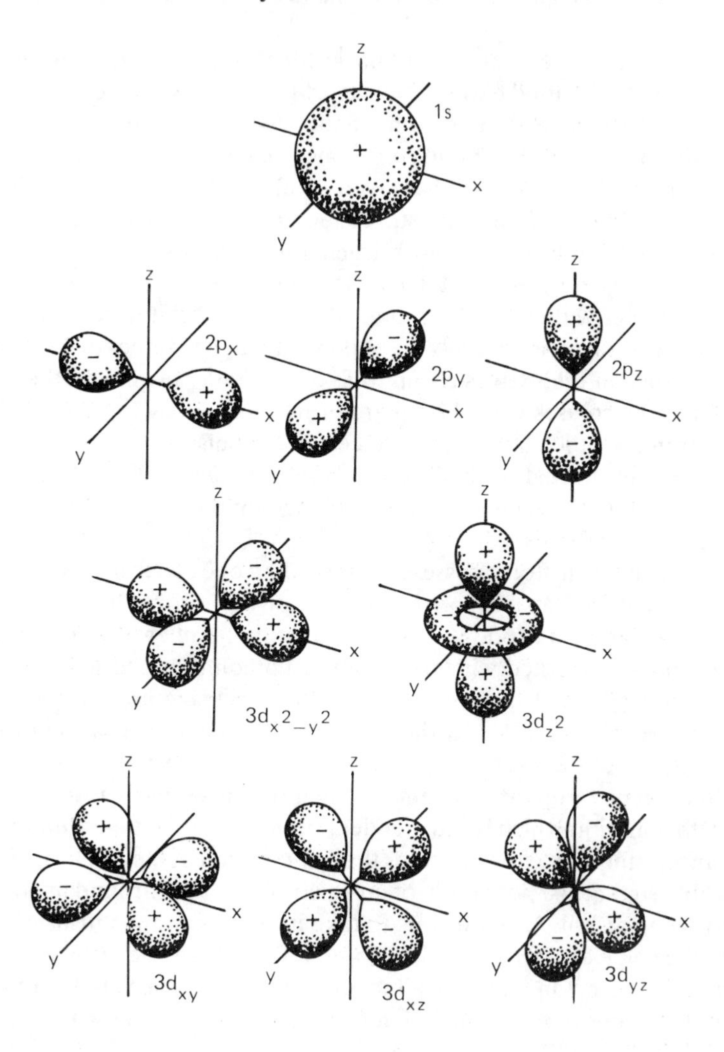

Fig. 1.2 — Diagrammatic representations of the spatial distribution of s, p and d atomic orbitals.

produced the solution of the wave equation for the hydrogen atom. **Interelectronic repulsion** must be considered. As negatively charged particles, electrons repel each other, and the understanding of the effects of such repulsion is fundamental to the understanding of a great amount of chemistry. The wave equation (1.11) can only be solved for polyelectronic systems by using the enormous iterative capacity of computers. The basis of the method is to guess the form of ψ for each orbital employed (taking hydrogen-like orbitals as the first guess) and to calculate the corresponding energy of the atomic system. The acceptable atomic orbitals as a solution of the wave equation are those which confer the minimum energy upon the system.

The solutions, in general, mimic those for the hydrogen atom, and the same nomenclature may be used to describe the orbitals of polyelectronic systems as have been described for the hydrogen atom. The shapes of their spatial distributions are like those for hydrogen. The major difference arises in the energies of the orbitals. The degeneracy of the levels with a given n value is lost. Orbitals with the same n value and with the same l value are still degenerate. Those with the same n value but with different values of l are no longer degenerate. This does not affect the 1s orbital, which is singly degenerate, but it does affect the 2s and 2p orbitals. The 2p orbitals have a higher energy than the 2s orbitals, the three 2p orbitals retaining their three-fold degeneracy. Likewise the sets of orbitals with higher values of n split into s, p, d, ... subsets which in themselves retain their degeneracy. In general the energy of an atomic orbital decreases with the square of the nuclear charge, and superimposed upon this trend are the effects of interelectronic repulsion.

It is important to realize that the loss of degeneracy, together with the general decrease in energy as Z increases, causes changes in the order of energies of various subsets of atomic orbitals.

When dealing with atoms possessing more than one electron it is vital to consider the electron-holding capacity of the orbitals of that atom. In order to do this it becomes necessary to introduce a fourth quantum number — s, **the spin quantum number**. This is concerned with the quantized amount of energy possessed by the electron independent of that concerned with its passage around the nucleus, the latter energy being controlled by the value of l. The electron has an intrinsic energy which is associated with the term **spin**. This is unfortunate since it may give rise to the impression that electrons are spinning on their own axes much as the moon spins on its axis with a motion which is independent of its orbital motion around the earth.

The uncertainty principle indicates that observation of the position of an electron is impossible so that the possibility of an electron spinning around its own axis must be left to the imagination! However, to take into account the intrinsic energy of an electron, the value of s must be either $\frac{1}{2}$ or $-\frac{1}{2}$. Essentially the intrinsic energy of the electron may interact in a quantized manner with that associated with the angular momentum represented by l, such that the only permitted interactions are $l+s$ and $l-s$.

The simple conclusion that the maximum number of electrons which may occupy any orbital is two comes from recognition of the **Pauli exclusion principle**. This is the cornerstone in understanding the chemistry of the elements since their electronic configurations are based upon it and their chemistry is intimately concerned with their electronic configurations. The exclusion principle may be stated as: 'No two electrons in an atom may possess **identical** sets of values of the four quantum numbers'. The consequences of the exclusion principle are (i) to restrict the number of electrons per orbital to a maximum of two and (ii) to restrict the number of any particular orbital (defined by its n, l, and m values) to one per atom.

Consider the 1s orbital; $n = 1$, $l = 0$ and $m = 0$ — there are no possibilities for changes in these values — any electron in a 1s orbital must be associated with them. One electron could have a value of $s = \frac{1}{2}$ whilst another electron must have the alternative value of $s = \frac{1}{2}$. The two electrons occupying the same orbital must have opposite 'spins'. Since there are no other combinations of the values of the four quantum numbers it is concluded that only two electrons may occupy the 1s orbital and that there can only be one 1s orbital in any one atom. Similar conclusions are valid for all other orbitals. The application of the Pauli exclusion principle provides the necessary framework for the understanding of the electronic configurations of the elements.

1.4 THE ELECTRONIC CONFIGURATIONS OF THE ELEMENTS — THE STRUCTURE OF THE PERIODIC CLASSIFICATION OF THE ELEMENTS

The modern form of the periodic classification of the elements, as approved by the International Union of Pure and Applied Chemistry, is shown in Fig. 1.3. The atomic number, Z, of each element is shown. There are eighteen **groups** according to modern convention.

The quantum rules define the different types of atomic orbitals which may be used for electron occupation in atoms. The Pauli exclusion principle defines the number of each type of orbital and limits each orbital to a maximum electron occupancy of two. Experimental observation, together with some sophisticated calculations, indicates the energies of the available orbitals for any particular atom. The electronic configuration — the orbitals which are used to accommodate the appropriate number of electrons — may be decided by the application of what is known as the **aufbau** (building up) **principle**, which is that electrons in the ground state of an atom occupy the orbitals of lowest energy such that the total electronic energy is minimized.

In the ground state of the hydrogen atom, the electronic configuration is that in which the only electron occupies the 1s orbital, written as: $1s^1$, the number of electrons occupying the orbital being indicated by the superscript.

For element number two (helium), there are two possible configurations which should be considered; $1s^2$ and $1s^1 2s^1$. There are two factors which decide which of the two configurations is of lower energy. These are the difference in energy between the 2s and 1s orbitals and the magnitude of the interelectronic repulsion energy in the $1s^2$ case. If the 2s–1s energy gap is larger than the interelectronic repulsion energy between the two electrons in the 1s orbital, the two electrons will pair up in the lower orbital. The point may be made by some calculations.

A combination of equations (1.8) and (1.9) (the term Z^2, where Z is the atomic number, is included so that the equation applies to hydrogen-like atoms — of nuclear charge, $+Ze$, and a single electron), gives the equation:

$$E = -I_H Z^2/n^2 \tag{1.15}$$

The energy of the 1s orbital in the helium atom is given by putting $Z = 2$, and $n = 1$, in equation (1.15) giving:

$$E(1s_{He}) = -4I_H \tag{1.16}$$

The energy of the 2s orbital of helium is given by putting $n = 2$, so that:

THE PERIODIC CLASSIFICATION OF THE ELEMENTS

1	2	3	4	5	6	7	8	9	10	11	12	13	14	15	16	17	18
1 H																	2 He
3 Li	4 Be											5 B	6 C	7 N	8 O	9 F	10 Ne
11 Na	12 Mg											13 Al	14 Si	15 P	16 S	17 Cl	18 Ar
19 K	20 Ca	21 Sc	22 Ti	23 V	24 Cr	25 Mn	26 Fe	27 Co	28 Ni	29 Cu	30 Zn	31 Ga	32 Ge	33 As	34 Se	35 Br	36 Kr
37 Rb	38 Sr	39 Y	40 Zr	41 Nb	42 Mo	43 Tc	44 Ru	45 Rh	46 Pd	47 Ag	48 Cd	49 In	50 Sn	51 Sb	52 Te	53 I	54 Xe
55 Cs	56 Ba	71 Lu	72 Hf	73 Ta	74 W	75 Re	76 Os	77 Ir	78 Pt	79 Au	80 Hg	81 Tl	82 Pb	83 Bi	84 Po	85 At	86 Rn
87 Fr	88 Ra	103 Lr	104 Unq	105 Unp	106 Unh	107 Uns	108 Uno	109 Une									

57 La	58 Ce	59 Pr	60 Nd	61 Pm	62 Sm	63 Eu	64 Gd	65 Tb	66 Dy	67 Ho	68 Er	69 Tm	70 Yb
89 Ac	90 Th	91 Pa	92 U	93 Np	94 Pu	95 Am	96 Cm	97 Bk	98 Cf	99 Es	100 Fm	101 Md	102 No

Fig. 1.3 — The periodic classification of the elements.

$$E(2s_{He}) = -I_H \tag{1.17}$$

Ignoring the effect of interelectronic repulsion, it is obvious that the energy of the $1s^2$ configuration ($-8I_H$) is lower than that of the $1s^1 2s^1$ configuration ($-5I_H$). An estimate of the interelectronic repulsion energy in the $1s^2$ configuration allows a conclusion as to which of the two possible configurations is appropriate to the ground state of the helium atom. The first ionization energy of the helium atom — the energy required to cause the ionization:

$$He(g) \rightarrow He^+(g) + e^-(g) \tag{1.18}$$

is observed to be 2370 kJ mol^{-1} (rather than $4I_H$ which is 5248 kJ mol^{-1}) and the second ionization energy — that needed to cause the change:

$$He^+(g) \rightarrow He^{2+}(g) + e^-(g) \tag{1.19}$$

is observed to be 5248 kJ mol^{-1} (exactly equal to $4I_H$).

The discrepancy between the calculated and observed values of the first ionization energy of He gives an estimate of the interelectronic repulsion energy (the equations used are for hydrogen-like atoms possessing a single electron). The difference between the hydrogen-like calculated value and the observed value for the first ionization energy of the helium atom gives an estimate of the magnitude of the interelectronic repulsion energy of $5248 - 2370 = 2878$ kJ mol^{-1}, an amount which is less than the 2s–1s energy gap which is given by $3I_H$ (3936 kJ mol^{-1}).

There is no doubt that the ground state of the helium atom has the configuration: $1s^2$. It must be emphasized that **electrons only pair up in the same orbital when it is the lowest energy option**.

The third element, lithium ($Z = 3$), has a full 1s orbital, with the third electron entering the 2s orbital giving the configuration: $1s^2 2s^2$. Beryllium ($Z = 4$) has the configuration: $1s^2 2s^2$, the interelectronic repulsion energy in $2s^2$ being lower than the 2p–2s energy gap.

The fifth electron in the boron ($Z = 5$) atom enters one of the three-fold degenerate 2p orbitals. Since these are truly degenerate it is not proper to specify which of the three orbitals is singly occupied, although some texts choose the $2p_x$ orbital for alphabetical reasons. Boron has the electronic configuration: $1s^2 2s^2 2p^1$. The carbon configuration is of considerable interest since it is the first example of electrons occupying degenerate orbitals. From what has been considered previously it is a simple matter to conclude that the two electrons occupying the 2p orbitals should occupy separate orbitals where interelectronic repulsion is less than if they doubly occupy a single orbital.

What is not so obvious is that the two electrons, occupying two different 2p orbitals, should have identical values of the spin quantum number, s. The reason for this is that electrons with parallel spins (identical s values) have zero probability of occupying the same space — a kind of Pauli restriction — whereas two electrons with opposed spins (different s values) have a finite chance of occupying the same region of space within the atom and, in consequence, the interelectronic repulsion is greater than if their spins were parallel. This is the rationalization of what are known as **Hund's rules**. These may be stated in the following way.

In filling a set of degenerate orbitals the number of unpaired electrons is maximized, and that such unpaired electrons will possess parallel spins. An alternative statement appears in section 1.6.

The electronic configuration of the carbon atom is $1s^2 2s^2 2p^2$ or, if the detailed content of the 2p orbitals is being discussed, it may be written as: $1s^2 2s^2 2p_x^1 2p_y^1$, the choice of *x* and *y* being merely alphabetical.

By similar arguments it is easily concluded that with the nitrogen atom the lowest energy electronic configuration is $1s^2 2s^2 2p_x^1 2p_y^1 2p_z^1$ or $1s^2 2s^2 2p^3$ for general purposes.

In the oxygen, fluorine and neon atoms the extra electrons doubly occupy the appropriate number of 2p orbitals since pairing is the lowest energy option — the 3s–2p gap being greater than any interelectronic repulsion energy involved.

The atoms of the elements Li, Be, B, C, N, O, F and Ne form the first short **period** of the Periodic Classification of the Elements. The second short period contains the elements Na, Mg, Al, Si, P, S, Cl and Ar, which have electronic configurations which are that of neon ($1s^2 2s^2 2p^6$) plus those concerned with the regular filling of the 3s and 3p orbitals as has been described for that of the 2s and 2p orbitals. Identical arguments apply and the elements of the second short period are arranged under their counterparts in the first short period with identical 'outer electronic configurations', except for the change in value of *n* from 2 to 3. The term 'outer' is a reference to the value of *n* and is related to the greater diffuseness of orbitals as *n* increases in value — the orbitals become larger. It is the nature of the outer electronic configuration which determines the chemistry of an element and it is for this reason that more emphasis is placed upon it rather than the arrangement of all the electrons of an atom. It is only by considering the total number of electrons that the outer electronic configuration may be determined.

The next elements are potassium and calcium in which the outer electrons occupy the next lowest energy orbital, which is the 4s since it has a lower energy than the 3d orbitals. In the next ten elements (scandium to zinc). the five 3d orbitals are progressively occupied. The filling is in accordance with Hund's rules with two irregularities which are described below.

The outer electronic configurations of the elements Sc, Ti and V are:

Sc. $4s^2 3d^1$
Ti. $4s^2 3d^2$
V. $4s^2 3d^3$

but that of chromium is:

Cr. $4s^1 3d^5$.

The explanation of this irregularity is that the 3d level is now of lower energy than the 4s and that the gap between the two sets of orbitals is small enough to prevent electron pairing in the 3d orbitals. Regularity returns with the next four elements:

Mn.	$4s^2 3d^5$
Fe.	$4s^2 3d^6$
Co.	$4s^2 3d^7$
Ni.	$4s^2 3d^8$.

With the next element, copper, the 3d level is sufficiently lower than the 4s to ensure complete pairing up in the 3d orbitals, leaving the sole unpaired electron in the 4s orbital:

Cu. $4s^1 3d^{10}$.

The next element is zinc which, with the configuration $4s^2 3d^{10}$, completes the first series of **transition** elements.

The next orbitals to be used in building up the elements are the 4p set, which are filled in a regular fashion in the elements gallium to argon, thus finishing the first long period of the periodic table. The first two elements (K and Ca) are arranged so that they come below Na and Mg and form parts of groups 1 and 2 respectively of the periodic table. The first set of transition elements form the first members respectively of groups 3 to 12, and the elements from Ga to Kr are placed under those from Al to Ar as members of groups 13 to 18.

The filling of the 5s, 4d and 5p orbitals accounts for the elements of the second long period (Rb and Sr in the s block, Y to Cd in the second transition series or d block, and In to Xe in the p block). As was the case with the 3d orbitals, the filling of the 4d set is not regular owing to the closeness of the energies of the 5s and 4d orbitals. The irregularities are not the same as in the the first set of transition elements. The outer electronic configurations of the second series of transition elements are:

Y	$5s^2 4d^1$
Zr	$5s^2 4d^2$
Nb	$5s^1 4d^4$
Mo	$5s^1 4d^5$
Tc	$5s^1 4d^6$
Ru	$5s^1 4d^7$
Rh	$5s^1 4d^8$
Pd	$4d^{10}$
Ag	$5s^1 4d^{10}$
Cd	$5s^2 4d^{10}$

In Y and Zr the 5s orbital is sufficiently lower in energy than the 4d level to enforce the production of the $5s^2$ pairing. In the elements from Nb to Rh the 4d orbitals are marginally lower in energy than the 5s orbital and the two sets of orbitals behave as though they are degenerate. In the elements Pd, Ag and Cd the 4d energy is distinctly lower than that of the 5s orbital so that complete pairing occurs in Pd, and then the 5s filling follows in Ag and Cd.

More irregularities are to follow in the formation of the third long period. The next orbital of lowest energy to be used is the 6s whose filling accounts for the outer electronic configurations of Cs and Ba. The next orbitals to be used are the 5d and 4f sets whose energies are very nearly identical, but vary with the nuclear charge. The outer electronic configurations of the next fifteen elements are shown in Table 1.3.

Table 1.3 — Outer electronic configurations of the lanthanide elements

Element, symbol	6s	5d	4f
Lanthanum, La	2	1	0
Cerium, Ce	2	0	2
Praseodymium. Pr	2	0	3
Neodymium, Nd	2	0	4
Promethium, Pm	2	0	5
Samarium, Sm	2	0	6
Europium, Eu	2	0	7
Gadolinium, Gd	2	1	7
Terbium, Tb	2	0	9
Dysprosium, Dy	2	0	10
Holmium, Ho	2	0	11
Erbium, Er	2	0	12
Thulium, Tm	2	0	13
Ytterbium, Yb	2	0	14
Lutetium, Lu	2	1	14

The 5d orbital is singly occupied at the beginning, in the middle and at the end of the series of elements in Table 1.3, which is where it has very similar energy to the 4f set of orbitals. For the other elements, the 4f energy is sufficiently lower than that of the 5d for the latter not to be used. The elements La to Yb form the f block and are the fourteen elements concerned with the filling of the 4f orbitals with its completion giving lutetium ($6s^2 5d^1 4f^{14}$) which is placed as the first member of the third transition series.

The next nine elements from hafnium to mercury complete the third transition series, although there are irregularities of filling the 5d orbitals since the 5d and 6s energies are close together. The outer electronic configurations of these elements are:

Hf	$6s^2 5d^2$
Ta	$6s^2 5d^3$
W	$6s^2 5d^4$
Re	$6s^2 5d^5$
Os	$6s^2 5d^6$
Ir	$6s^2 5d^7$

— a regular filling of the 5d orbitals whose energies are higher than that of the 6s orbital, but after Ir the 5d energy becomes lower than that of the 6s and the small energy gap causes the irregularities in the next two elements:

Pt	$6s^1 5d^9$
Au	$6s^1 5d^{10}$

and the filling of the 6s orbital is completed in mercury:

Hg	$6s^2 5d^{10}$

The regular filling of the 6p orbitals accounts for the outer electronic configurations of the elements from thallium to radon, thus completing the third long period with its integral f block elements.

The fourth long period is incomplete but, as far as it goes, mirrors the third long period. The first two elements, francium and radium, have outer electronic configurations $7s^1$ and $7s^2$ respectively. The energies of the next orbitals to be filled (6d and 5f) are similar and give rise to irregularities dependent upon which of them is the lower in energy and by how much. The accepted outer electronic configurations are:

Ac	$7s^2 6d^1$
Th	$7s^2 6d^2$
Pa	$7s^2 6d^1 5f^2$
U	$7s^2 6d^1 5f^3$
Np	$7s^2 6d^1 5f^4$
Pu	$7s^2 \quad 5f^6$

— the irregularities persist up to plutonium, and after that there is the presumption that there is a regular filling of the 5f orbitals, much like that of the 4f orbitals, with the exception of there being a reappearance of a single 6d electron in the case of curium. The trans-plutonium elements are intensely radioactive and have short lifetimes so that detailed studies of their electronic configurations have not been carried out and the published ones are speculative.

Similarly to the lanthanide elements, the actinides (actinium to nobelium) are placed as a set of fourteen elements as a separate f block, with element 104, Lr, being the first member of the fourth transition series. The fourth transition series is unfinished and the element with the highest Z value, which has so far been synthesized, is number 109.

The periodic table consists of eighteen groups corresponding to the filling of the ns, $(n-1)$d and np orbitals for the values of n up to seven, with the f block elements situated separately, there are two short periods for $n = 1$ and 2, since there are no d orbitals at that stage. Hydrogen is placed as the first element in group one, helium being the first element in group eighteen.

The majority of elements in any particular group have identical outer electronic configurations (apart from having varying values of the principal quantum number, n). Because of the various irregularities in orbital filling it is not possible to state that **all** elements in a group have identical configurations. The historical basis of the periodic classification was in terms of the members of any particular group sharing very similar chemical properties. As has been seen, it is not possible for there to be a completely regular filling of orbitals, which would allow any member of a group to have the same outer electronic configuration. This is usually not the case for any particular oxidation state of the elements of one group. These tend to have identical outer electronic configurations. The outer configurations of the elements Ni, Pd and Pt are s^2d^8, s^0d^{10} and s^1d^9, respectively, but their +2 ions have the common d^8 configuration.

1.5 THE PERIODICITY OF THE IONIZATION ENERGIES, THE SIZES AND THE ELECTRONEGATIVITIES OF THE ELEMENTS

This section deals with the variation, along the periods and down the groups of the periodic table, of the ionization energies, sizes and electronegativity coefficients of the elements.

Ionization energies

The strict definition of the first ionization energy of an atom is the minimum energy required to convert one mole of the gaseous atom (in its ground electronic state) into its unipositive ion:

$$A(g) \rightarrow A^+(g) + e^- \tag{1.20}$$

The first ionization energies of hydrogen, helium and the s and p block elements of periods one to three are plotted in Fig. 1.4. There is a pattern of the values for the

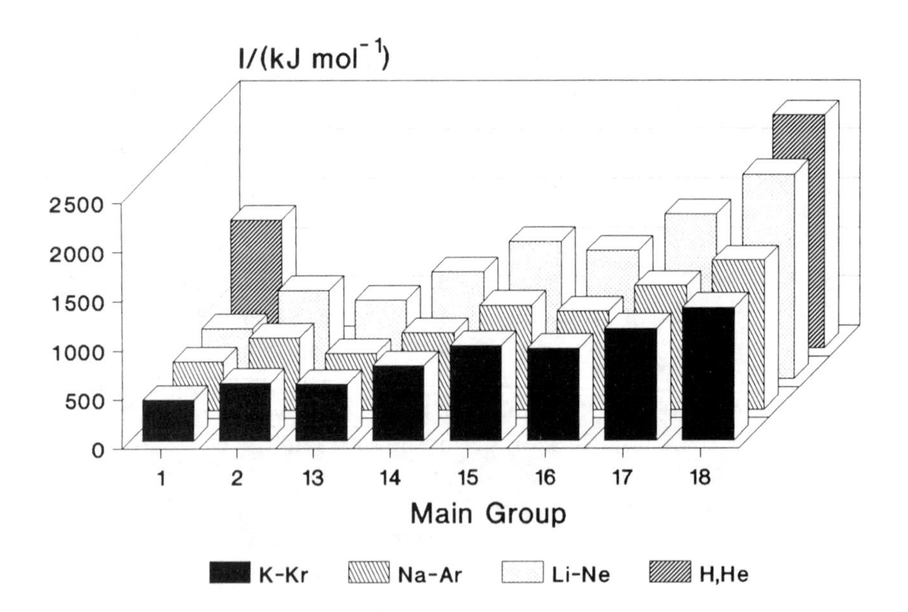

Fig. 1.4 — First ionization energies of the main group elements from hydrogen to krypton.

elements Li to Ne (the first short period) which is repeated for the elements Na to Ar (the second short period) and which is repeated again for the elements K, Ca and Al to Kr (the s and p block elements of the first long period). In addition to this periodicity it is obvious that the ionization energy decreases down any of the groups. Both observations are explicable in terms of the electronic configurations of the elements.

Variation of ionization energy along a period

Consider the variation of the first ionization energies of the elements of the first short period. The electron which is removed in the ionization process is the one (or is one of those) with the highest energy — the one requiring the least energy to remove from the attractive influence of the atomic nucleus. The value of the first ionization energy of the lithium atom is 519 kJ mol^{-1} and corresponds to the change in electronic configuration $1s^2 2s^1$ to $1s^2$.

The increase to 900 kJ mol^{-1} in the case of the beryllium atom is due to the increase in nuclear charge. The electron most easily removed is one of the pair in the 2s orbital. In the case of the boron atom, in spite of an increase in nuclear charge, there is a decrease in the first ionization energy to 799 kJ mol^{-1}. This is because the electron removed is from a 2p orbital — higher in energy than the 2s level.

There follows a general increase in the first ionization energies of the carbon (1090 kJ mol^{-1}) and nitrogen (1400 kJ mol^{-1}) atoms. The electrons are removed from the 2p orbitals, and the increase is due to the increase in nuclear charge. With oxygen there is a slight decrease to 1310 kJ mol^{-1} in spite of an increase in nuclear charge. In the cases of B, C and N the electron removed in the ionization process is the sole resident of one of the 2p orbitals. In oxygen there are two 2p orbitals which are singly occupied and one which is doubly occupied. Consideration of the higher interelectronic repulsion associated with a pair of electrons in the same orbital leads to the conclusion that it is one of the paired-up electrons in the 2p level of oxygen which is most easily removed. Consideration of Hund's rules leads to the conclusion that all three 2p electrons in the resulting O^+ ion possess parallel spins. The interelectronic-repulsion-assisted removal of paired-up 2p electrons applies to the cases of the next two elements, F and Ne, the increases being due to the increases in nuclear charge.

The general pattern of the variation of the first ionization energies of the elements of the first short period is repeated for the respective elements of the s and p blocks of the subsequent periods of the periodic system.

In any such series there are two variable quantities: the nuclear charge and the number of electrons of the atoms considered. It is possible to eliminate the effect of the nuclear charge (but not the effects of changes in the effectiveness of the nuclear charge) by considering the successive ionization energies of the neon atom, for example. Fig. 1.5 is a plot of the first eight successive ionization energies of the neon atom and, as would be expected from a reduction in the number of electrons being attracted by the constant nuclear charge, the values exhibit a general increase.

The first three ionizations arise from doubly occupied 2p orbitals by the same reasoning as is given above for the first ionization of the oxygen atom. All three ionizations are assisted by the large interelectronic repulsion associated with the

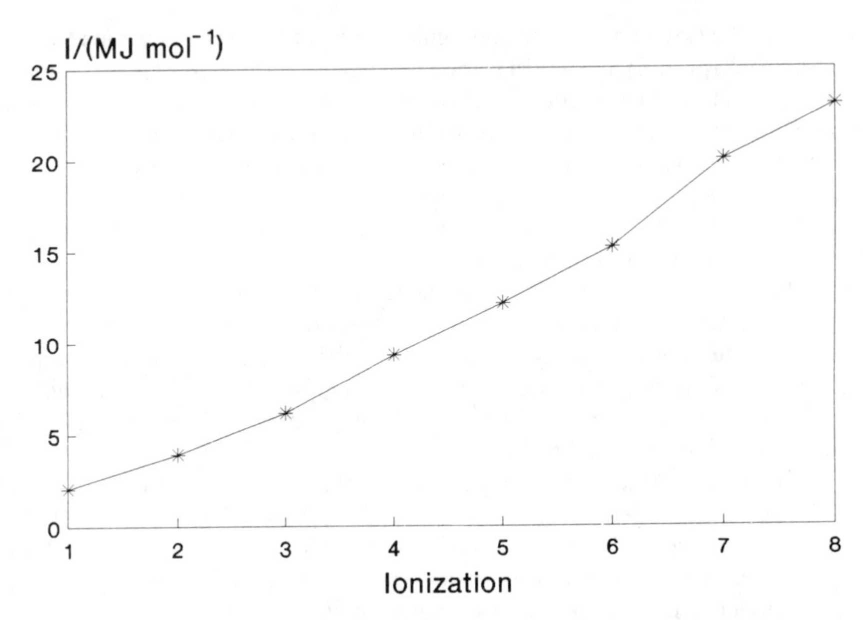

Fig. 1.5 — Successive ionization energies of the neon atom.

double occupation of orbitals. As may be seen from the extrapolation of the line joining the first three points in Fig. 1.5 the next three electrons are considerably more difficult to remove than the first three. There is an increasing effectiveness of the nuclear charge with the additional difficulty of electron removal due to the absence of the high interelectronic repulsion assistance enjoyed by the first three electrons to be removed.

The second discontinuity shown in Fig. 1.5 is associated with the ionization of the 7th and 8th electrons, which are very much more difficult to remove because they originate in the 2s orbital of the neon atom.

Variation of ionization energy down a group

In general there is a decrease in the first ionization energies of the atoms of any particular group of the periodic table as the nuclear charge increases. This may be seen from the data plotted in Fig. 1.4. As any group is descended, the orbital from which the electron is ionized has a progressively larger value of n, and has the highest energy for each particular atom in the group. This essentially offsets the effect of the increasing nuclear charge. It is as though the outermost electrons are shielded from the effect of the nuclear charge by the full inner sets of orbitals.

Variations in atomic size across periods and down groups

The size of an atom is not a simple concept. There are three different modes by which sizes may be assigned to any particular atom. These are as follows.

(i) Atomic radius

The atomic radius of an element is considered to be half the interatomic distance between identical (singly bonded) atoms. This may apply to iron in its metallic state, in which case the quantity may be regarded as the metallic radius of the iron atom, or to a molecule such as Cl_2. The difference between the two examples is sufficient to demonstrate that some degree of caution is necessary when comparing the atomic radii of different elements. It is best to limit such comparisons to elements with similar types of bonding — metals for example. Even that restriction is subject to the drawback that the metallic elements exhibit at least three different crystalline arrangements with possibly different coordination numbers (the number of nearest neighbours for any one atom).

(ii) Covalent radius

The covalent radius of an element is considered to be one half of the covalent bond distance of a molecule such as Cl_2 (equal to its atomic radius in this case) where the atoms concerned are participating in single bonding. Covalent radii for participation in multiple bonding are also quoted. In the case of a single bond between two different atoms, the bond distance is divided up between the participants by subtracting the covalent radius of one of the atoms, whose radius is known, from it. A set of mutually consistent values is now generally accepted and, since the vast majority of the elements take part in some form of covalent bonding, the covalent radius is the best quantity to consider for the study of general trends. Only atoms of the Group 18 elements (except Kr and Xe) do not have covalent radii assigned to them because of their general inertness with respect to the formation of molecules. The use of covalent radii for comparing the sizes of atoms is subject to the reservation that its magnitude, for any given atom, is dependent upon the oxidation state of that element.

(iii) Van der Waals radius

The Van der Waals radius of an element is half the distance between two atoms of an element which are as close to each other as is possible without being formally bonded by anything except Van der Waals **intermolecular forces**. Such a quantity is used for the representation of the size of an atom with no chemical bonding tendencies — the Group 18 elements. That for krypton, for instance, is half of the distance between nearest neighbours in the solid crystalline state, and is equal to the atomic radius. Van der Waals radii of atoms and molecules are of importance in discussions of the liquid and solid states of molecular systems, and in the details of some molecular structures where two or more groups attached to the same atom may approach each other.

Fig. 1.6 shows how the covalent radii vary across periods and down groups of the periodic table. Across periods there is a general reduction in atomic size, whilst down any group the atoms become larger. These trends are consistent with the understanding gained from the study of the variations of the first ionization energies of the elements. As the ionization energy is a measure of the effectiveness of the nuclear charge in attracting the extra-nuclear electrons, it might be expected that an increase in nuclear effectiveness would lead to a reduction in atomic size. The trends in atomic size in the periodic system are almost the exact opposite to those in the first ionization energy.

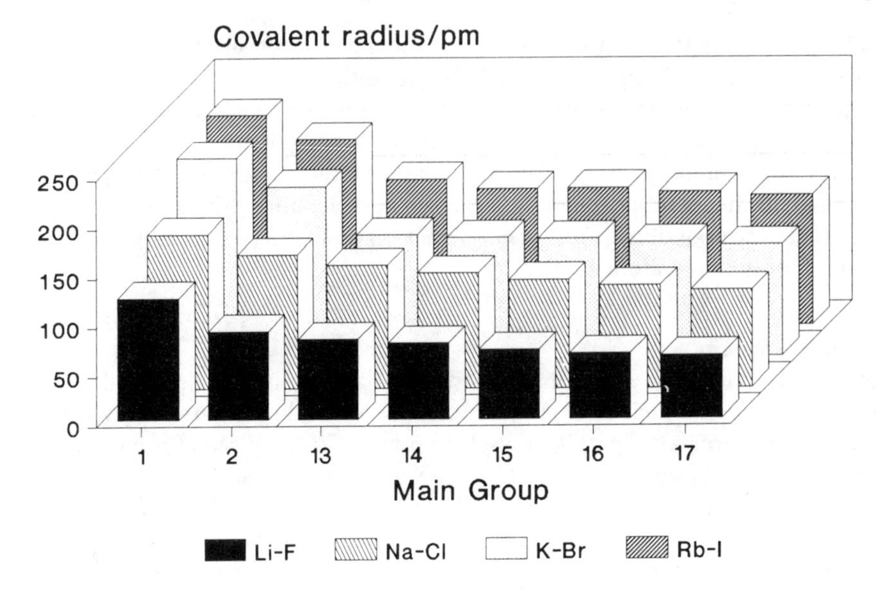

Fig. 1.6 — Covalent radii of some main group elements.

Variations in electronegativity across periods and down groups

The concept of electronegativity is derived from such experimental observations that the elements, fluorine and chlorine, are highly electronegative in that they exhibit a very strong tendency to become negative ions. The elements of Group 1, on the other hand, are not electronegative and are better described as being electropositive — they have a strong tendency to form positive ions. A scale of **electronegativity coefficients** would be useful in allowing a number to represent the tendency of an element in a molecule of attracting electrons to itself. The establishment of such a scale has involved the powers of two Nobel Prize winners (Pauling and Mulliken), and after many other efforts there are still doubts about the currently accepted values and about their usefulness. The difficulties are exemplified by briefly describing the contributions of Pauling and Mulliken and then discussing the extension of their values to the majority of elements (the ones that possess a covalent radius).

The Pauling electronegativity scale

Pauling noted that the bond dissociation energy of a diatomic molecule, AB, was greater than the average of those of the molecules, A_2 and B_2. He proposed that the geometric mean of the bond dissociation energies of the molecules, A_2 and B_2, represented the strength of a pure covalent bond between the atoms A and B, and that any extra strength possessed by the A−B bond was due to the difference in the electronegativity coefficients of the two atoms. If x is used to represent the electronegativity coefficient of an atom, and Δ for the difference between the observed dissociation energy of the bond A−B and its value when the bond is purely covalent, then Δ is given by:

$$\Delta = D(\mathrm{AB}) - \{D(\mathrm{A_2}) \times D(\mathrm{B_2})\}^{1/2} \qquad (1.21)$$

and Pauling found that if

$$|x_A - x_B| = \Delta^{1/2} \tag{1.22}$$

(here |...| represents the positive value of the expression within the vertical lines) the values for the xs were additive. By suitably adjusting the units of the quantities used in the calculation (he used electron-volts, 1 eV being equal to 96.487 kJ mol^{-1}), and by assigning the value of 2.1 to x_H, he was able to estimate values of x for a small number of elements for which the thermochemical data were available. The data were scarce and more values of x were calculated with every advance in thermochemical knowledge.

The Mulliken electronegativity scale

Mulliken adopted a different approach to the problem than that of Pauling. The basis of his argument is altogether more logical and easier to follow. He considered the two possibilities for electron transfer reactions between the elements A and B in the gas phase:

$$A(g) + B(g) \rightarrow A^+(g) + B^-(g) \qquad \text{(I)} \tag{1.23}$$
$$A(g) + B(g) \rightarrow A^-(g) + B^+(g) \qquad \text{(II)} \tag{1.24}$$

Which way the electron would be most likely to be transferred would depend on the relative electronegativities of A and B. If B should be more electronegative than A ($x_B > x_A$) then the more likely reaction would be (I) and the change in internal energy, ΔU(I), would be smaller (i.e. less positive, more negative) than that for the alternative reaction, ΔU(II):

$$\Delta U(\text{I}) < \Delta U(\text{II}) \tag{1.25}$$

The reactions (I) and (II) involve the removal of an electron from one element and its placement on the other—the change in internal energy being the difference between the ionization energy of one atom and the electron affinity of the other. The above inequality then becomes:

$$I_A - E_B < I_B - E_A \tag{1.26}$$

and collecting terms in A at one side and those in B at the other side of the inequality sign gives:

$$I_A + E_A < I_B + E_B \tag{1.27}$$

Comparison of the inequality (1.27) with that proposed for the x values:

$$x_A < x_B \tag{1.28}$$

suggests that there is a relationship between the x value for an element and the sum of the ionization energy and the electron affinity values for that element:

$$x \propto I + E \tag{1.29}$$

Mulliken found that x values calculated from available data were very similar to those obtained by Pauling (when a common conversion factor was used) and was extended to elements for which the Pauling-type data were unavailable. The scale was restricted by the availability of E values.

The Allred-Rochow electronegativity scale

This scale is the one which is now generally accepted, and is based on the concept that the electronegativity of an element is related to the force of attraction experienced by an electron at a distance from the nucleus by the covalent radius of the particular atom. According to Coulomb's law this force is given by:

$$F = Z_{\text{eff}} e^2 / r_{\text{cov}}^2 \tag{1.30}$$

where Z_{eff} is the effective atomic number and e is the electronic charge. The effective atomic number is considered to be the difference between the actual atomic number, Z, and a shielding factor, S, which is estimated by the use of Slater's rules.

The best fit with the accepted Pauling/Mulliken values for electronegativity coefficients is the equation:

$$x = 3590 {}^* Z_{\text{eff}} / r_{\text{cov}}^2 + 0.744 \tag{1.31}$$

(with r_{cov} expressed in picometres).

The values given in modern versions of the periodic table and in data books are the Allred–Rochow values rounded off (usually) to one decimal place. To attempt to represent x more accurately would be imprudent after consideration of the difficulties in obtaining any values at all! The method of calculation offers the possibilities of assigning x values to the different oxidation states of the same element — it would be expected, for instance, that the electronegativities of manganese in its oxidation states of 0, II, III, IV, V, VI and VII would be different, and providing that the necessary covalent radii are known, the x values are easily calculated. As the element becomes smaller (as the oxidation state increases) the x value rises. It would be expected that the smaller and more highly charged an atom becomes, the higher would be its attraction for electrons.

Fig. 1.7 shows the variation of Allred–Rochow electronegativity coefficients for singly bonded elements along periods and down groups of the periodic table. In general the value of the electronegativity coefficient increases across the periods and decreases down the groups. That is precisely the opposite of the trends in covalent radii but similar to the trends in first ionization energy (Fig. 1.4). This latter

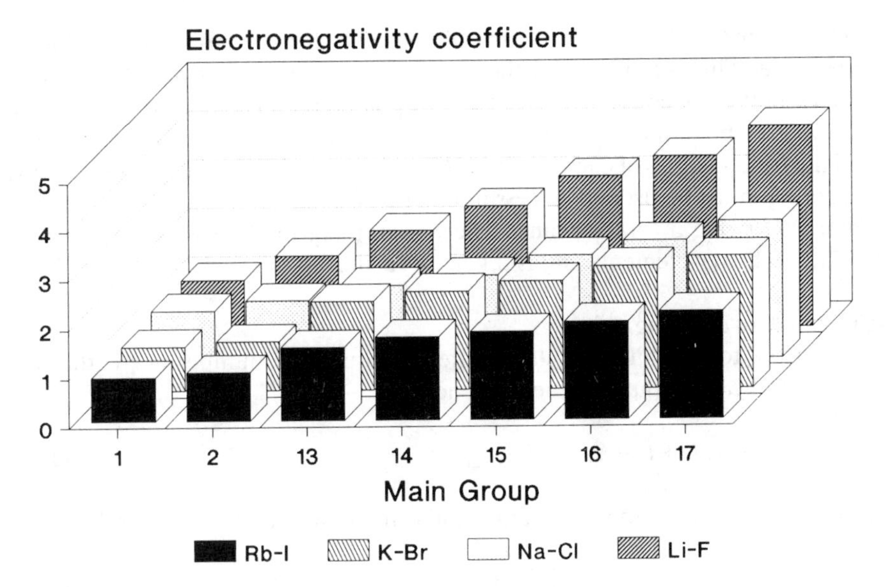

Fig. 1.7 — Allred–Rochow electronegativity coefficients for some main group elements.

conclusion is no surprise, as it is to be expected that elements with a high tendency to attract electrons possess high first ionization energies.

1.6 ELECTRONIC CONFIGURATIONS AND ELECTRONIC STATES

The term 'electronic state' is used when referring to a very precise method of describing the distribution of electrons within a given electronic configuration. It is necessary because the electronic configuration as such is not sufficiently precise for some purposes and because there are many instances when one configuration represents more than one **state**. The electronic state is representative of the precise manner in which atomic orbitals are occupied and takes into account the various spin possibilities for a given configuration.

That of the carbon atom has been dealt with in some detail. The general expression of its electronic configuration is $1s^2 2s^2 2p^2$ and the 'ground state' has been decided to be that in which the two 2p electrons possess the same value of the spin quantum number, *s*.

Electronic states of the carbon atom ($2p^2$)

More quantum numbers are required to define the possible states arising from the $2p^2$ configuration of the carbon atom. This is because of the quantized nature of the interaction of the two electrons' orbital and spin momenta to give a resultant total orbital angular momentum (L) and a resultant total spin angular momentum (S). The method of interaction described in this section is known as Russell–Saunders coupling. It depends upon there being strong interactions between the individual l

values to give an L value and strong interactions between the individual s values to give an S value. This is followed by the quantized interaction between the L and S values. The method of coupling of orbital and spin momenta applies to the majority of cases. If the above interactions, $l \Leftrightarrow l$ and $s \Leftrightarrow s$, are not strong relative to those between the l and s values of each electron, $l \Leftrightarrow s$, then the alternative method, known as j–j coupling, must be used. The $l \Leftrightarrow s$ interactions give appropriate j values which then interact to give a total J value. The j–j coupling method is generally more appropriate to heavy elements and is taken no further in this text.

Russell–Saunders coupling

The rule for the total orbital angular momentum quantum number (L) is that it is made up from the individual l values in general as:

$$L = l_1 + l_2, l_1 + l_2 - 1, \ldots, |l_1 - l_2| \tag{1.32}$$

There is a similar rule for the total spin angular momentum quantum number (S):

$$S = s_1 + s_2, s_1 + s_2 - 1 = |s_1 - s_2| \tag{1.33}$$

Rather than use the rules as written in equations (1.32) and (1.33) it is better and more certain to use rules which refer to the permitted l and s values with respect to an externally applied magnetic field and use their combinations to decide upon the permitted values of L and S. The rules for L and S values with respect to an external magnetic field are:

$$L = M_L, M_L - 1, \ldots, 0, \ldots, -(M_L - 1), -M_L \tag{1.34}$$

and

$$S = M_S, M_S - 1, \ldots, 0, \ldots, -(M_S - 1), -M_S \tag{1.35}$$

where the upper case Ms refer to totals of lower case ms in both equations:

$$M_L = \sum m_l \tag{1.36}$$

and

$$M_S = \sum m_s \tag{1.37}$$

the l and s subscripts being present to distinguish the orbital and spin magnetic quantum number values.

By using these rules it is possible to derive the electronic states associated with the $2p^2$ electronic configuration with certainty. It is necessary to write down all the possible **microstates** appropriate to the two 2p electrons. This is done by writing down the permitted combinations of the possible values of m_l and m_s, each different combination representing a microstate. There are fifteen ways of distributing two indistinguishable electrons between three orbitals given that there is a choice of two values for the spin quantum number. With three orbitals and two spin values there are six ways of assigning the first electron. That leaves only five ways of assigning the second electron, the combinations of the two assignments being thirty in total. This has to be divided by two to take account of the indistinguishability of the two electrons, giving a final total of fifteen microstates. These are specified in Table 1.4.

Table 1.4 — The microstates of the $2p^2$ configuration

	Value of m_l				
Microstate no.	1	0	−1	Σm_l	Σm_s
1	⇅			2	0
2		⇅		0	0
3			⇅	−2	0
4	↑	↑		1	1
5	↓	↓		1	−1
6	↑		↑	0	1
7	↓		↓	0	−1
8		↑	↑	−1	1
9		↓	↓	−1	−1
10	↑	↓		1	0
11	↓	↑		1	0
12	↑		↓	0	0
13	↓		↑	0	0
14		↑	↓	−1	0
15		↓	↑	−1	0

All of the microstates have equal probability. The problem is to sort them out into proper electronic states. This means finding those sets of microstates which represent individual electronic states. This means finding sets of values of Σm_l which are consistent with there being a valid value of L in accordance with equation (1.34), and the best way of doing this is to look for the highest value of Σm_l in Table 1.4 — this corresponds to microstate number 1, with a value of 2. If equation (1.34) is to be satisfied, there should be values of Σm_l of 1, 0, −1 and −2, all with the same value of $\Sigma m_s = 0$ (since that is the value associated with the Σm_l value of 2). There are such values present in microstates numbers 10 and 11 ($\Sigma m_l = 1$), numbers 2, 12 and 13 ($\Sigma m_l = 0$), numbers 14 and 15 ($\Sigma m_l = -1$) and number 3 ($\Sigma m_l = -2$), all of which have zero total spin ($\Sigma m_s = 0$). Only one microstate is required for each value of Σm_l, and the duplicate and triplicate microstates can and do contribute to other states present. What is certain is that the M_L (Σm_l) values of 2, 1, 0, −1 and −2, all with an M_S (Σm_s) value of zero, are present and satisfy the rules for there being a particular electronic state arising from the $2p^2$ configuration. Before isolating the other possible states it is necessary to deal with the symbolism used in the description of electronic states.

For any electronic state the value of L (the maximum Σm_l value) is used to assign a letter to the state in the same way in which l is used to assign a descriptive letter to an orbital. The coding is just the same as that used for the values of l except that upper case letters are used for the L-value codes. Some values of L and the associated code letters are given in Table 1.5.

Table 1.5 — Coding letters for some L values

Value of L	Code letter
0	S
1	P
2	D
3	F
4	G

Table 1.5 is a portion of an infinite series of L values and the coding follows alphabetical order after F. The L value of the electronic state already identified for the $2p^2$ configuration is a D state. The five microstates forming the D state are associated with a single value of Σm_s, and accordingly it is to be described as a **singlet** state. A more formal method of arriving at this conclusion is to calculate the **multiplicity** of the state by using the equation:

$$\text{Multiplicity} = 2S + 1 \tag{1.38}$$

in which S is given by the maximum value of Σm_s for any particular state. The multiplicity is indicated as a left-hand superscript to the code letter for the state. Thus the state under discussion is to be written as: ^{1}D.

There is a quantized interaction between the total angular orbital momentum and the total spin angular momentum, which may be expressed as a rule for yet another quantum number, J, the resultant total angular momentum quantum number. The rule for J is:

$$J = L + S, L + S - 1, \ldots, |L - S| \tag{1.39}$$

The multiplicity of a state indicates the number of different J values it possesses.

For the ^{1}D state under discussion, since $S = 0$, there is only one J value which is equal to the value of L ($= 2$). There are as many J values for a state as its multiplicity. The J value of a state is indicated by a right-hand subscript on the code letter, in the current example as: $^1D_2$. This symbolic description of the electronic state is called the **term symbol** for the particular state.

Returning to the other microstates of the $2p^2$ configuration, the next highest value of Σm_l is 1 and the associated values of Σm_s are 1, 0 and -1, which are indicative of a triplet state. If S is equal to 1, then the multiplicity is three (by equation (1.38)), with the values of Σm_s as indicated in the previous sentence. The value of Σm_l indicates the presence of a P state, but to confirm this and that it is a triplet state it is essential to identify the nine microstates associated with such a state. There are nine microstates which form the ^3P state because the three values of Σm_l of 1, 0 and -1 (for a P state $L=1$) are in turn associated with the three values of Σm_s of 1, 0 and -1 (for a triplet state $S=1$). The appropriate microstates are identified in Table 1.6.

Table 1.6 — The microstates of Table 1.5 contributing to the ^3P state

		Values of Σm_l		
		1	0	-1
Values	1	4	6	8
of	0	10, 11	2, 12, 13	14, 15
Σm_s	-1	5	7	9

As may be seen from Table 1.6 there is an excess of microstates over the nine actually required, but that is also the situation with regard to the ^1D state. For instance the $\Sigma m_l = 1$, $\Sigma m_s = 0$ combination arises in microstates numbers 10 and 11 of Table 1.5. One is needed for the ^1D state and one for the ^3P state, and since they are identical in value they both contribute to both states in the form of the linear combinations, $\psi_{10} + \psi_{11}$, and $\psi_{10} - \psi_{11}$.

The ^3P state has three permitted J values: 2, 1 and 0 by the operation of the J rule expressed by equation (1.39). Fully written down these are: $^2P_2$, $^3P_1$ and $^3P_0$.

The ^1D and ^3P states require fourteen microstates for their complete description so that there is one microstate from the fifteen of Table 1.5 which is unaccounted for. This is the one (of a possible three) which has $\Sigma m_l = \Sigma m_s = 0$ and which can only be a singlet-S state: ^1S. There is only one J value (zero) so that the full term symbol for the state is $^1S_0$.

By this laborious procedure it is possible to conclude that the $2p^2$ **configuration** has associated with it the ^1D, ^3P and ^1S **states**. It is now possible to re-state Hund's rules in a more exact and understandable form. They may be formulated as indicating that the **electronic ground state** is:

(i) that which has the highest multiplicity and if there are more than one state in that category then the one of lowest energy is:
(ii) that with the highest value of L.

There is an additional rule concerning the values of J:

(iii) for states of identical L values, the lowest in energy is that with the lowest J value if the electron shell is less than half full, and is the one with the highest value of J if the electron shell is more than half full. The ground states of those atoms with a half-full shell are S states ($L = 0$) with only one J value, whatever the multiplicity may be. In such a case this subrule is irrelevant.

The ground state of the carbon atom is now revealed to be the ^3P state and its make-up from nine microstates — a more complex but more accurate description than by considering it as 'two singly occupied 2p orbitals with the two electrons possessing parallel spins'.

The formulation and accurate descriptions of electronic states and the assignment of term symbols is of great importance in the understanding of the electronic spectra of atoms and molecules (dealt with in Chapter 6), and is of particular importance in the interpretation of the spectra of complexes containing transition metal ions (dealt with in Chapter 8).

In the case of the carbon atom, the ^1D and ^1S states are, respectively, 95.7 and 251 kJ mol^{-1} higher in energy than the ^3P ground state. These figures exemplify the differences in interelectronic effects in the three cases and help to emphasize the importance of distinguishing between configurations and states.

1.7 CONCLUDING SUMMARY

This chapter contains a non-mathematical treatment of atomic structure leading to an understanding of the following topics.

(1) The line emission spectrum of the hydrogen atom and its relevance to the development of the concept of the quantization of electron energies.
(2) Atomic orbitals — their definition by the quantum rules and an appreciation of their spatial distributions.
(3) The general structure of the periodic classification of the elements and its dependence upon the Pauli exclusion principle which defines the available orbitals and limits their occupancy.
(4) The electronic configurations of the elements and their dependence upon Hund's rules, which in turn are dependent upon interelectronic repulsion.
(5) The variations of ionization energies, covalent radii and electronegativity coefficients (i) down groups and (ii) across periods of the periodic classification which are related to the electronic configurations of the elements.
(6) The difference between electronic configurations and electronic states and the relationship between them.
(7) The method of deciding which states are generated by an electronic configuration, and the derivation of term symbols.

2

Molecular symmetry and group theory

2.1 INTRODUCTION

The theories of (i) atomic and molecular orbitals, (ii) the electronic configurations and states of atoms and molecules, (iii) the spectroscopy of atoms and molecules and (iv) the shapes and vibrational modes of molecules, are all heavily dependent upon aspects of group theory.

This chapter deals with the necessary group theoretical ideas as applied to molecular symmetry. The applications in the areas mentioned above are demonstrated extensively in the remainder of this book.

2.2 MOLECULAR SYMMETRY AND GROUP THEORY

A **group** consists of a set of **elements** which are related to each other according to certain rules. The particular kind of elements which are relevant to the symmetries of molecules are **symmetry elements**. With each symmetry element there is an associated **symmetry operation**. The necessary rules are referred to when appropriate.

Elements of symmetry and symmetry operations

There are seven elements of symmetry which are commonly possessed by molecular systems. An element of symmetry is possessed by a molecule if, after the associated symmetry operation is carried out, the atoms of that molecule are not perceived to have moved. The molecules are in an **equivalent** configuration. The individual atoms may have moved but only to positions previously occupied by identical atoms.

The seven elements of symmetry, their notations and their related symmetry operations are given in Table 2.1.

1. The identity, E

The symmetry element, known as the identity and symbolized by E (or in some texts by I), is possessed by all molecules independently of their shape. The related

Table 2.1 — Elements of symmetry and their associated operations

Symmetry element	Symbol	Symmetry operation
Identity	E	Leave the molecule alone
Proper axis	C_n	Rotate the molecule by 360°/n around the axis
Horizontal plane	σ_h	Reflect the molecule through the plane which is perpendicular to the major axis
Vertical plane	σ_v	Reflect the molecule through a plane which contains the major axis
Dihedral plane	σ_d	Reflect the molecule through the plane which bisects two C_2 axes
Improper axis	S_n	Rotate the molecule by 360°/n around the improper axis and then reflect the molecule through the plane which is perpendicular to the improper axis
Inversion centre or centre of symmetry	i	Invert the molecule through the inversion centre

symmetry operation of leaving the molecule alone seems too trivial a matter to have any importance. The importance of E is that it is essential, for group theoretical purposes, for a group to contain it. For example, it expresses the result of performing some operations twice — the double reflexion of a molecule in any particular plane of symmetry, for instance. Such action restores every atom of the molecule to its original position so that it is equal to the performance of the operation of leaving the molecule alone — expressed by E.

2. Proper axes of symmetry, C_n

A proper axis of symmetry, denoted by C_n, is an axis around which a molecule is rotated by 360°/n to produce an equivalent configuration. The trigonally planar molecule, BF_3, may be set up so that the molecular plane is contained by the xy Cartesian plane (that containing the x and y axes) and so that the z Cartesian axis passes through the centre of the boron nucleus as is shown in Fig. 2.1. If the molecule is rotated around the z axis by 120° (360°/3), an equivalent configuration of the molecule is produced. The boron atom does not change its position, and the fluorine atoms exchange places depending upon the direction of the rotation. The rotation described is the symmetry operation associated with the C_3 axis of symmetry, and the demonstration of its production of an equivalent configuration of the BF_3 molecule is what is required to indicate that the C_3 proper axis of symmetry is possessed by that molecule.

There are other proper axes of symmetry possessed by the BF_3 molecule. The three lines joining the boron and fluorine nuclei are all contained by C_2 axes (from hereon the term 'proper' is dropped unless it is absolutely necessary to remove possible

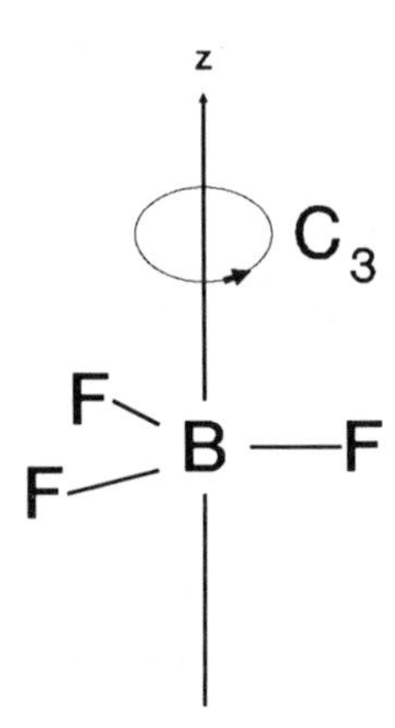

Fig 2.1 — The C_3 proper axis of the boron trifluoride molecule.

confusion) as may be seen from Fig. 2.2. The associated symmetry operation of rotating the molecule around one of the C_2 axes by $360°/2 = 180$ produces an equivalent configuration of the molecule. The boron atom and one of the fluorine atoms do not move whilst the other two fluorine atoms exchange places. There are, then, three C_2 axes of symmetry possessed by the BF_3 molecule.

The value of the subscript, n, in the symbol, C_n, for a proper axis of symmetry, is known as the **order** of that axis. The axis (and there may be more than one) of highest order possessed by a molecule is termed the **major** axis. The concept of the major axis is important in distinguishing between horizontal and vertical axes of symmetry. It is also important in the diagnosis of whether a molecule belongs to a C group or a D group — terms which will be defined at a later stage. As is the case with the C_3 axis of BF_3, the axis of symmetry coincides with one of the Cartesian axes (z), but that is because of the manner in which the diagram in Fig. 2.1 was drawn. It is a **convention** that the major axis of symmetry should be coincident with the Cartesian z axis. It is not necessary for there to be any coincidences between the axes of symmetry of a molecule and the Cartesian axes, but it is of considerable convenience if there is at least one coincidence.

Planes of symmetry, σ

There are three types of plane of symmetry, all denoted by the Greek lower case sigma, σ, and *all* of them are such that reflexion of the molecule through them produces equivalent configurations of that molecule. Reflexion through the plane is the symmetry operation associated with a plane of symmetry.

3. Horizontal planes, σ_h

A horizontal plane of symmetry (denoted by σ_h) is one that is perpendicular to the major axis of symmetry of a molecule. The molecular plane of the BF_3 molecule is an example of a horizontal plane - it is perpendicular to the C_3 axis. Reflexion in the horizontal plane of the BF_3 molecule has no effect upon any of the four atoms. A better example is the PCl_5 molecule (shown in Fig. 2.3) which, since it is trigonally

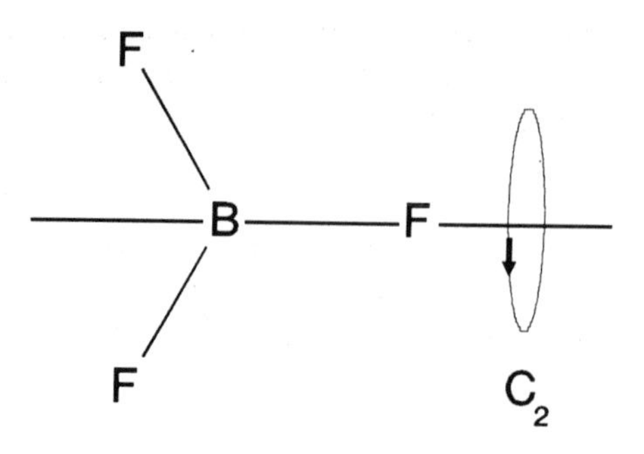

Fig 2.2 — One of the C_2 axes of the boron trifluoride molecule.

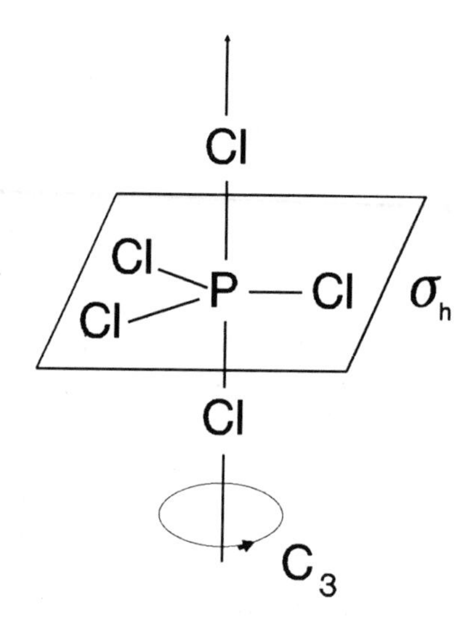

Fig 2.3 — The C_3 axis and the horizontal plane in PCl_5.

bipyramidal, possesses a C_3 axis (again arranged to coincide with the z axis) and a horizontal plane (the xy plane as it is set up) which contains the phosphorus atom and the three chlorine atoms of the trigonal plane. Reflexion through the horizontal plane causes the apical (out-of-plane) chlorine atoms to exchange places in producing an equivalent configuration of the molecule.

4. *Vertical planes, σ_v*

A vertical plane of symmetry (denoted by σ_v) is one which contains the major axis. The BF_3 and PCl_5 molecules both possess three vertical planes of symmetry. These contain the C_3 axis and, in the BF_3 case, the boron atom and one each of the fluorine atoms respectively. In the PCl_5 case the three vertical planes, σ_v, contain the C_3 axis, the phosphorus atom, the two apical chlorine atoms and one each of the chlorine atoms of the trigonal plane respectively.

5. Dihedral planes, σ_d

A dihedral plane of symmetry (denoted by σ_d) is one which bisects two C_2 axes of symmetry. In addition it contains the major axis and so is a special type of vertical plane. An example is shown in Fig. 2.4, which contains a diagram of the square

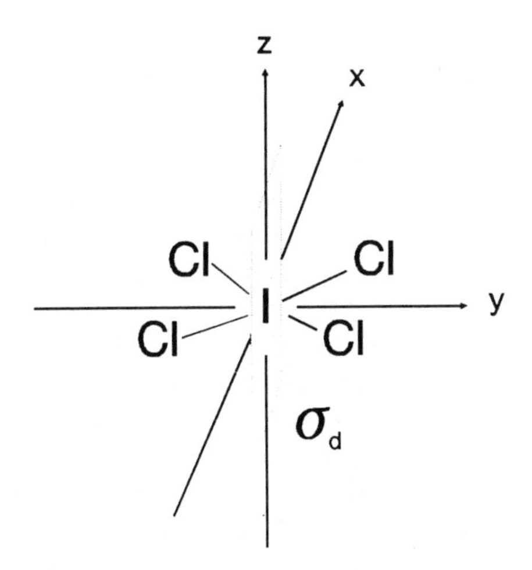

Fig 2.4 — One of the dihedral planes of the ICl_4^- ion.

planar ICl_4^- ion. As may be seen from Fig. 2.4 the ICl_4^- ion has a major axis which is a C_4 axis. The C_4 axis is *also* a C_2 axis in that rotations around it of 90° and 180° both produce equivalent configurations of the molecule. A rotation through 270° also produces an equivalent configuration but that is equivalent to a rotation through 90° in the opposite direction and so does not indicate an extra type of axis of symmetry. The molecular plane contains two C_2' axes (the prime is used to distinguish them from the C_2 axis which is coincident with the C_4 axis) both of which contain the iodine atom and two diametrically opposed chlorine atoms. The two C_2' axes are contained respectively by the two vertical planes which also contain the major axis. There are, in addition, two C_2'' axes which are contained by the horizontal plane, are perpendicular to the major axis, and also bisect the respective Cl–I– Cl angles. The double primes serve to distinguish these axes from the C_2 and C_2' axes. The C_2'' axes are contained by two extra planes of symmetry which are termed dihedral planes since they bisect the two C_2' axes. The dihedral planes contain the major axis.

6. Improper axes of symmetry, S_n

If rotation about an axis by 360°/n followed by reflexion through a plane perpendicular to the axis produces an equivalent configuration of a molecule, then the molecule contains an improper axis of symmetry. Such an axis is denoted by S_n, the associated symmetry operation having been described. The C_3 axis of the PCl_5 molecule is also an S_3 axis. The operation of S_3 on PCl_5 causes the apical chlorine atoms to exchange places.

The operation of reflexion through a horizontal plane may be regarded as a special case of an improper axis of symmetry of order one — S_1. The rotation of a molecule around an axis by 360°produces an identical configuration ($C_1 = E$) and the reflexion in the horizontal plane is the only non-trivial part of the operations associated with the S_1 improper axis. This may be symbolized as:

$$S_1 = C_1 \times \sigma_h = E \times \sigma_h = \sigma_h \tag{2.1}$$

7. *The inversion centre or centre of symmetry, i*

An inversion centre (denoted by *i*) is possessed by a molecule which has pairs of identical atoms which are diametrically opposed to each other about the centre. Any particular atom with the coordinates x, y, z must be partnered by an identical atom with the coordinates $-x$, $-y$, $-z$, if the molecule is to possess an inversion centre. That condition must apply to all atoms in the molecule which are off-centre. The ICl_4^- ion possesses an inversion centre, as does the dihydrogen molecule, H_2. The BF_3 molecule (trigonally planar) does not possess an inversion centre. The symmetry operation associated with an inversion centre is the inversion of the molecule in which the diametrically opposed atoms in each such pair exchange places. The inversion centre is another special case of an improper axis in that the operations associated with the S_2 element are exactly those which produce the inversion of the atoms of a molecule containing *i*. Reference to Fig. 2.5 shows the effects of carrying

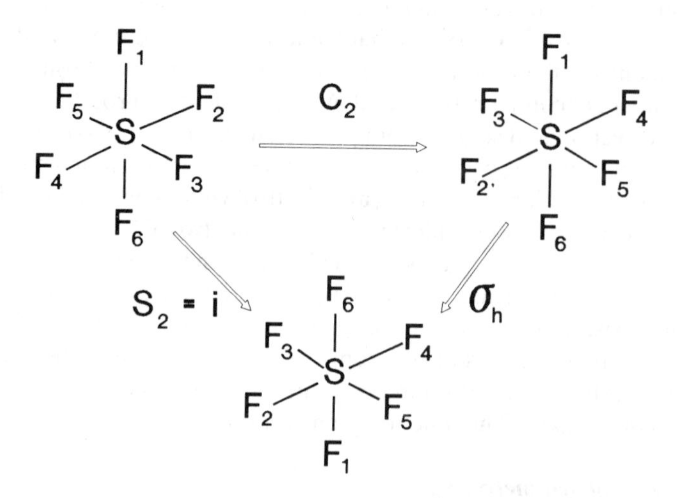

Fig 2.5 — The inversion of the SF_6 molecule by two paths.

out the operation, C_2, followed by σ_h, and that the fluorine atoms in each of the three diametrically opposed pairs in the octahedral SF_6 molecule have been exchanged with each other — all of which is what is understood to be the inversion of the molecule. This is symbolized as:

$$S_2 = C_2 \times \sigma_h = i \tag{2.2}$$

The subscript numbering of the fluorine atoms in Fig. 2.5 is done to make clear the atomic movements which take place as a result of the application of the symmetry operations.

2.3 POINT GROUPS

The symmetry elements which may be possessed by a molecule are defined above and exemplified. The next stage is to decide which, and how many, of these elements are possessed by particular molecules so that the molecules can be assigned to **point groups**. A point group consists of all the elements of symmetry possessed by a molecule and which intersect at a point. Such elements represent a group according to the rules to be outlined.

A simple example is the water molecule, which is a bent triatomic system. Its symmetry elements are easily detected. There is only one proper axis of symmetry, which is that which bisects the bond angle and contains the oxygen atom. It is a C_2 axis, and the associated operation of rotating the molecule about the axis by 180° results in the hydrogen atoms exchanging places with each other. The demonstration of the effectiveness of the operation is sufficient for the diagnosis of the presence of the element.

Fig. 2.6 shows the molecule of water set up so that the C_2 axis is coincident with

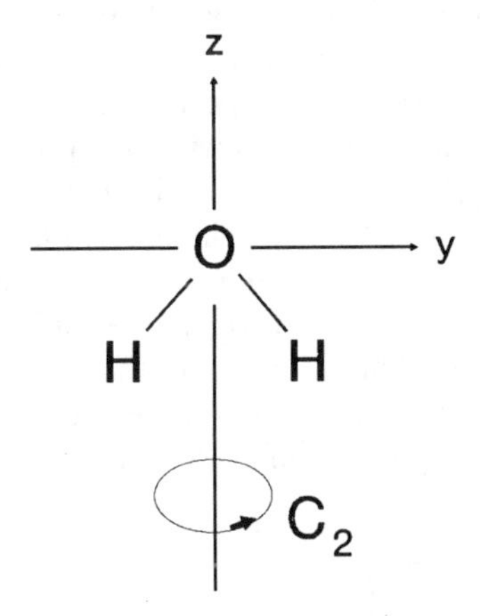

Fig 2.6 — The water molecule in the *yz* plane.

the *z* axis, as is conventional. The molecular plane is setup to be in the *yz* plane so that the *x* axis is perpendicular to the paper, as is the xz plane. There are two vertical planes of symmetry — the *xz* and the *yz* planes — and these are designated as σ_v and

σ_v' respectively, the prime serving to distinguish between the two. The only other symmetry element possessed by the water molecule is the identity, E.

The four symmetry elements form a group, which may be demonstrated by introducing the appropriate rules. The rules are exemplified by considering the orbitals of the atoms present in the molecule. Such a consideration also develops the relevance of group theory in that it leads to an understanding of which of the atomic orbitals are permitted to combine to form molecular orbitals.

The electrons which are important for the bonding in the water molecule are those in the **valence shell**: $2s^2 2p^4$. It is essential to explore the **character** of the 2s and 2p orbitals, and this is done by deciding how each orbital **transforms** with respect to the operations associated with each of the symmetry elements possessed by the water molecule.

The character of the 2s orbital of the oxygen atom

The character of an orbital is symbolized by a number which expresses the result of any particular operation on its wave function. In the case of the 2s orbital of the oxygen atom (whose ψ values are everywhere positive) there is no change of sign of ψ with any of the four operations: E, $C_2\,(z)$, $\sigma_v\,(xz)$ and $\sigma_v'(yz)$. These results may be written down in the form:

	E	C_2	σ_v	σ_v'
2s	1	1	1	1

the 1s indicating that the 2s orbital is **symmetric** with respect to the individual operations. Such a collection of characters is termed a **representation**, this particular example being the totally symmetric representation.

The character of the $2p_x$ orbital of the oxygen atom

The $2p_x$ orbital of the oxygen atom is symmetric with respect to the identity, but is **antisymmetric** with respect to the C_2 operation. This is because of the spatial distribution of ψ values in the 2p orbitals with one positive lobe and one negative lobe. The operation, C_2, causes the positive and negative regions of the $2p_x$ orbital to exchange places with each other. This sign change is indicated as a character of -1 in the representation of the $2p_x$ orbital as far as the C_2 operation is concerned. The $2p_x$ orbital is unchanged by reflexion in the xz plane but suffers a sign change when reflected through the yz plane. The collection of the characters of the $2p_x$ orbital with respect to the four symmetry operations associated with the four elements of symmetry forms another representation of the group currently being constructed:

	E	C_2	σ_v	σ_v'
$2p_y$	1	-1	-1	1

The character of the $2p_y$ orbital of the oxygen atom

The $2p_y$ orbital of the oxygen atom changes sign when the C_2 operation is applied to it, and when it is reflected through the xz plane, but is symmetric with respect to the molecular plane, yz. The representation expressing the character of the $2p_y$ orbital is another member of the group:

	E	C_2	σ_v	σ'_v
$2p_y$	1	-1	-1	1

The character of the $2p_z$ orbital of the oxygen atom

The $2p_z$ orbital of the oxygen atom has exactly the same set of characters as the 2s orbital — it is another example of a totally symmetric representation.

The multiplication of representations

At this stage it is important to use one of the rules of group theory which states that the product of any two representations of a group must also be a member of that group. This rule may be used on the examples of the representations for the $2p_x$ and $2p_y$ orbitals as deduced above. The product of two representations is obtained by multiplying together the individual characters for each symmetry element of the group. The normal rules of arithmetic apply so that the representation of the product of those of the two 2p orbitals under discussion is given by:

	E	C_2	σ_v	σ'_v
$2p_x$	1	-1	1	-1
$2p_y$	1	-1	-1	1
$2p_x.2p_y$	1	1	-1	-1

The new representation is, by the rules, a member of the group and is also the only possible addition to the set of representations so far deduced for the molecular shape under consideration. This statement may be checked by trying other double products amongst the four representations. The four representations may be collected together and be given symbols, the collection of characters being termed a **character table**:

	E	C_2	σ_v	σ'_v
A_1	1	1	1	1
A_2	1	1	-1	-1
B_1	1	-1	1	-1
B_2	1	-1	-1	1

The symbols used for the representations are those proposed by Mulliken. The A representations are those which are symmetric with respect to the C_2 operation, and the Bs are antisymmetric to that operation. The subscript 1 indicates that a representation is symmetric with respect to the σ_v operation, the subscript 2 indicating antisymmetry to it. No other indications are required since the characters in the σ_v' column are decided by another rule of group theory. This rule is: the product of any two columns of a character table must also be a column in that table. It may be seen that the product of the C_2 characters and those of σ_v give the contents of the σ_v' column.

The representations deduced above must be described as **irreducible representations**. They cannot be simpler, but there are other representations (examples follow) which are reducible in that they are sums of irreducible ones. Character tables, in general, are a list of the irreducible representations of the particular group, and as in the ones shown in Appendix I of this book they have an extra column which indicates the representations by which various orbitals **transform**, and, symbolized by R_x, R_y and R_z , the transformation properties of **rotations** about those respective axes.

In the example under discussion the 2s and $2p_z$ orbitals transform as a_1 , the $2p_x$ orbital transforms as b_1 and the $2p_y$ orbital transforms as b_1. In this context 'transform' refers to the behaviour of the orbitals with respect to the symmetry operations associated with the symmetry elements of the particular group. It should be noticed that lower case Mulliken symbols are used to indicate the irreducible representations of *orbitals*. The upper case Mulliken symbols are reserved for the description of the symmetry properties of *electronic states*.

Symmetry properties of the hydrogen 1s orbitals

Neither of the 1s orbitals of the hydrogen atoms, taken separately, transforms within the group of irreducible representations deduced for the water molecule. The two 1s orbitals must be taken together as one or the other of two **group orbitals**. A more formal treatment of the group orbitals which two 1s orbitals may form is dealt with in Chapter 3.

Whenever there are two or more identical atoms linked to a central atom their wavefunctions must be combined in such a way as to demonstrate their indistinguishability. This is achieved by making linear combinations of the wavefunctions of the ligand atoms. For the two 1s orbitals of the hydrogen atoms in the water molecule their wavefunctions may be combined to give:

$$h_1 = 1s_A + 1s_B \tag{2.3}$$

and

$$h_2 = 1s_A - 1s_B \tag{2.4}$$

where h_1 and h_2 are the wavefunctions of the group orbitals and $1s_A$ and $1s_B$ represent the 1s wavefunctions of the two hydrogen atoms, A and B. The two group orbitals are shown diagrammatically in Fig. 2.7. By inspection they may be shown to transform as a_1 and b_2 representations respectively. The h_1 orbital behaves exactly

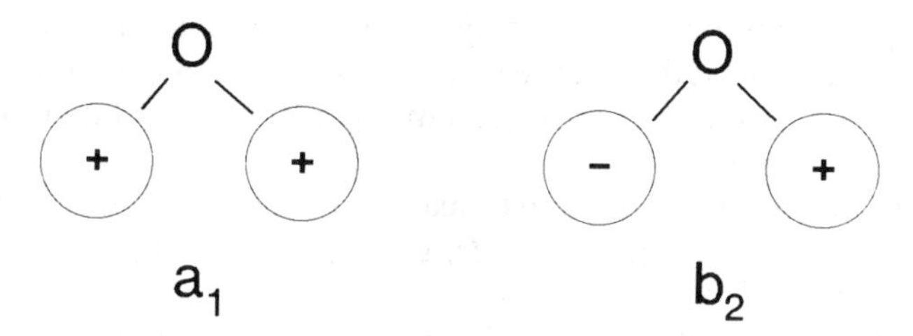

Fig 2.7 — The two group orbitals of the hydrogen atoms in the water molecule.

like the 2s and $2p_z$ orbitals of the oxygen atom, and the h_2 orbital behaves exactly like the $2p_y$ orbital of the oxygen atom, with respect to the four symmetry operations of the point group.

2.4 LABELLING OF POINT GROUPS AND CHARACTER TABLES

It is clearly necessary to label the various point groups to which molecules may belong. The labelling system used is to use a letter which is related to the major axis, and to use the value of n (the order of the major axis) together with a letter indicating the plane of symmetry (h, v or d) of highest importance for descriptive purposes as subscripts. The system was suggested by Schoenflies, and the labels are known as Schoenflies symbols.

The normal method of deciding the point group of a molecule is that described below, and which, after practice with examples, is accurate.

Assignment of point groups to molecules.

There are three shapes which are important in chemistry and are easily recognized by the number of their faces, all of which consist of equilateral triangles. They are the tetrahedron (four faces), the octahedron (six faces) and the icosahedron (twenty faces). The first two of these shapes are extremely common in chemistry, whilst the third shape is of some importance in boron chemistry and some clusters. The three special shapes are associated with point groups and their character tables, and are labelled T_d, O_h and I_h respectively.

The point group to which a molecule belongs may be decided by the answers to four main questions.

(1) Does the molecule belong to one of the special point groups, T_d, O_h or I_h? If it does, the point group has been identified.

(2) If the molecule does not belong to one of the special groups, it becomes necessary to identify the major axis (or axes).

The major axis is the one with the highest order — the highest value of n — and is designated C_n. In a case where there is more than one axis which could be classified as major (of equal values of n), it is conventional to regard the axis placed along the z axis as being the major one. In the trivial cases where there is only a C_1 axis, the point group of the molecule is C_1, unless there is either a plane of symmetry or an inversion centre present, indicating the point groups C_s or C_i respectively.

The main question is, are there n C_2 axes perpendicular to the major axis, C_n? If there are, the molecule belongs to a D_n group; otherwise the molecule belongs to a C_n group.

(3) The third question applies only to D_n groups: is there a horizontal plane of symmetry present? If σ_h is present, the molecule belongs to the point group D_{nh}. If there is no σ_h but there are $n\sigma_v$ present, the molecule belongs to the point group, D_{nd}. If no σ_v are present then the point group of the molecule is D_3.

(4) The fourth question applies only to C_n groups: is there a σ_h present? If there is, the molecule belongs to the point group, C_{nh}. If there is no σ_h but there are $n\sigma_v$ present, the molecule belongs to the point group, C_{nv}. If no σ_v are present, but there is an S_{2n} improper axis present, the molecule belongs to the S_{2n} point group. If there is no S_{2n} the point group of the molecule is C_n.

The above procedure will identify the point groups of the vast majority of molecules for which there is any reason so to do.

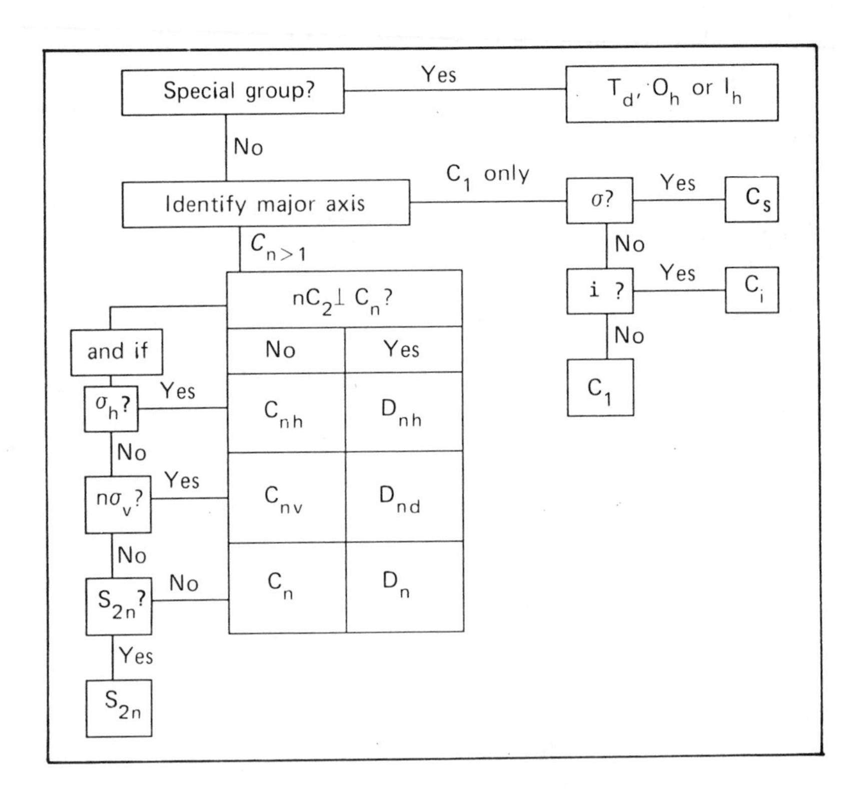

Fig 2.8 — A flow sheet for the assignment of molecules to point groups.

The procedure may be written in the form of a flow sheet which is shown in Fig. 2.8 and allows the assignment of molecules to point groups. The most frequently used character tables are to be found in Appendix I.

2.5 CONCLUDING SUMMARY

This chapter contains an introduction to group theory and to symmetry elements and operations.

The reader should be able to (i) identify elements of symmetry possessed by a molecule, (ii) assign a molecule to its correct point group, and (iii) select the appropriate character table.

3

Covalent bonding in diatomic molecules

3.1 INTRODUCTION

This chapter is concerned with the application of the symmetry concepts of Chapter 2 to the production of molecular orbitals of a range of diatomic molecules. **Molecular orbital theory** is developed in the discussion of the bonding of three sets of compounds. The basis of the theory is described in the treatment of the dihydrogen molecule-ion, H_2^+, and the dihydrogen molecule. It is further developed in a section on the homonuclear diatomic molecules, A_2, where A is a member of the first short period of the periodic classification (Li–Ne). The third section is concerned with some heteronuclear diatomic molecules where differences in electronegativity can be important.

3.2 COVALENT BONDING IN H_2^+ AND H_2

The simplest covalent bond is that which occurs in the molecule-ion, H_2^+, which consists of two protons and one electron. Even so, it represents a quantum mechanical three-body problem, and solutions of the wave equation are obtained by iterative methods. The molecular orbitals derived from the combination of two 1s atomic orbitals serve to describe the electronic configurations of the species: H_2^+, H_2, He_2^+ and He_2.

Production of Molecular Orbitals

The basic concept of **molecular orbital theory** is that molecular orbitals may be constructed from a set of contributing atomic orbitals such that the molecular wavefunctions consist of **linear combinations of atomic orbitals**. In the case of the combination of two 1s hydrogen atomic orbitals to give two molecular orbitals, the two linear combinations are those already proposed for the two hydrogen atoms of the water molecule and given by equations (2.3) and (2.4). They are rewritten below in a modified fashion so that atomic wavefunctions are represented by ψ and molecular wavefunctions by ϕ:

$$\phi_1 = \psi_A + \psi_B \tag{3.1}$$

$$\phi_2 = \psi_A - \psi_B \tag{3.2}$$

where ψ_A and ψ_B are the two hydrogen atomic 1s wavefunctions of atoms A and B respectively.

The H_2^+ and H_2 molecules belong to the $D_{\infty h}$ point group. The two 1s atomic orbitals, individually, do not transform within the $D_{\infty h}$ point group, but together their character may be elucidated. This is done by considering each of the symmetry elements of the $D_{\infty h}$ group in turn, and writing down under each element **the number of orbitals unaffected** by the associated symmetry operation:

	E	C_∞^ϕ	σ_v	i	S_∞^ϕ	C_2
$\psi_A + \psi_B$	2	2	2	0	0	0

If the two 1s orbitals are left alone by the E operation they will be unaffected by it and hence the number 2 is written down in the above table. Rotation by any angle, ϕ, around the C_∞ axis does not affect the orbitals — hence the second 2 appears as the character of the two 1s orbitals. The third 2 appears because the two orbitals are unaffected by reflexion in any of the infinite number of vertical planes which contain the molecular axis. The operation of inversion affects both orbitals in that they exchange places with each other and so a zero is written down in the i column. Likewise an S_∞ operation causes the orbitals to exchange places and a zero is written in that column. There are an infinite number of C_2 axes which pass through the inversion centre and are perpendicular to the molecular axis. The associated operation of rotation through 180° around any C_2 axis causes the 1s orbitals to exchange places with each other so that there is a final zero to be placed in the table above.

The logic underlying the above derivation of the character of the two 1s orbitals arises from a consideration of the **transformation matrix** for each symmetry operation. There are two vectors, υ_1 and υ_2, which may be considered to represent the displacement of the two hydrogen atoms from the centre of the molecule. These may, or may not, be altered by any symmetry operation. If the C_∞ operation is carried out, the transformation matrix is produced as follows:

$$\upsilon_1 = \upsilon_1 + 0\upsilon_2 \tag{3.3}$$

$$\upsilon_2 = 0\upsilon_1 + \upsilon_2 \tag{3.4}$$

i.e. the matrix is

$$\begin{vmatrix} 1 & 0 \\ 0 & 1 \end{vmatrix}$$

which has a value of 2 for the sum of the numbers on the top-left to bottom-right diagonal and represents the character of the two 1s hydrogen orbitals with respect to the C_∞ operation. Similarly the matrix representing the effects of the i operation is generated from the equations:

$$\upsilon_1 = 0\upsilon_1 + \upsilon_2 \tag{3.5}$$

$$\upsilon_2 = \upsilon_1 + 0\upsilon_2 \tag{3.6}$$

and is

$$\begin{vmatrix} 0 & 1 \\ 1 & 0 \end{vmatrix}$$

the top-left to bottom-right diagonal, giving a zero value for the character.

The method of writing down the number of atoms (or orbitals if that is what is being considered) unaffected by the operation is a quick way of arriving at the result. There are some complications which can arise from such methods, and examples are dealt with when they arise.

The sequence of numbers arrived at represent the character of the two 1s orbitals with respect to $D_{\infty h}$ symmetry. Such a combination of numbers is not to be found in the $D_{\infty h}$ character table — it is an example of a **reducible representation**. Its reduction to a sum of irreducible representations is, in this instance, a matter of realizing that the sum of the σ_g^+ and σ_u^+ characters represents the character of the two 1s orbitals:

	E	C_∞^ϕ	σ_v	i	S_∞^ϕ	C_2
σ_g^+	1	1	1	1	1	1
σ_u^+	1	1	1	−1	−1	−1
$\sigma_g^+ + \sigma_u^+$	2	2	2	0	0	0

Lower case letters are used for the symbols representing the symmetry properties of orbitals. Greek letters are used to symbolize the irreducible representations of the $D_{\infty h}$ point group with g or u subscripts, and with + and − signs as superscripts. The g and u subscripts refer to the character in the i column — g (German **g**erade meaning even) indicating symmetrical behaviour and u (**u**ngerade meaning odd) indicating antisymmetrical behaviour with respect to inversion. The + and − signs refer respectively to symmetry and antisymmetry with respect to reflexion in one of the vertical planes. The 2s in the E and some of the other columns are indications of doubly degenerate representations.

By referring to the diagrams in Fig. 2.7 it may be seen that the h_1 orbital, now referred to as σ_1, transforms as σ_g^+, and that the h_2 orbital, now referred to as σ_2, transforms as σ_u^+. The only differences between the h_1, h_2 pair of orbitals and the σ_1, σ_2 pair are that they are closer together in the H_2^+ and H_2 cases, and that they have different symmetry symbols because they are participating in molecules belonging to different point groups.

That two molecular orbitals are produced from the two atomic orbitals is an important part of molecular orbital theory — a law of conservation of orbital numbers. The two molecular orbitals differ in energy, both from each other and from the energy of the atomic level. To understand how this arises it is essential to consider the **normalization** of the orbitals. Normalization is the procedure of arranging for the integral over all space of the square of the orbital wavefunction to be unity. This is expressed by the equation:

$$\int_0^\infty \phi^2 \, d\tau = 1 \tag{3.7}$$

where $d\tau$ is a volume element (equal to $dx.dy.dz$).

The probability of unity expresses the certainty of finding an electron in the orbital. It must be the case that by transforming atomic orbitals into molecular ones that no loss, or gain, in electron probability should occur. For equation (3.7) to be valid, a **normalization factor**, N, must be introduced into equations (3.1) and (3.2), which then become:

$$\phi_1 \; (\sigma_g^+) = N_1(\psi_A + \psi_B) \tag{3.8}$$

and

$$\phi_2 \; (\sigma_g^+) = N_2(\psi_A + \psi_B) \tag{3.9}$$

where N_1 and N_2 are normalization factors.

To determine the value of N_1 in equation (3.8) the expression for ϕ_1 must be placed into equation (3.7), giving:

$$N_1^2 \int_0^\infty (\psi_A + \psi_B)^2 \, d\tau = 1 \tag{3.10}$$

and expanding the square term in the integral gives:

$$N_1^2 \int_0^\infty (\psi_A^2 + 2\psi_A\psi_B + \psi_B^2) \, d\tau = 1 \tag{3.11}$$

which may be written as three separate integrals:

$$N_1^2\left(\int_0^\infty \psi_A^2 \, d\tau + 2\int_0^\infty \psi_A\psi_B \, d\tau + \int_0^\infty \psi_B^2 \, d\tau\right) = 1 \tag{3.12}$$

The assumption that the atomic orbital wavefunctions are separately normalized leads to the conclusion that the first and third integrals in equation (3.12) are both equal to unity. The second integral is known as the **overlap integral**, symbolized by S. This allows equation (3.12) to be simplified to:

$$N_1^2\,(1 + 2S + 1) = 1 \tag{3.13}$$

which gives a value for N_1 of:

$$N_1 = 1/(2 + 2S)^{1/2} \tag{3.14}$$

A similar treatment of equation (3.9) gives a value for N_2 of:

$$N_2 = 1/(2 - 2S)^{1/2} \tag{3.15}$$

The next stage in the full description of the molecular orbitals is to calculate their energies. This is done by considering the Schrödinger equation for molecular wavefunctions:

$$H\phi = E\phi \tag{3.16}$$

If both sides are premultiplied by ϕ (this is essential for equations containing operators such as H) this gives:

$$\phi H\phi = \phi E\phi = E\phi^2 \tag{3.17}$$

there being no difference between $E\phi^2$ and $\phi E\phi$ since E is not an operator. Equation (3.17) may be integrated over all space to give:

$$\int_0^\infty \phi H\phi \, d\tau = E\int_0^\infty \phi^2 \, d\tau = E \tag{3.18}$$

since the integral on the right-hand side is equal to unity for normalized orbitals. Equation (3.18) may be used to calculate the energy of the molecular orbital, ϕ_1, by substituting its value from equation (3.8):

$$E_{\phi_1} = N_1^2\int_0^\infty (\psi_A + \psi_B)H(\psi_A + \psi_B) \, d\tau$$

$$= N_1^2\left(\int_0^\infty \psi_A H \psi_A \,d\tau + \int_0^\infty \psi_B H \psi_B \,d\tau + 2\int_0^\infty \psi_A H \psi_B \,d\tau\right)$$
$$= N_1^2(\alpha + \alpha + 2\beta)$$
$$= 2N_1^2(\alpha + \beta) \quad (3.19)$$

The first two integrals are entirely concerned with atomic orbitals, ψ_A and ψ_B, respectively and have identical values (since they refer to identical orbitals) which are put equal to α, which is a quantity known as the **Coulomb integral**. In essence it is the energy of an electron in the 1s orbital of the hydrogen atom and equal to the ionization energy of that atom. The third integral is really the sum of two identical integrals (again because ψ_A and ψ_B are identical) and is put equal to 2β, where β is called the **resonance integral**. β represents the extra energy gained by an electron, over that it possesses in any case by being in the 1s atomic orbital of the hydrogen atom, when it occupies the molecular orbital, ϕ_1. Because the electron is more stable in σ_1 than it is in ψ_A or ψ_B, ϕ_1 is called a **bonding** molecular orbital. Occupancy of ϕ_1 leads to the stabilization of the system.

If the value of N_1 from equation (3.14) is substituted into equation (3.19), the expression for the energy of the bonding orbital becomes:

$$E_{\phi_1} = \frac{\alpha + \beta}{1 + S} \quad (3.20)$$

A similar treatment of ϕ_2 produces the equation:

$$E_{\phi_2} = \frac{\alpha - \beta}{1 - S} \quad (3.21)$$

which shows it to have a higher energy than ϕ_1, and which is also higher than that of ψ_A (or ψ_B) — it is therefore called an **anti-bonding molecular orbital**. Electrons in anti-bonding orbitals are less stable than in the atomic orbitals from which the molecular orbital was constructed. Such anti-bonding electrons contribute towards a weakening of the bonding of the molecule, or sometimes to complete dissociation of the molecule.

Fig. 3.1 shows the relative energies of the atomic orbitals, ψ_A and ψ_B, and the molecular orbitals, ϕ_1 and ϕ_2, which are involved in the production of H_2^+ and H_2. Both α and β are negative quantities on an energy scale with the ionization limit as the reference zero. The electronic configuration of the H_2^+ molecule-ion may be written as ϕ_1^1, or in symmetry symbols as $(\sigma_g^+)^1$. That of the dihydrogen molecule is ϕ_1^2, or $(\sigma_g^+)^2$, provided that the energy gap between ϕ_2 and ϕ_1 is sufficiently large to force the electrons to pair up in the bonding orbital. This must be so because the configuration $\phi_1^1\phi_2^1$ would lead to dissociation into the separate hydrogen atoms. The occupation of the bonding orbital by a pair of electrons is the simplest example of a **single covalent bond**.

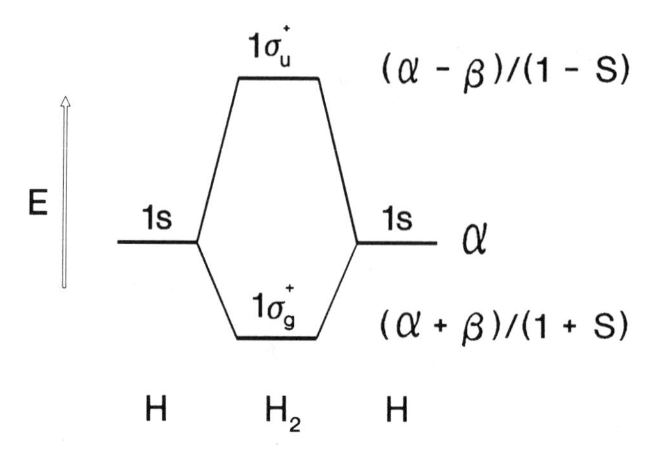

Fig 3.1 — Molecular orbital diagram for the dihydrogen molecule.

The molecular orbitals ϕ_1 and ϕ_2 may be used to describe the electronic configurations of the helium molecule-ion, He_2^+, and the dihelium molecule, He_2. The former is a three-electron case so that two electrons pair up in the bonding orbital, leaving one unpaired electron to occupy the anti-bonding orbital, $\phi_1^2\phi_2^1$ being the electronic configuration of He_2^+. The single anti-bonding electron offsets some of the bonding effect of the pair of electrons in the bonding orbital to give a bond with a strength about equal to that of the bond in H_2^+.

The electronic configuration of dihelium would be $\phi_1^2\phi_2^2$, resulting in zero bonding. The dihelium molecule does not exist. The bond dissociation energies of H_2^+, H_2 and He_2^+ are 264, 436 and 297, respectively, and are consistent with the expectations from molecular orbital theory.

3.3 ENERGETICS OF THE BONDING IN H_2^+ AND H_2

It is helpful in the understanding of covalent bond formation to consider the energies due to the operation of attractive and repulsive forces in H_2^+ and H_2, and to estimate the magnitude of the interelectronic repulsion energy in the dihydrogen molecule. Fig. 3.2 shows the Morse curves for H_2^+ and H_2, which are plots of their potential energies against internuclear distance. Under standard conditions both molecules exist in their zero-point vibrational states, i.e. in the lowest vibrational energy level. The dissociation limits for both species are identical — the complete separation of the two atoms, which is taken as an arbitrary zero of energy. The difference between the zero of energy and the zero-point vibrational energy in both cases represents the bond dissociation energies, respectively, of H_2^+ and H_2.

To obtain an accurate assessment of the interelectronic repulsion energy of the H_2 molecule it is essential to carry out calculations in which the hydrogen nuclei are a constant distance apart. The following calculations are for an internuclear distance of

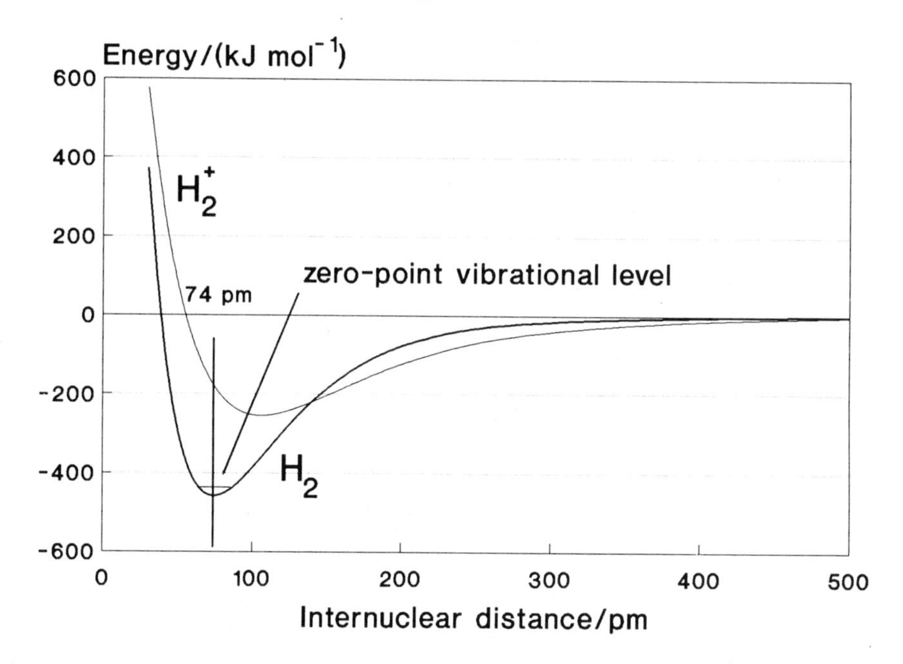

Fig 3.2 — Morse curves for the dihydrogen molecule and the dihydrogen molecule-ion.

74 pm for both molecules, which is the equilibrium internuclear distance in the dihydrogen molecule.

3.4 THE H_2^+ MOLECULE-ION

There are only two forces operating in H_2^+: the attractive force between the nuclei and the single electron, and the repulsive force between the two nuclei. The interproton repulsion energy may be calculated from Coulomb's law:

$$E_{\text{p–p}} = Ne^2/4\pi\varepsilon_0 r \tag{3.22}$$

where e is the charge on the proton, r is the interproton distance, ε_0 is the permittivity of free space and N is the Avogadro number.

For two protons separated by 74 pm the force of repulsion between them causes an increase in energy of:

$$\begin{aligned} E_{\text{p–p}} &= \frac{6.022 \times 10^{23} \times (1.602 \times 10^{-19})^2}{8.854 \times 10^{-12} \times 74 \times 10^{-12} \times 4\pi} \\ &= 1877 \text{ kJ mol}^{-1} \end{aligned} \tag{3.23}$$

compared to the infinite separation of H^+ and H (the arbitrary zero of energy). From the Morse curve for H_2^+ in Fig. 3.2 it may be estimated that if H^+ and H are brought from infinite separation to an interproton distance of 74 pm there is a stabilization of 180 kJ mol^{-1}. This represents the resultant energy of the system with both forces operating. It means that the attractive force operating between the electron and the two protons produces a stabilization which is in excess of 1877 kJ mol^{-1} by 180 kJ mol^{-1}, so that the quantity known as the **electronic binding energy** is calculated to be 1877 + 180 = 2057 kJ mol^{-1}. The interrelationship of these energies is shown in the diagram of Fig. 3.3. Notice that the actual dissociation energy is relatively small

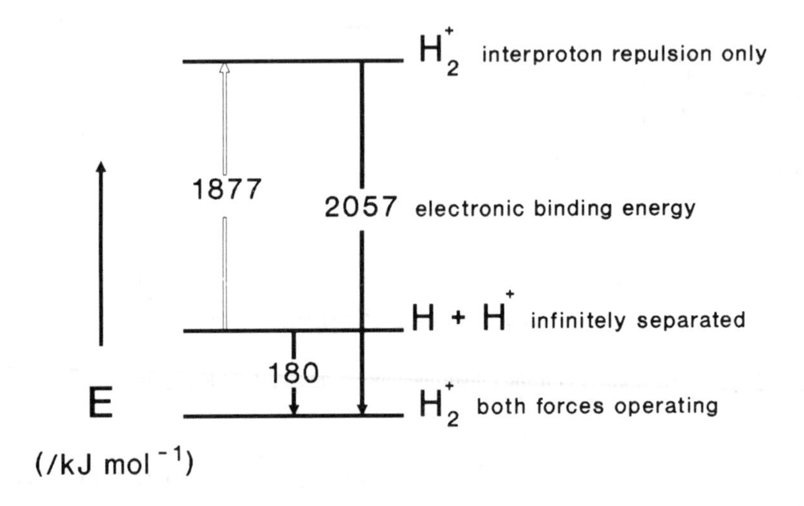

Fig 3.3 — Energetics of the formation of the dihydrogen molecule-ion.

compared to the energies representing the effects of the attractive and repulsive forces operating in the H_2^+ system.

3.5 THE DIHYDROGEN MOLECULE

In the dihydrogen molecule there are three forces operating:

(i) interproton repulsion,
(ii) proton–electron attraction and
(iii) interelectronic repulsion.

The force of interproton repulsion produces a destabilization of the dihydrogen system equal to that of the H_2^+ molecule-ion since the interproton distance is taken to be 74 pm. The resultant stabilization of all three forces is equal to the bond dissociation energy of H_2, which is 436 kJ mol^{-1}. For a comparison with H_2^+ the electronic binding energy may be calculated as: 1877 + 436 = 2313 kJ mol^{-1}. It is 12% greater than that in H_2^+ indicating that two bonding electrons are only marginally better than one at binding the two nuclei together. The reason for this is that, with two electrons present, there is a substantial destabilization of the system as a result of the interelectronic repulsion.

The magnitude of this may be calculated by assuming that the electronic binding energy per electron is as calculated for the H_2^+ system (2057 kJ mol^{-1}). For two electrons the stabilization from the electronic binding energy is $2 \times 2057 = 4114$ kJ mol^{-1} and that amount is offset by the interelectronic repulsion energy so that the resultant is 2313 kJ mol^{-1} to give $4114 - 2313 = 1801$ kJ mol^{-1} as representing the interelectronic repulsion energy in H_2. Notice that all three energy quantities in H_2 are large in comparison with the resultant bond dissociation energy. The above calculations are represented diagrammatically in Fig. 3.4.

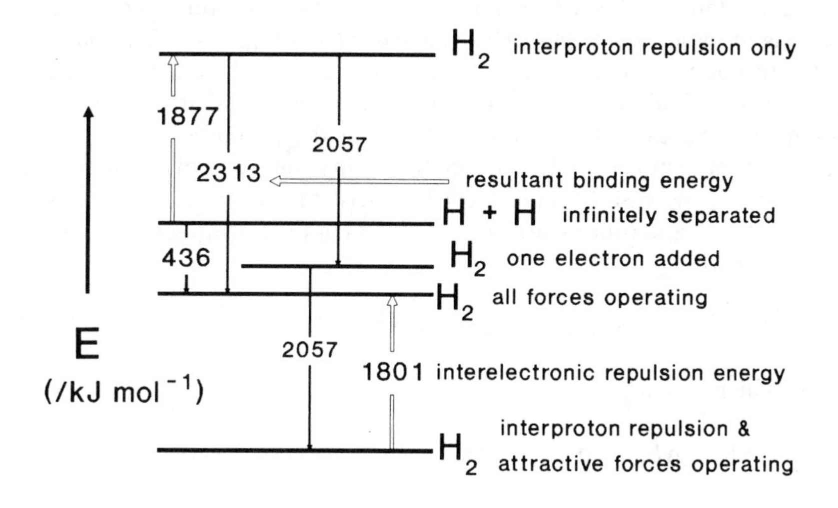

Fig 3.4 — Energetics of the formation of the dihydrogen molecule.

It is of interest to calculate the magnitude of the interelectronic repulsion energy in the hydride ion, H^-, which possesses an ionic radius of 208 pm. The electron affinity of the proton is -1312 kJ mol^{-1} (numerically equal to the ionization energy of the hydrogen atom) and represents the energy released when an electron enters the 1s orbital of the hydrogen atom. The electron affinity of the hydrogen atom is considerably smaller than that of the proton and is -71 kJ mol^{-1}. Since the electron enters the same 1s orbital the difference between the two electron affinities gives an estimate of the interelectronic repulsion energy of $-71 + 1312 = 1241$ kJ mol^{-1}. This is appreciably smaller than the value calculated for the two electrons occupying the bonding orbital of the dihydrogen molecule. The increased size of the hydride ion is one reason for this, the other being that in H_2 there are two attracting protons which draw the electrons closer to each other in spite of their like charges. The other case for which interelectronic repulsion energy has been estimated is for the helium atom (Chapter 1) where the value was calculated to be 2878 kJ mol^{-1}. Such a high value is due to the double charge on the single nucleus which has a greater attractive effect upon the two electrons than do the two singly charged, and spaced out, nuclei in the case of H_2.

3.6 SOME EXPERIMENTAL OBSERVATIONS

In the proper development of a scientific theory it is to be expected that theory and experimental observations are to be consistent with one another. It is essential to obtain experimental observations by which ideas such as molecular orbital theory may be tested. As put forward by Popper it is only by trying to invalidate theories that experimentation serves to refine and improve such ideas.

One very helpful method of obtaining experimental measurement of the energies of electrons in molecules is **photoelectron spectroscopy** (p.e.s.). The basis of the method is to bombard an atomic or molecular species with radiation of sufficient energy to cause its ionization. If the quantum energy of the radiation is high enough, ionizations may be caused from one or other of the permitted levels within the bombarded atom or molecule. In addition to causing ionization of the target species, the radiation (in the case of molecules) bombardment may cause changes in the vibrational, ΔE_{vib}, and rotational, ΔE_{rot}, energies of the resulting positive ion. This may be written as:

$$M(g) + h\upsilon \rightarrow M^{+}(g) + e^{-}(\text{photoelectron}) \quad (3.24)$$

the energy balance being:

$$h\upsilon = I_M + \Delta E_{vib} + \Delta E_{rot} + \text{K.E.} \quad (3.25)$$

The kinetic energies (K.E.) of the photoelectrons are measured by the use of a modification of a conventional β-ray spectrometer as used in the study of β-ray (electron) emissions from radioactive nuclei. The photoelectron spectrum of a molecule is presented as a plot of the count rate (the intensity of the photoelectrons detected) against the electron energy. The p.e.s. for the dihydrogen molecule is shown in Fig. 3.5.

The energy of the quantum used to cause the ionization is 21.22 eV (1 eV = 96.487 kJ mol^{-1}) so that the ionization energy of the dihydrogen molecule is 21.22 minus the energy of the photoelectron (5.8 eV), which gives 15.42 eV as the first ionization energy of H_2. The additional peaks in the photoelectron spectrum represent ionizations of the molecule, with additional energy being used to vibrationally and rotationally excite the product H_2^+ molecule-ion. The rotational fine structure is not observed but contributes to the width of the vibrational bands that are observed.

The second peak in the dihydrogen p.e.s. corresponds to the energy required for its ionization plus that required to give H_2^+ one quantum of vibrational excitation. The difference in energy between the first two peaks is a measure of the magnitude of one quantum of vibrational excitation energy of H_2^+. The difference amounts to 0.3 eV or 29 kJ mol^{-1}. It is useful to compare such a value with the energy of one

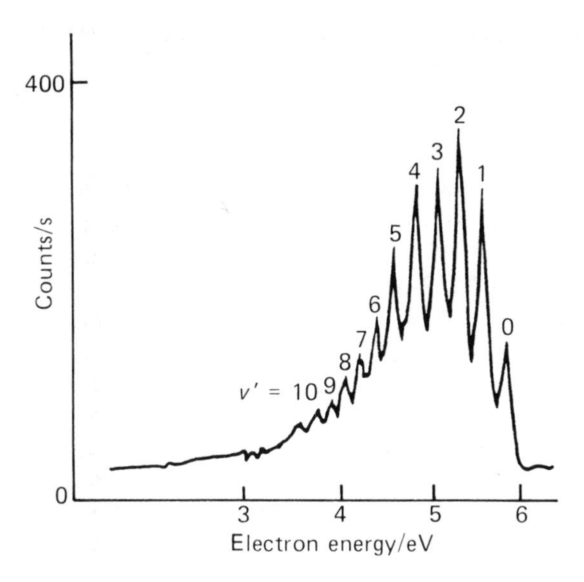

Fig 3.5 — The photoelectron spectrum of the dihydrogen molecule. (Reprinted with permission, from D. W. Turner and D. P. May, *J. Chem. Phys.*, **45** 471 (1966).

quantum of vibrational excitation of the dihydrogen molecule which is 49.8 kJ mol^{-1}.

The frequency (and, in consequence, the energy) of the vibration of two atoms bonded together is related to the bond strength (see Chapter 6) so it may be concluded that the removal of an electron from a dihydrogen molecule causes the remaining bond (in H_2^+) to be considerably weaker than that in the parent molecule. This is a confirmation of the bonding nature of the electron removed in the ionization process. Diagnosis of the bonding, non-bonding or anti-bonding nature of electrons in molecules may be made from a study of the effects of their ionization upon the vibrational frequency of the resulting positive ion. The removal of a non-bonding electron would have very little effect upon the bond strength and the vibrational frequencies of the neutral molecule, and its unipositive ion would be expected to be similar. The vibrational frequency of the ion would be greater than that of the neutral molecule if the electron removed originally occupied an anti-bonding orbital.

Experimental confirmation of the energy of the anti-bonding level in the dihydrogen molecule comes from the observation of its **absorption spectrum** in the far ultraviolet region. Dihydrogen absorbs radiation of a wavelength of 109 nm which is equivalent to a quantum energy of 1052 kJ mol^{-1}. In terms of an electronic transition the process of absorbing the quantum is:

$$\phi_1^2 \text{ (ground state)} \xrightarrow{h\upsilon} \phi_1^1\phi_2^1 \text{ (excited state)} \tag{3.26}$$

which would lead to the dissociation of the molecule into two hydrogen atoms. The energy of the transition should not be equated to the difference in energy between the two molecular orbitals, ϕ_1 and ϕ_2, but rather to a difference in energy between the ground and excited electronic states of H_2. In the ground state there is considerable interelectronic repulsion since the ϕ_1 orbital is doubly occupied. In the excited state there is much less repulsion between the electrons, which occupy separate orbitals, so that the difference in energy, $\phi_2 - \phi_1$, will be greater than the actual difference in energy between the two electronic states; $E_{\text{excited}} - E_{\text{ground}} < \phi_2 - \phi_1$.

3.7 HOMONUUCLEAR DIATOMIC MOLECULES OF THE FIRST ROW ELEMENTS

The extension of molecular orbital theory to the homonuclear diatomic molecules of the first short period elements, A_2, involves the arrangement of the 2s and 2p orbitals as group orbitals, and their classification within the $D_{\infty h}$ point group to which the molecules belong. The molecular axis of an A_2 molecule is arranged to be coincident with the z axis by convention. It is necessary to look at the 2s and 2p orbitals separately.

Classification of the 2s orbitals of A_2 molecules

The classification of the two 2s orbitals of an A_2 molecule is very similar to that of the two 1s orbitals of dihydrogen, and is not be repeated here. The two combinations of the 2s orbitals are:

$$\phi_{2\sigma_g^+} = (\psi_{2s_A} + \psi_{2s_B}) \tag{3.27}$$

and

$$\phi_{2\sigma_u^+} = (\psi_{2s_A} - \psi_{2s_B}) \tag{3.28}$$

the A and B subscripts referring to the two atoms contributing to the molecule. It is assumed that all wavefunctions are assumed to be normalized even though the normalization factors are omitted.

There is a bonding combination (equation (3.27)) which transforms within the $D_{\infty h}$ point group as a $2\sigma_g^+$ irreducible representation, the prefix 2 being assigned because of the bonding combination of the 1s orbitals having the same symmetry (and termed $1\sigma_g^+$). Likewise the anti-bonding combination (equation (3.28)) is termed $2\sigma_u^+$.

Classification of the 2p orbitals of A_2 molecules

The character of the reducible representation of the 2p orbitals of A_2 molecules may be obtained by writing down, under each symmetry element of the $D_{\infty h}$ group, the number of such orbitals which are unchanged by each symmetry operation. This produces:

	E	C_∞^ϕ	σ_v	i	S_∞^ϕ	C_2
$6 \times 2p$	6	$2 + 4\cos\phi$	2	0	0	0

This result requires some explanation, particularly with regard to the character of the 2p orbitals with respect to the C_∞ operation. The two $2p_z$ orbitals, lying along the C axis with their positive lobes overlapping, are unaffected by the associated operation, and account for the 2 in the character column. The term 4 cos ϕ, arises because of the two $2p_x$ and two $2p_y$ orbitals, which are perpendicular to the C_∞ axis. Although a rotation through m degrees around that axis does not move any of the orbitals to another centre, it does alter their disposition with regard to the xz and yz planes.

If the angle, ϕ, was chosen to be 180°, for instance, it would have the effect of inverting the $2p_x$ and $2p_y$ orbitals, and it would be necessary to place -1 for each orbital as their characters (note that $\cos 180 = -1$). To take into account all possible values of ϕ it is essential to express the character of each orbital as the cosine of the angle of rotation, ϕ. Effectively this implies that the character of each orbital is represented by the resolution of the orbital on to the plane it occupied before the symmetry operation was carried out. This ensures that for a rotation through 180° the character of a $2p_x$ or $2p_y$ orbital will be -1; in effect such an orbital, whilst not moving from its original position, changes the signs of its ψ values.

The character with respect to reflexion in one of the infinite number of vertical planes requires some explanation. It is best to choose a particular vertical plane such as that represented by the xz plane. Reflexion in **any** of the vertical planes has no effect upon the two $2p_z$ orbitals, which gives 2 as their character. Reflexion of the two $2p_x$ orbitals in the xz plane does not change them in any way — their character is 2. The reflexion of the two $2p_y$ orbitals in the xz plane causes their ψ values to change sign, and because they are otherwise unaffected, their character is -2. The resultant character of the six 2p orbitals, with respect to the operation, σ_v, is given by $2+2-2=2$.

The reducible representation of the six 2p orbitals may be seen, by inspection of the $D_{\infty h}$ character table, to be equivalent to the sum of the irreducible representations:

$$6 \times 2p = \sigma_g^+ + \sigma_u^+ + \pi_g + \pi_u \tag{3.29}$$

The table below demonstrates the truth of equation (3.29):

	E	C_∞^ϕ	σ_v	i	S_∞^ϕ	C_2
δ_g^+	1	1	1	1	1	1
σ_u^+	1	1	1	-1	-1	-1
π_g	2	$2\cos\phi$	0	2	$-2\cos\phi$	0
π_u	2	$2\cos\phi$	0	-2	$2\cos\phi$	0
$\sigma_g^+ + \sigma_u^+ + \pi_g + \pi_u$	6	$2+4\cos\phi$	2	0	0	0

It is important to realize which orbital combinations are represented by the above irreducible representations. This is best achieved by looking at the diagrams in Fig. 3.6 for the overlaps represented by the equations:

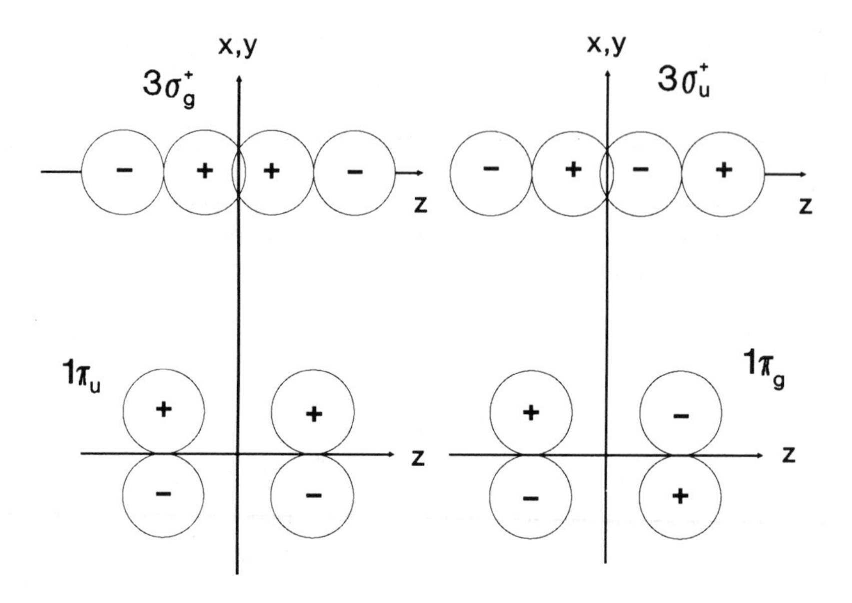

Fig 3.6 — Orbital overlaps between 2p atomic orbitals in the formation of some diatomic molecules.

$$\phi_{3\sigma_g^+} = (2p_{zA} + 2p_{zB}) \tag{3.30}$$

$$\phi_{3\sigma_u^+} = (2p_{zA} - 2p_{zB}) \tag{3.31}$$

$$\phi_{1\pi_u} = (2p_{xA} + 2p_{xB});\ (2p_{yA} + 2p_{yB}) \tag{3.32}$$

$$\phi_{1\pi_g} = (2p_{xA} - 2p_{xB});\ (2p_{yA} - 2p_{yB}) \tag{3.33}$$

The convention used throughout this text to express the form of a molecular orbital is to use a plus sign to indicate a bonding combination and a minus sign to indicate an anti-bonding combination.

The molecular orbital $\phi_{3\sigma_g^+}$, is bonding and is the third highest energy σ_g^+ orbital — hence the prefix 3. The $\sigma_{3\sigma_g^+}$ orbital is the anti-bonding combination of the two $2p_z$ orbitals, and the third highest energy σ_u^+ orbital. The π_u and π_g orbitals are both doubly degenerate, the π_u combination being bonding, the π_g combination being anti-bonding. They are both prefixed by the figure 1 since they are the lowest energy orbitals of their type.

In the forthcoming discussion of the bonding of the A_2 molecules, the orbitals will be referred to by their symmetry symbols with the appropriate numerical prefixes.

Fig. 3.7 is a diagram of the relative energies of the molecular orbitals of the A_2

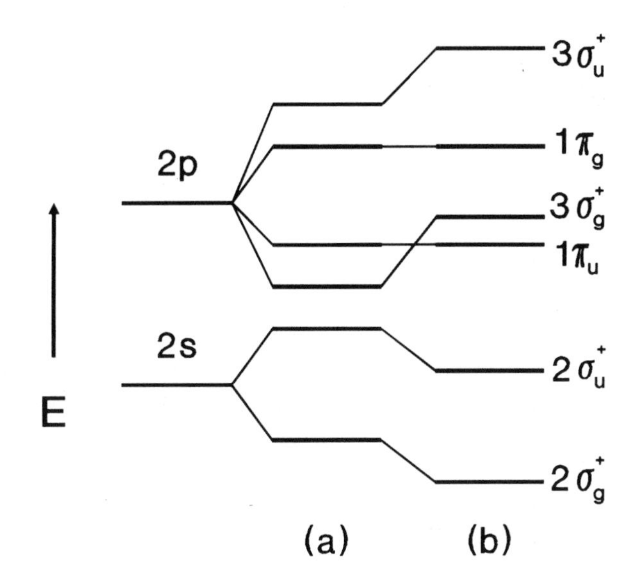

Fig 3.7 — The molecular orbital diagram for homonuclear diatomic molecules of the first short period.

molecules, together with those of the atomic orbitals from which they were constructed. The 'sideways' overlap involved in the production of π orbitals is by no means as effective as the 'end-on' overlap which characterizes the production of σ orbitals. For a given interatomic distance the overlap integral for σ-type overlap is generally higher than that for π-type overlap between two orbitals. The consequence of this is that the bonding stabilization and the anti-bonding destabilization associated with π orbitals are significantly less than those associated with σ orbitals. This accounts for the differences in energy, shown in Fig. 3.7(a), between the σ and π orbitals which originate from the 2p atomic orbitals of the A_2 atoms. The order of energies of the orbitals of A molecules is dependent upon the assumption that the energy difference between the 2p and 2s atomic orbitals is sufficient to prevent significant interaction between the molecular orbitals.

If two molecular orbitals have identical symmetry, such as the $2\sigma_g^+$ and $3\sigma_g^+$ orbitals, they may interact by the formation of linear combinations. The resulting combinations still have the same symmetry (and retain the nomenclature) but the lower orbital is stabilized at the expense of the upper one. Such interaction is also possible for the $2\sigma_u^+$ and $3\sigma_u^+$ orbitals.

The **extent** of such interaction is determined by the energy gap between the two contributors. If the energy gap between the 2p and 2s atomic orbitals is small enough the interaction between the $2\sigma_g^+$ and $3\sigma_g^+$ molecular orbitals may be so extensive as to cause the upper orbital ($3\sigma_g^+$) to have an energy which is greater than that of the $1\pi_u$ set. Such an effect is shown in Fig. 3.7(b). The magnitude of the 2p–2s energy gap varies along the elements of the first short period as is shown in Fig. 3.8. The energy

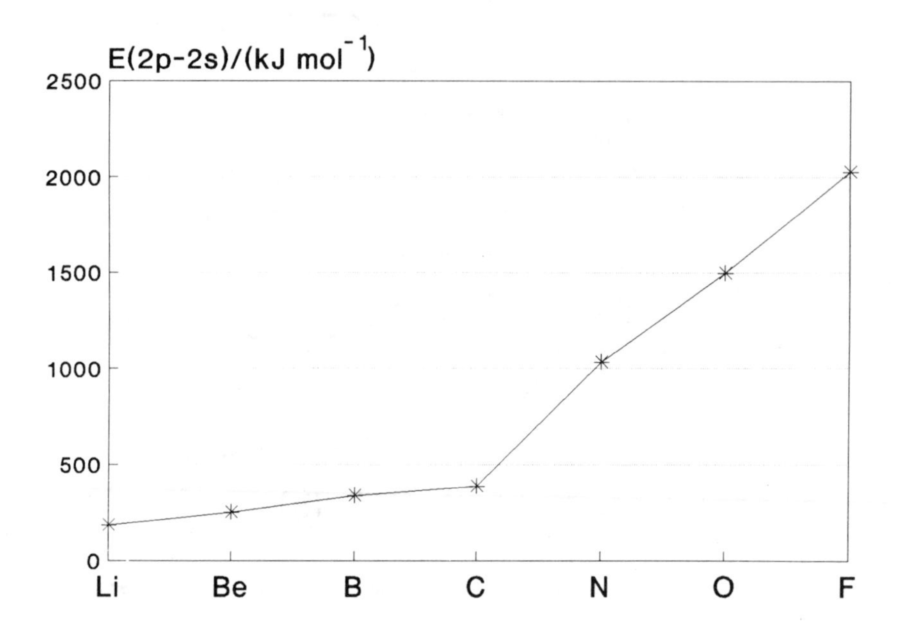

Fig 3.8 — Variation of the 2p–2s energy gap along the first short period.

gaps in the elements lithium to nitrogen are sufficiently small to make significant $2\sigma_g^+ - 3\sigma_g^+$ interaction possible such that Fig. 3.7(b) is relevant in determining the electronic configurations of the molecules, Li_2, Be_2, B_2, C_2 and N_2. Fig. 3.7(a) is to be used to determine the electronic configurations of the molecules, O_2, F_2 and Ne_2, since the 2p–2s energy gaps in O, F and Ne are sufficiently large to preclude significant molecular orbital interaction.

The interaction of molecular orbitals of the same symmetry has important consequences for all systems where it occurs, and more examples will be referred to in later chapters. It is possible to carry out the mixing of the original atomic orbital (known as hybridization) before the m.o.s are formed. Some examples of this approach are included in later chapters. Both approaches give the same eventual result for the contributions of atomic orbitals to the molecular ones.

Electronic configurations of homonuclear diatomic molecules

The electronic configurations of the homonuclear diatomic molecules of the elements of the first short period, and some of their ions, are given in Table 3.1.

Table 3.1

Molecule/ion	$2\sigma_g^+$	$2\sigma_u^+$	$3\sigma_g^+$	$1\pi_u$	$1\pi_g$	$3\sigma_u^+$
Li_2	2					
Be_2	2	2				
B_2	2	2		2		
C_2	2	2		4		
N_2	2	2	2	4		
N_2^+	2	2	1	4		
N_2^-	2	2	2	4	1	
O_2	2	2	2	4	2	
O_2^+	2	2	2	4	1	
O_2^-	2	2	2	4	3	
O_2^{2-}	2	2	2	4	4	
F_2	2	2	2	4	4	
Ne_2	2	2	2	4	4	2

In all cases the four electrons occupying the $1\sigma_g^+$ and $1\sigma_u^+$ orbitals are not indicated. The interactions of the 1s orbitals of the atoms under discussion are minimal and, although they may be regarded as occupying the molecular orbitals previously indicated, they are virtually non-bonding. In any case the slight bonding character of the two electrons occupying $1\sigma_g^+$ is cancelled by the slight anti-bonding nature of the two electrons in $1\sigma_g^+$. In some texts the four electrons are indicated symbolically by KK — a reference to the 'K shell' — now recognized as the 1s atomic orbital.

Dilithium, Li_2

This molecule exists in the gas phase and has a bond dissociation energy of 107 kJ mol^{-1} and a bond length of 267 pm. The weak, and very long, bond is understandable in terms of the two electrons in $2\sigma_g^+$ being the only ones having bonding character. The bond may be described as having a **bond order** of one. The bond order is defined as a half of the resultant excess of the number of bonding electrons above the number of anti-bonding electrons. The four 1s electrons have no resultant bonding effect and yet contribute considerably to the interelectronic repulsion energy. The electron affinity of the lithium atom is 59.8 kJ mol^{-1}, which indicates that the nuclear charge of $+3e$ is not very effective in attracting more electrons. The atom also has a very low ionization energy (513 kJ mol^{-1}), which is another indication of the low effectiveness of the nuclear charge. That has been discussed in Chapter 1 in terms of there being a considerable amount of interelectronic repulsion between the three electrons possessed by the lithium atom. Another approach is to use the concept of the nuclear charge being **shielded** by the various occupied atomic orbitals so as to reduce its effectiveness in attracting extra electrons. For bonding to be achieved, the attraction between the two shielded nuclei and the two bonding electrons must outweigh the two repulsive interactions — internuclear and interelectronic. The efficient shielding of the lithium nuclei by their $1s^2$ 'core' configurations contributes to the weakness of the bond in the Li_2 molecule. Solid lithium does not contain discrete molecules and has a metallic lattice. The bonding in metals is discussed in Chapter 5.

Diberyllium, Be_2

Since the electronic configuration of the diberyllium molecule, Be_2, would be $(2\sigma_g^+)^2$ $(2\sigma_u^+)^2$, with two bonding electrons being counterbalanced by two anti-bonding electrons, it is not surprising that the molecule does not exist with that configuration. It would possess a bond order of zero. There is a possibility that excited states could exist transiently where one of the anti-bonding electrons had been excited to the $1\pi_u$ or $3\sigma_g^+$ bonding levels.

Diboron, B_2

The diboron molecule, B_2, has a transient existence in the vapour of the element, and it is known that its bond dissociation energy is 291 kJ mol^{-1}, the bond length being 159 pm. The bond is stronger and shorter than that in Li_2. The first ionization energy of the boron atom (800 kJ mol^{-1}) indicates that the nuclear charge is considerably more effective than that of the lithium atom and a somewhat stronger (and shorter) bond is to be expected for B_2 as compared to that in Li_2. The two pairs of sigma electrons have a zero resultant bonding effect, leaving the stability of the bond to the two electrons which occupy the $1\pi_u$ orbitals. It is of interest that, since the $1\pi_u$ level is doubly degenerate (so that Hund's rules apply to their filling), the two orbitals are singly occupied and the two π_u electrons have parallel spins. The bonding in B_2 consists of two 'half-π' bonds if the term 'bond' is understood to indicate a pair of bonding electrons. The bonding may be described in terms of the bond order being unity.

Evidence which is consistent with the above description of the bonding in B_2 is the observation that the molecule is **paramagnetic** — the property associated with an unpaired electron (or with more than one unpaired electron with parallel spins). In the B_2 case, the evidence for the presence of unpaired electrons comes from the observation of its **electron spin resonance** (esr) spectrum. The detailed theory of electron spin resonance spectra is not dealt with in this book. The essentials of the method depend upon there being a difference in the energy of an unpaired electron when subjected to a magnetic field. The electron spin is aligned either in the same direction as the applied field or against it. The difference in energy between the two quantized alignments corresponds to the energies of radiofrequency photons. No esr signal is obtained from paired-up electrons since neither of the electrons may change its spin without violating the Pauli exclusion principle.

Dicarbon, C_2

This molecule exists transiently in flames and has a bond dissociation energy of 590 kJ mol^{-1} and a bond length of 124 pm. The bond order is 2 since both the orbitals of the bonding $1\pi_u$ set are filled. The atoms are held together by two π bonds — a very unusual example. The carbon nucleus is more effective than that of the boron atom and, combined with there being twice as many resultant bonding electrons, serves to produce a much more stable molecule (with respect to the constituent atoms) than in the case of B_2. The average bond energy for C–C σ-type bonds is generally accepted to be 348 kJ mol and that for C=C σ,π bonds is 612 kJ mol^{-1}. These figures are consistent with the view expressed above that π-type bonding is weaker than σ-type. The extra bond energy of the C=C double bond is given by $612-348=264$ kJ mol^{-1}, which figure should be compared to the 348 kJ mol^{-1} for the strength of the C–C single σ bond.

Dinitrogen, N_2, and the ions, N_2^+ and N_2^-

In the dinitrogen molecule, N_2, two electrons occupy the $3\sigma_g^+$ orbital, and the bond order is three - one sigma pair plus the two pi pairs. The electronic configuration is consistent with the very high bond dissociation energy of 942 kJ mol^{-1} and the short bond length of 109 pm. The molecule is chemically inert to oxidation and reduction, although it does easily form some complexes when it acts as a ligand as, for example, in $[Ru(NH_3)_5N_2]^{2+}$. It undergoes reaction with dihydrogen only under conditions of high temperature and pressure in the presence of a catalyst. Some bacteria possess the capability of reducing dinitrogen at ambient temperature. The great strength of the bond, in dinitrogen, is associated with the presence of an excess of six bonding electrons together with the greater effectiveness of the nuclear charge compared to that of carbon.

The ionization of the molecule to give the N_2^+ ion causes the bond order to be reduced to 2.5, with consequent weakening (bond dissociation energy = 841 kJ mol^{-1}) and lengthening (bond length = 112 pm) of the bond as compared to that in N_2. The electron removed in the ionization comes from the $3\sigma_g^+$ orbital, which, because of the interaction with the $2\sigma_g^+$ orbital, is only moderately bonding. The effects upon the bond strength and length are, therefore, relatively slight.

The N_2^- ion is produced by adding an electron to the anti-bonding $1\pi_g$ level and, as does the N_2^+ ion, has a bond order of 2.5. Since the $1\pi_g$ level is anti-bonding, with no off-setting effects, the addition of an electron causes the bond in N_2^- to be significantly weaker (bond dissociation energy = 765 kJ mol^{-1}) and longer (bond length = 119 pm) than the one in N_2.

Dioxygen, O_2, and the ions, O_2^+, O_2^- and O_2^{2-}

The dioxygen molecule, O_2, has the dinitrogen configuration (except that the $1\pi_u$ level is higher in energy than the $3\sigma_g^+$ orbital) with an extra two electrons which occupy the $1\pi_g$ level. Since $1\pi_g$ is doubly degenerate, the orbitals are singly occupied by electrons with parallel spins (Hund's rules). The molecule is paramagnetic as would be expected, and since the additional electrons occupy anti-bonding orbitals, the bond order decreases to two as compared with the N_2 molecule. In consequence the bond dissociation energy (494 kJ mol^{-1}) is considerably lower, and the bond length (121 pm) is considerably larger, than in the N_2 case.

The ionization of an electron from the highest energy level ($1\pi_g$) of the O_2 molecule produces the positive ion, O_2^+, which has a bond order of 2.5 (comparable to that of N_2^+). The bond dissociation energy of O_2^+ is 644 kJ mol^{-1} and its bond length is 112 pm. The nuclear charge of the oxygen is not as effective as that of nitrogen, and the anti-bonding electron in O_2^+ has a very significant weakening effect.

In the superoxide ion, O_2^-, there are three electrons occupying the $1\pi_g$ orbitals. The bond order is 1.5, which is consistent with its observed bond dissociation energy of 360 kJ mol^{-1} and bond length of 132 pm.

The peroxide ion, O_2^{2-}, has a filled set of $1\pi_g$ orbitals and the bond is weaker (bond dissociation energy = 149 kJ mol^{-1}) and longer (bond length = 149 pm) than that of O_2^-. The bond order of the O_2^{2-} ion is 1.0.

Difluorine, F_2

The difluorine molecule, F_2, has an identical electronic configuration to that of the peroxide ion. The bond order is 1.0, and the bond dissociation energy of 155 kJ mol^{-1} and bond length of 144 pm being very similar to the values for O_2^{2-}.

The species with any occupancy of the $1\pi_g$ (anti-bonding) level (O_2^+, O_2, O_2^-, O_2^{2-} and F_2) exhibit considerable chemical reactivity.

Dineon, Ne_2

The dineon molecule, Ne_2, with all its molecular orbitals filled would not be expected to exist with that particular electronic configuration.

3.8 SOME HETERONUCLEAR DIATOMIC MOLECULES

The heteronuclear diatomic molecules, nitrogen monoxide, NO, carbon monoxide, CO, and hydrogen fluoride, HF, are dealt with in this section. They belong to the point group, $C_{\infty v}$, and possess a C_∞ axis of symmetry and an infinite number of vertical planes all containing that axis. The orbitals of the molecules, NO and CO, are similar to those of the A_2 molecules of the previous section but have different terminology. In addition to the different symmetry there are effects due to the participating atoms having different electronegativities, the energies of the combining atomic levels not being identical.

Nitrogen monoxide (nitric oxide), NO

The molecular orbital diagram for the nitrogen monoxide molecule is shown in Fig. 3.9. The orbitals are produced from the same pairs of atomic orbitals as in the cases of the homonuclear diatomic molecules of section 3.7. The first ionization energies of nitrogen (1400 kJ mol^{-1}) and oxygen (1314 kJ mol^{-1}) are quite similar so the atomic orbitals match up reasonably well. The terminology is different, and because of the absence of an inversion centre there are no g or u subscripts, and this alters the numbering of the σ and π orbitals. The two pairs of 1s electrons (or KK) form the $1\sigma^+$ and $2\sigma^+$ orbitals, the other σ^+ orbitals following on in order of increasing energy. The two sets of π orbitals become 1π (bonding) and 2π (anti-bonding) respectively. The electronic configuration of the nitrogen monoxide molecule is thus: KK $(3\sigma^+)^2(4\sigma^+)^2(1\pi)^4(5\sigma^+)^2(2\pi)^1$. The bond order is 2.5, consistent with the bond dissociation energy (626 kJ mol) and bond length (114 pm). The molecule is chemically very reactive and is paramagnetic because of the single unpaired anti-bonding electron. The bond order in the NO^+ ion is 3.0 in the absence of any electrons in the 2π level.

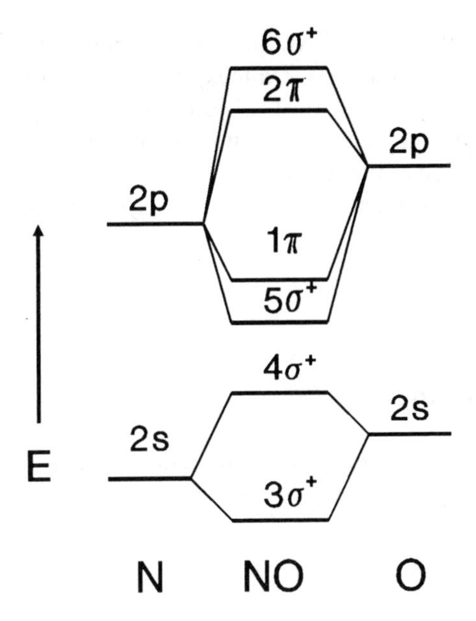

Fig 3.9 — Molecular orbital diagram for nitrogen monoxide.

In the NO case (ignoring the $1s^2$ pairs) there are four orbitals possessing the same symmetry, σ^+, and which can therefore mix to an extent which depends upon energy differences. It would be expected that all four orbitals would possess 2s and 2p contributions although it is not possible to quantify those in qualitative molecular orbital theory. The 2p–2s gaps in nitrogen and oxygen are relatively large so that the 2p–2s admixtures in the σ^+ m.o.s would not be expected to approach equivalence.

Carbon monoxide

As the first ionization energies imply, the energies of the atomic orbitals of carbon (first I.E. = 1086 kJ mol^{-1}) and oxygen (first I.E. = 1314 kJ mol^{-1}) do not match up very well. When the 2p–2s energy gaps are compared (C, 386 and O, 1544 kJ mol^{-1}) it is obvious that there is a great mismatch between the respective 2s levels. That of the oxygen atom is too low to interact with the 2s atomic orbital of the carbon atom in any significant manner. The 2s atomic orbital of the oxygen atom is to be regarded as a virtually non-bonding orbital ($3\sigma^+$). The small 2p–2s energy gap in the carbon atom facilitates the mixing, or hybridization, of its 2s and $2p_z$ orbitals (assuming, as is conventional, that the molecular axis is coincident with the z axis). The two orbitals participate in σ^+ orbitals of the molecule, and may mix. Because of the relatively small 2p–2s energy gap, they do mix, the new orbitals being written as:

$$h_1 = \psi_{2s}(C) + \psi_{2p_z}(C) \quad (3.34)$$

and

$$h_2 = \psi_{2s}(C) - \psi_{2p_z}(C) \quad (3.35)$$

The hybrid orbitals, h_1 and h_2, are shown diagrammatically in Fig. 3.10. They have two unequal lobes of oppositely signed ψ values, the large positive lobe of h_1 being directed at the oxygen atom, with that of h_2 pointing in the opposite direction. It would be difficult for h_2 to play any significant part in the bonding to the oxygen atom, and it is to be regarded as being non-bonding ($5\sigma^+$). The interaction of $h_1(C)$ and the $2p_z$ orbital of the oxygen atom gives a bonding combination:

$$\phi(4\sigma^+) = h_1(C) + \psi_{2p_z}(O) \qquad (3.36)$$

and an anti-bonding combination:

$$\phi(6\sigma^+) = h_1(C) - \psi_{2p_z}(O) \qquad (3.37)$$

The molecular orbitals are completed by the interaction of the two sets of $2p_x$ and $2p_y$ orbitals to give doubly degenerate 1π (bonding) and 2π (anti-bonding) levels:

$$\phi(1\pi) = \psi_{2p_{x,y}}(C) + \psi_{2p_{x,y}}(O) \qquad (3.38)$$

and

$$\phi(2\pi) = \psi_{2p_{x,y}}(C) - \psi_{2p_{x,y}}(O) \qquad (3.39)$$

The molecular orbital diagram is given in Fig. 3.11. The electronic configuration of the CO molecule is thus:

$$KK\ (3\sigma^+)^2(4\sigma^+)^2(1\pi)^4(5\sigma^+)^2$$

the bond order is 3, being consistent with the high bond dissociation energy of 1090 kJ mol^{-1} and the short bond length of 113 pm. Carbon monoxide is a relatively inert chemical substance, but it does have an extensive involvement with the lower oxidation states of the transition elements with which it forms a great many carbonyl complexes, in which it acts as a ligand. The bonding of CO to a transition metal involves the use of the otherwise non-bonding electron pair in the $5\sigma^+$ orbital. The vacant 2π orbital is also important in the bonding of CO to transition metals, the details being dealt with in Chapter 8.

The cyanide ion, CN^-, is isoelectronic with carbon monoxide and has an extensive chemistry of involvement with transition metals but, unlike CO, it exhibits a preference for the positive oxidation states of the elements. This is because of its negative charge.

Hydrogen fluoride, HF

The molecule of hydrogen fluoride, HF, belongs to the $C_{\infty v}$ point group. The hydrogen atom uses its 1s atomic orbital to make bonding and anti-bonding combinations with the $2p_z$ orbital of the fluorine atom (by convention the z axis is made to coincide with the molecular axis). Because of the different values of the first ionization energies of the two elements (hydrogen, 1312 and fluorine, 1681 kJ

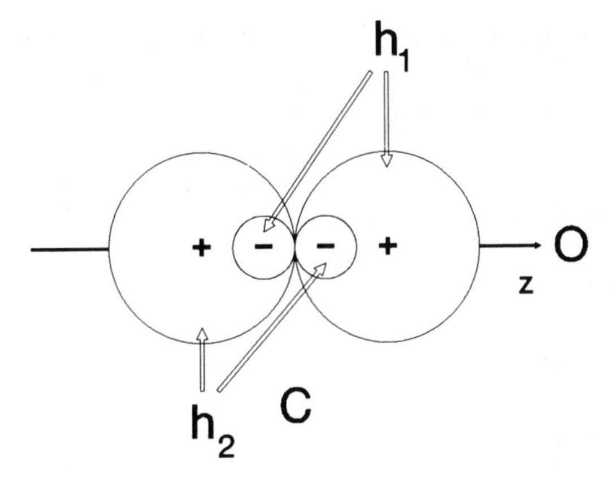

Fig 3.10 — The two sp hybrid orbitals of the carbon atom in carbon monoxide.

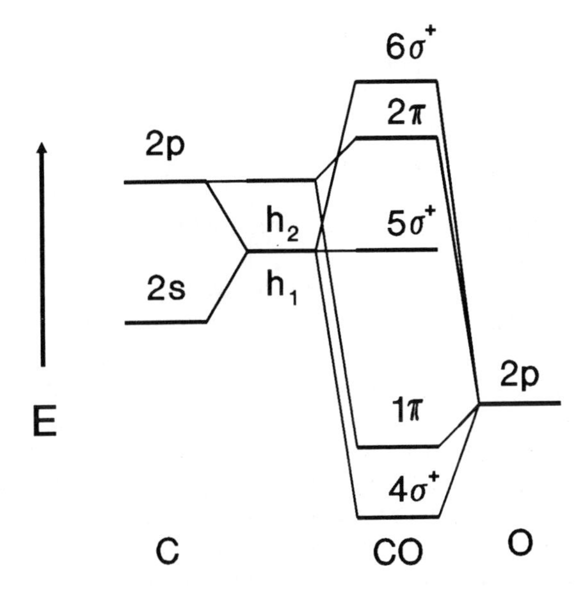

Fig 3.11 — Molecular orbital diagram for carbon monoxide.

mol^{-1}) the molecular orbital diagram (shown in Fig. 3.12)) is considerably 'asymmetric' with the lower, bonding orbital ($3\sigma^+$), having a major input from the fluorine $2p_z$ orbital. The higher, anti-bonding orbital ($4\sigma^+$) has its major contribution from the hydrogen 1s orbital. Such inequality in contributing to molecular orbitals may be expressed by including a factor, λ, in the equations for the linear combinations of the 1s(H) and $2p_z$ (F) wavefunctions:

$$\phi(3\sigma^+) = \psi_{1s}(H) + \lambda\psi_{2p_z}(F) \tag{3.40}$$

and

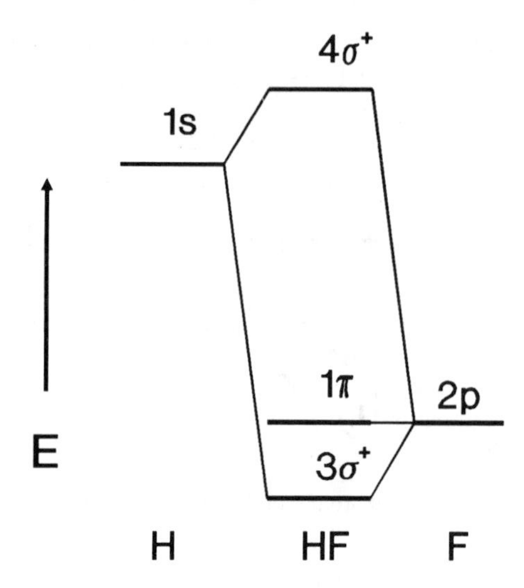

Fig 3.12 — Molecular orbital diagram for hydrogen fluoride.

$$\phi(4\sigma^+) = \lambda\psi_{1s}(H) - \psi_{2p_z}(F) \tag{3.41}$$

Because the squares of wavefunctions are proportional to electron probabilities, the 'share' of the two bonding electrons experienced by the fluorine atom is given by the factor, $\lambda^2/(1+\lambda^2)$. That experienced by the hydrogen atom is $1/(1+\lambda^2)$. The HF molecule possesses a **dipole moment** of 0.43, which is experimental evidence for such a charge separation. The hydrogen 'end' of the molecule is, in effect, positive to the extent of $0.43e$, whilst the fluorine 'end' is negative to the extent of $-0.43e$. Another viewpoint is that the 'covalent' bond is really 43% **ionic**.

The other orbitals in Fig. 3.12 are the non-bonding $2\sigma^+$ (the 2s atomic orbital of the fluorine atom), and the non-bonding 1π (the $2p_x$ and $2p_y$ atomic orbitals of the fluorine atom). The $1s^2$ pair of fluorine electrons ($1\sigma^+$) is omitted from the diagram.

The discrepancy in matching up of the 1s(H) and $2p_z$(F) orbitals is $1681 - 1312 = 369$ kJ mol^{-1}, and leads to a bond with considerable ionic character. The difference in energy between the orbitals of the sodium atom (3s ionization energy = 496 kJ mol^{-1}) and the fluorine atom ($2p_z$ ionization energy = 1681 kJ mol^{-1}) amounts to 1185 kJ mol and leads to the conclusion that covalent bond formation is impossible between Na and F. If the elements are to combine at all, it has to be by an alternative method (**ionic bonding**, which is dealt with in Chapter 5).

3.9 CONCLUDING SUMMARY

This chapter contains a discussion of some important aspects of molecular orbitals.

(1) The theory is applied to H_2^+ and H_2, and the importance of interelectronic repulsion energy is stressed. The interelectronic repulsion energies in He, H^-

and H_2 are estimated and compared. The theory is extended to He_2^+ and He_2. The labelling of molecular orbitals with appropriate symmetry symbols is dealt with.

(2) The measurement and interpretation of photoelectron spectra are described, and the photochemical decomposition of H_2 is discussed.

(3) The various bonding and anti-bonding combinations of s and p orbitals to form the molecular orbitals of the homonuclear diatomic molecules of the elements from Li to Ne are dealt with, leading to an understanding of the bond strengths and bond lengths of such molecules. The labelling of the orbitals is included, and interaction between orbitals of identical symmetries discussed.

(4) The effects of electronegativity differences in some heteronuclear diatomic molecules are discussed, leading to the extreme case where the ionic bond is a better description of bonding than is covalency.

4

Bonding and shapes of polyatomic molecules

4.1 INTRODUCTION

This chapter is concerned with the bonding in polyatomic molecules. Molecular orbital theory is applied, not only to the cohesion of atoms, but to the factors responsible for the determination of bond angles and molecular shapes. A method of predicting molecular shape (the valence shell electron pair repulsion theory, VSEPR) is described and criticized. Molecular orbital theory is extended and applied to the production of diagrams which correlate the orbital energies of different possible extreme geometries of a molecule.

The discussions in this chapter are restricted to (i) triatomic molecules, (ii) some examples of polyatomic molecules with as many as eight atoms, and (iii) the HF_2^- ion as an example of hydrogen bonding.

4.2 TRIATOMIC MOLECULES

Triatomic molecules may be linear or bent. The shape adopted by any particular molecule is that which is consistent with the minimization of its **total** energy. Two approaches to the problems of molecular shape are described; the valence shell electron pair repulsion theory and molecular orbital theory.

1. The valence shell electron pair repulsion (VSEPR) theory

The VSEPR theory is based upon the original ideas of Sigwick and Powell, and was extended by Gillespie and Nyholm. It is concerned with the minimization of the repulsions between the pairs of electrons in the valence shell. The term 'valence shell' refers to the orbitals of the **central** atom of the molecule which could possibly be involved in bonding. The basic operation of the theory is to add up the number of electrons in the valence shell of the central atom together with suitable contributions from the ligand atoms — one electron per bond formed. This sum is then divided by two to give the number of pairs of electrons. These electron pairs repel each other to give a spatial distribution which is that corresponding to the minimization of the repulsive forces.

This method of predicting molecular shape affords a very rapid and easy-to-use set of arguments which almost always produces an answer which is consistent with observation. The arguments are open to some doubts and, because the theory gives the right answer for the shape of the molecule, there is a temptation to believe that some knowledge of the bonding has been gained. This is not true, as is demonstrated in various sections of this book.

2. Molecular orbital (M.O.) theory

The basis of this has already been described, and its extension to the treatment of triatomic molecules forms a major part of this chapter. It gives a very satisfactory description of the shapes and the bonding of molecules in general, and is consistent with observations of photoelectron and absorption spectra.

Valence shell electron pair repulsion theory

As has been pointed out above, the shape of a molecule (or the geometry around any particular atom connected to at least two other atoms) is dependent upon the minimization of the repulsive forces operating between the pairs of valence electrons. There is an important restriction upon the type of electron pairs in that they must be 'sigma' pairs. Any 'pi' or 'delta' pairs must be discounted. The terms 'sigma', 'pi' and 'delta' refer to the type of overlap undertaken by the contributory atomic orbitals in producing the molecular orbitals.

The overlap of two s orbitals on separate atoms and that of two p orbitals along the bond axis give rise to sigma orbitals. These are cylindrically symmetrical with respect to the bond axis. The sigma-type overlap of two p orbitals is shown in Fig. 3.6. The sideways overlap of two p orbitals gives rise to π-type orbitals as is also shown in Fig. 3.6. Two d_{z^2} orbitals overlapping along the z axis give rise to σ-type orbitals.

Fig. 4.1 demonstrates the types of overlap for $d_{xz} - d_{xz}$ (in the xz plane) and $d_{xy} - d_{xy}$ (along the z axis). The 'sideways' (two lobes) overlap by the d_{xz} orbitals is π-type, the four-lobe overlap by the two d_{xy} orbitals being termed δ-type. The Greek letters are used loosely for such orbitals — strictly they should be used for molecules possessing either $D_{\infty h}$ or $C_{\infty v}$ symmetry, but they are used generally to describe the type of overlap. Only sigma-type pairs are counted up in the VSEPR approach. This often means that some assumptions about the bonding have to be made before the theory may be applied.

Table 4.1 contains the basic shapes adopted by the indicated numbers of sigma (σ) electron pairs.

Fig. 4.2 shows representations of the basic shapes assumed by the various electron pair distributions.

The application of VSEPR theory to triatomic molecules is exemplified by considering water, carbon dioxide and xenon difluoride.

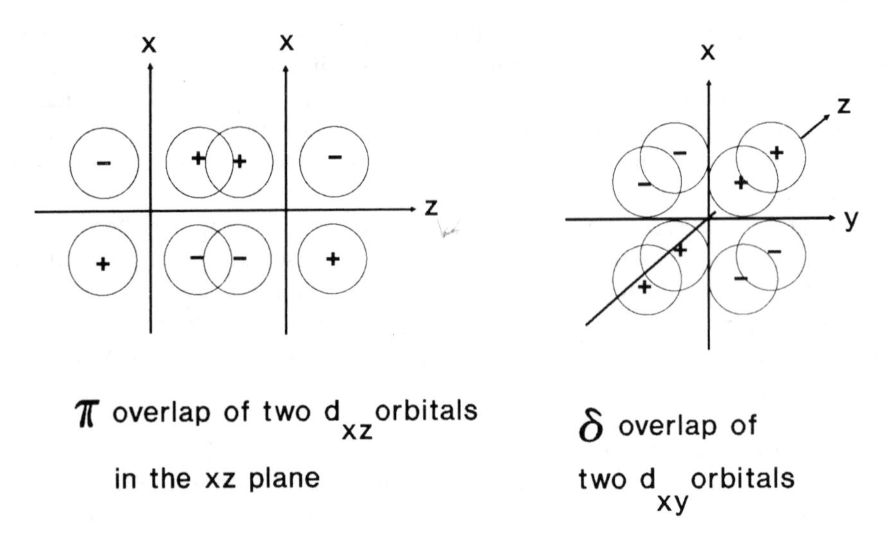

Fig. 4.1 — 'pi' and 'delta' overlap diagrams for d orbitals.

Table 4.1

Number of σ electron pairs	Shape of electron pair distribution
2	Linear
3	Trigonally planar
4	Tetrahedral
5	Trigonally bipyramidal
6	Octahedral

Water, H_2O

The isolated water molecule has a bond angle of 104.5°. The central oxygen atom has the valence shell electron configuration: $2s^22p^4$, with two of the 2p orbitals singly occupied. The hydrogen atoms supply one electron each to the valence shell of the central atom, which makes a total of eight σ electrons (since the 1s–2p overlaps are of the σ type). The four electron pairs are most stable when their distribution is tetrahedral. Two of the pairs are bonding pairs, the other two being non-bonding or

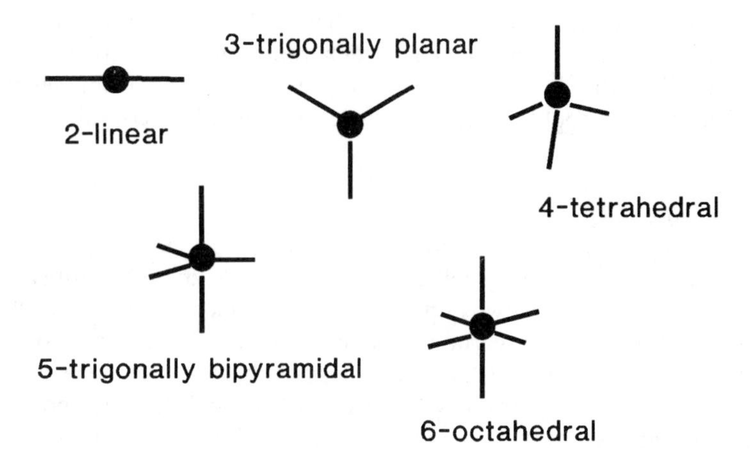

Fig. 4.2 — The basic distributions of 2, 3, 4, 5 and 6 'sigma' electron pairs corresponding to the minimalization of repulsions.

lone pairs. This picture of the water molecule would imply that the bond angle should be that of the regular tetrahedron; 109°28′. The theory predicts that the water molecule should be bent, and so it is. To refine the prediction of the bond angle it is possible to argue that the non-bonding pairs of electrons, which are more localized on the central oxygen atom than the bonding pairs, have a greater repulsive effect upon the bonding pairs than that exerted by the bonding pairs, upon each other. The result would be a squeezing together of the two bonding pairs making the bond angle somewhat less than the regular tetrahedral value. It is not possible to quantify this aspect of the theory. No further deductions should be made concerning the bonding in the water molecule.

It is dangerous to attempt to deduce anything more about the water molecule from this theory. There is a temptation to conclude that there are two identical electron pair bonds responsible for holding the three atoms together and that there are two identical lone pairs in the other two tetrahedral positions. When such erroneous conclusions are made, the errors are compounded by the attempt to fix the electron pair arrangement to coincide with those conclusions by mixing up the 2s and three 2p wavefunctions so that four identical sp^3 hybrid wavefunctions are directed to the vertices of a regular tetrahedron. Such an action may be carried out mathematically. It is a matter of taking the 2s orbital of the oxygen atom (a_1 in the C_{2v} point group to which the water molecule belongs) and the three 2p orbitals (b_1, b_2 and a_1 respectively), and mixing them to form four linear combinations. The character of the four hybrid orbitals is, of course, equal to the sum, $2a_1 + b_1 + b_2$, which is reducible to just those components. Although such an operation may be carried out it does nothing to further the understanding of the bonding in the water molecule. On the other hand, it contributes a great deal to the misunderstanding of the bonding! It is of use to those who picture real molecules in terms of the 'ball-and-stick' models which are used to demonstrate molecular shapes. The mistake which can be made with such models is to assume that each 'stick' represents an electron pair. Mulliken has said: 'I believe the chemical bond is not so simple as some people seem to think', and this quotation should always be borne in mind when considering molecular theories.

Carbon dioxide, CO_2

The carbon dioxide molecule is linear and belongs to the $D_{\infty h}$ point group. The carbon atom has the valence shell configuration of $2s^22p^2$. Ostensibly the carbon atom is divalent — there are two unpaired electrons which occupy two of the three 2p atomic orbitals. In order for carbon to be tetravalent and to arrange for there to be four unpaired electrons in its valence shell it is possible to cause the excitation of one of the 2s electrons to the previously unoccupied 2p orbital. It is then possible to feed two σ-type electrons into the 2s and one of the 2p orbitals, leaving the other two 2p orbitals to accept the two π electrons. Prior knowledge is used of the normal divalency of the oxygen atom and the impossibility of there being more than one σ bond between any pair of atoms — the spatial distribution of atomic orbitals decides that. The count of σ electron pairs comes to two, and quite correctly the theory predicts the shape of the CO_2 molecule to be linear, the two σ pairs repelling each other to a position of minimum repulsion. The two π pairs are used to complete the bonding picture. At that point the deductions should stop. Further (inadvisable) deductions include the hybridization of the 2s and $2p_z$ orbitals of the carbon atom to ensure that two identical electron pair bonds are directed at an angle of 180° from each other along the z axis, and the construction of two identical π bonds.

The excitation of an electron from 2s to 2p is an aid to the electron counting process — it does not actually happen in the formation of CO_2 or any other molecule. The mistake is often made that such manipulations have a connexion with reality!

There is an alternative way of operating the VSEPR theory which is less subject to misinterpretation. The application to CO_2 takes note that if the atoms were thought of as being in their common oxidation states: oxygen would be O(− 2) and that would make carbon C(+ 4), then C^{4+} would have **no** electrons in its valence shell. If two σ electrons were then added to the carbon(4 +) for each ligand atom (the two oxygen(− 2) species in this case) the carbon atom would then possess four σ electrons in its valence shell. These would form two pairs and would repel each other to the stable linear position that the CO_2 molecule usually adopts.

Xenon difluoride, XeF_2

The XeF_2 molecule is linear. VSEPR theory would recognize that the valence shell of the xenon atom has the configuration: $5s^25p^6$, and as such should be zero valent. There is the possibility of causing a 5p to 5d excitation to make the atom divalent. Addition of the two valency electrons from the fluorine atoms would make two bonding pairs and these, together with the remaining three pairs of non-bonding electrons, would contribute to the total of five electron pairs. The five pairs would assume a trigonally bipyramidal distribution, and it is logical for the fluorine atoms to be as far away from each other as possible to give a linear molecule. The alternative approach would divide the molecule up into Xe(+ 2) and two ligand fluoride ions.

The Xe(+ 2), $5s^25p^4$, would then accept two electron pairs from the fluorine atoms, one which would fill the third 5p orbital and the second would enter the next available level, 5d. The same (correct) conclusion would follow.

The VSEPR treatment of the XeF_2 molecule is open to the misinterpretation that the two identical Xe−F bonds are produced by electron pairs localized in two identical 5p–5d hybrid orbitals. The 5p to 5d excitation is to facilitate electron counting, and there should be no implication that it actually takes place.

Summary with regard to VSEPR theory

The VSEPR theory may be used to give fairly accurate conclusions concerning the **shapes** of molecules in their electronic ground states. It must not be used to indicate anything about the bonding of molecules. Such speculation is almost certain to be inaccurate. The theory gives no information or understanding about any electronically excited states of molecules.

Molecular orbital theory

The application of molecular orbital theory, even on a qualitative basis, is a more lengthy procedure than using VSEPR theory, but it produces so much more in understanding and is subject to rigorous testing by experimentation. It assumes only the atomic orbitals which are available for bonding and the number of electrons which have to be accommodated. The understanding of the bonding of a molecule arises from the proper application of symmetry theory as is demonstrated for the H_2O, CO_2 and XeF_2 molecules. Quantitative results may be obtained for quite complicated molecules using modern computer methods. The results of proper solutions of the wave equation for molecules are referred to when necessary.

Water

The groundwork for the application of molecular orbital theory to the water molecule has been carried out to a large extent in the section on group theory. The procedure is to identify the point group to which the molecule belongs. If the normal state of the molecule is to be treated, then that is a simple matter, providing that the shape is known. To demonstrate the power of molecular orbital theory, both bent (90°) and linear forms of the molecule are treated.

90° Water molecule

There is a very logical method for the derivation of the molecular orbital energy diagram for any molecule. It is applied to the 90° water molecule as follows.

(1) Identify the point group to which the molecule belongs.

The 90° form of the water molecule belongs to the C_{2v}. point group. There is only one axis of symmetry, C_v, and by convention this is arranged to coincide with the z axis. The position of the molecule with respect to the coordinate axes is as shown in Fig. 2.6.

(2) Classify the atomic orbitals in the valency shell of the central atom with respect to the point group of the molecule.

The classification of the orbitals of the oxygen atom is a matter of looking them up in the C_{2v} character table, a full version of which is included in Appendix I. The results are:

> the 2s(O) orbital transforms as a a_1 representation,
> the $2p_x$(O) orbital transforms as a b_1 representation,
> the $2p_y$(O) orbital transforms as a b_2 representation, and
> the $2p_z$(O) orbital transforms as another a_1 representation.

(3) Classify the valency orbitals of the ligand atoms with respect to the point group of the molecule and identify their group orbitals.

The two hydrogen 1s orbitals have the character shown in the following table — the individual characters are the number of orbitals unaffected by the particular symmetry operation carried out upon the two orbitals.

C_{2v}	E	C_2	$\sigma_v(xz)$	$\sigma'_v(yz)$
1s + 1s	2	0	0	2

This representation may be seen, by inspection of the C_{2v} character table, to be equivalent to the sum of the characters for the a_1 and b_1 irreducible representations. Those particular hydrogen group orbitals are very similar to the ones used to describe the bonding of the H_2 molecule — the nuclei are further apart in the water case. The a_1 hydrogen group orbital is H–H bonding, whilst the b_2 group orbital is H–H anti-bonding. Because of the greater distance between the hydrogen atoms in the water molecule, such bonding and anti-bonding interactions are not as great as those in the H_2 molecule. They are by no means negligible.

(4) Draw the molecular orbital diagram for the molecule.

Two considerations are of importance in drawing the molecular orbital diagram for a molecule. First, the relative energies of the orbitals must be borne in mind. The first ionization energies of the atoms involved in molecule formation give a good indication of how to position the atomic levels, and information about the magnitude of p–s energy gaps is useful. Second, and highly important, is the guiding principle that only atomic orbitals (single or group) belonging to the same irreducible representation may combine to give bonding and anti-bonding molecular orbitals.

It is also helpful to have knowledge of the photoelectron and absorption spectra to assist with the exact placement of the molecular levels. This information is not necessary for the production of a qualitative m.o. diagram. When such a diagram is seen to be consistent with the spectroscopic information it may be transformed into a reasonably scaled presentation. In addition it is very instructive to obtain the details of any proper quantum mechanical calculations which have been carried out on the particular molecule. Such calculations are possible for quite complicated molecules, and software packages exist which would allow the reader to carry out calculations of the molecular orbitals of systems of interest.

With regard to the water molecule it is important to realize that the first ionization energies of the hydrogen and oxygen atoms are almost identical (H, 1312; O, 1314 kJ

mol^{-1}), and that there is a large energy difference between the 2p and 2s levels of the oxygen atom (1544 kJ mol^{-1}). The latter piece of information indicates that the 2s(O) orbital does not participate in the bonding to a major extent. The 2s(O) and $2p_z$(O) orbitals do have identical symmetry properties and so have the possibility of mixing to give two hybrid orbitals but the large energy gap between them precludes very much interaction. That means that the main a_1 combination is between the $2p_z$(O) and h_1(H) orbitals. Combination occurs between the b_2 ($2p_y$(O)) and h_2(H) orbitals, with the b_1 ($2p_x$(O)) orbital remaining as a non-bonding m.o. The a_1 interaction produces a bonding orbital which is O−H bonding and which is also H−H bonding. The bonding orbital from the b_2 combination is O−H bonding but is H−H anti-bonding.

The above interactions are made obvious in the pictorial representation of the a_1 and b_2 orbital combination diagrams shown in Fig. 4.3. The molecular orbital energy

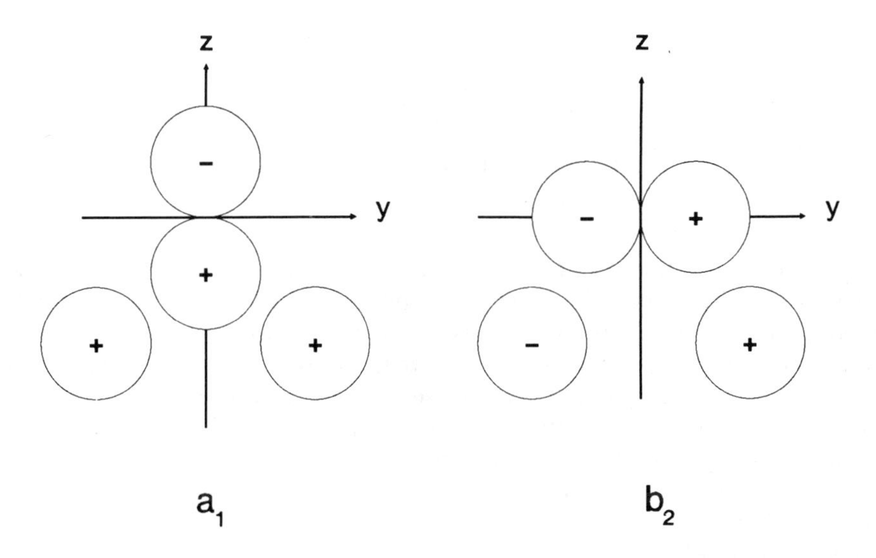

Fig. 4.3 — Overlap diagrams for the bonding orbitals of the water molecule.

diagram for the 90° water molecule is shown in Fig. 4.4. The $2a_1$ orbital gains some stability from slight interaction with the bonding $3a_1$ orbital which becomes destabilized to a similar extent. This causes the $3a_1$ orbital to have a higher energy than that of the $1b_2$ orbital. Such matters cannot be predicted by the qualitative approach used here, quantitative calculations being necessary. The 'tie lines' between the atomic and molecular orbitals in Fig. 4.4 (and all other m.o. diagrams in this text) do not take the mixing of molecular orbitals into account. The approach to m.o. diagrams in this book is to simplify them as far as possible and, wherever there is orbital mixing, the contributions from the various atomic orbitals are not always fully indicated by the tie lines.

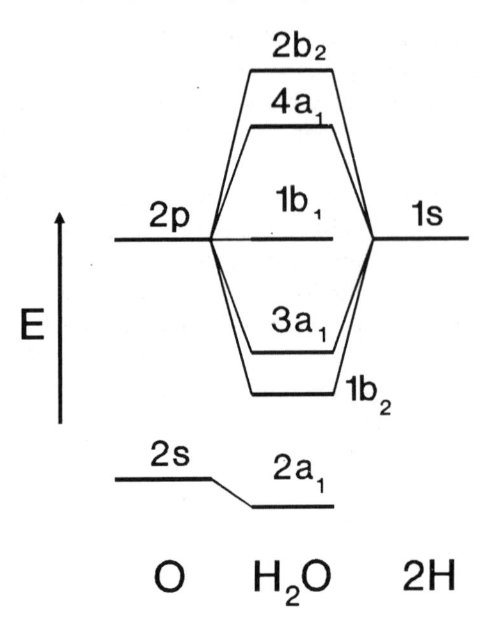

Fig. 4.4 — Molecular orbital diagram for the 90° water molecule.

One very important difference between VSEPR theory and molecular orbital theory should be noted. The molecular orbitals of the water molecule which are involved in the bonding are **three centre** orbitals. They are associated with all three atoms of the molecule. There are no **localized** electron pair bonds between pairs of atoms as VSEPR theory implies. The existence of three-centre orbitals (and multi-centre orbitals in more complicated molecules) is not only more consistent with symmetry theory — it allows for the reduction of interelectronic repulsion effects when more extensive, non-localized, orbitals are doubly occupied.

180° water molecule

(1) The linear water molecule belongs to the $D_{\infty h}$ point group. The classifications of atomic and group orbitals must be carried out using the $D_{\infty h}$ character table. The molecular axis, C_∞, is arranged to coincide with the z axis.

(2) The classification of the 2s and 2p atomic orbitals of the central oxygen atom.

The 2s(O) orbital transforms as σ_g^+,

the $2p_x$ and $2p_y$ orbitals transform as the doubly degenerate π_u representation and

the $2p_z$ orbital transforms as σ_u^+.

Throughout this text there is strict adherence to the use of the full symbols for all orbital symmetry representations. In most texts the '+' and '−' superscripts are omitted.

Unlike the 90° case, the 2s and $2p_z$ orbitals have different symmetry properties so there is no question of their mixing.

(3) Classification of the hydrogen group orbitals.
The H–H bonding group orbital, h_1, transforms as σ_g^+, and the H–H anti-bonding group orbital, h_2, transforms as σ_u^+. The detailed conclusions are dealt with in section 3.2.

(4) The molecular orbitals may now be constructed by allowing the orbitals of the oxygen and hydrogen atoms, belonging to the same representations, to combine. Thus the interaction between $\sigma_g^+(O)$ and h_1 is possible, by symmetry, but is very restricted by the large difference in energy between them. The major interaction is between $\sigma_g^+(O)$ and h_2 which possess very similar energies. The $2p_x$ and $2p_y$ orbitals of the oxygen atom remain as a doubly degenerate pair of π_u orbitals. The m.o. diagram for the linear water molecule is shown in Fig. 4.5.

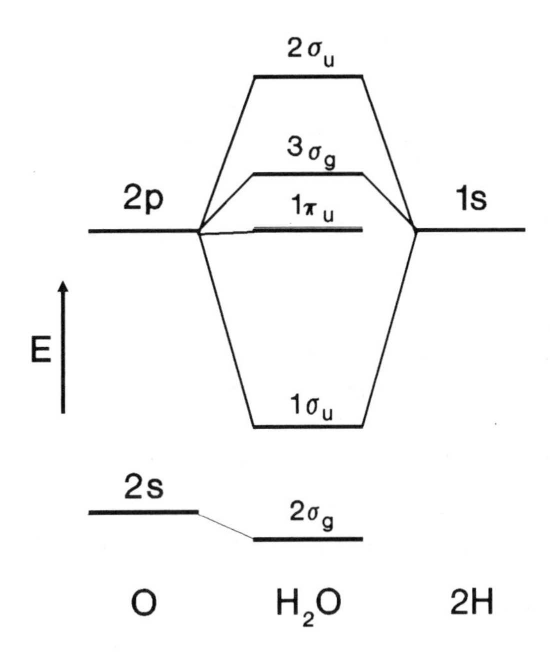

Fig. 4.5 — Molecular orbital diagram for the linear water molecule.

Comparison of the energies of 90° and 180° water molecules
In the water molecule there are eight valency electrons to be distributed in the lowest four molecular orbitals. In the 90° case this produces an electronic configuration which may be written as: 90° water: $2a_1^2 3a_1^2 1b_2^2 1b_1^2$ (ignoring the $1s^2$ pair of the oxygen atom which could be written as $1a_1^2$ or K). The $2a_1$ orbital is slightly bonding because of the interaction with the $3a_1$ orbital. The latter orbital has considerable bonding character, as has the $1b_2$, so that there are two bonding pairs of electrons responsible for the cohesion of the three atoms. The $1b_1$ orbital is non-bonding, there being no

hydrogen orbitals of that symmetry. The 90° molecule would have two similar, strongly bonding, electron pairs (in the $3a_1$ and $1b_2$ three-centre m.o.s) and two virtually non-bonding electron pairs (in the $1b_1$ and $2a_1$ orbitals) of very different energies.

The electronic configuration of the linear water molecule is $(2\sigma_g^+)^2\,(1\sigma_u^+)^2\,1\pi_u^4$ (the $1s^2$(O) pair of electrons occupy the $1\sigma_g^+$ molecular orbital). The $2\sigma_g^+$ pair of electrons are only weakly bonding, with the $1\sigma_u^+$ pair supplying practically the only cohesion for the three atoms, the other four electrons being non-bonding.

A comparison of the m.o. diagrams for the two forms of the water molecule is given in the correlation diagram of Fig. 4.6 and shows that the 90° angle confers the

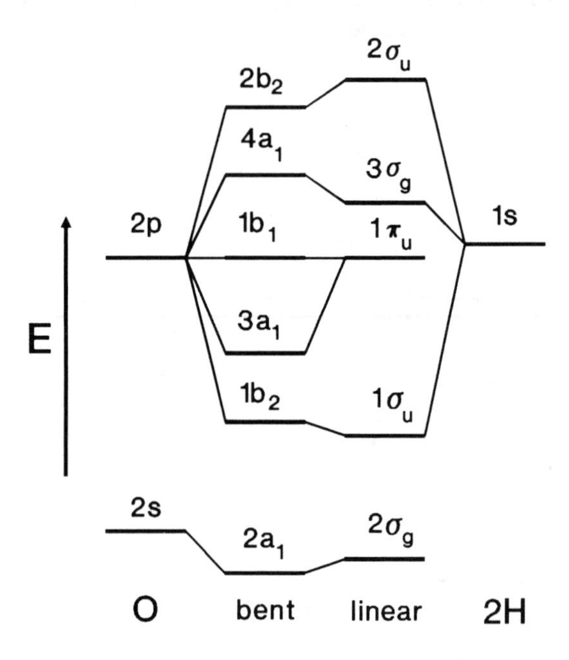

Fig. 4.6 — Correlation diagram for linear and 90° water molecules.

extra stability of two bonding pairs of electrons as opposed to the single pair of bonding electrons in the linear molecule.

It should be pointed out that neither interelectronic repulsions nor internuclear repulsions have been considered. The ignoring of interelectronic repulsions is not serious since the orbitals used in the two forms of the molecule are extremely similar. The internuclear repulsion in the 90° form would obviously be larger than in the linear case and contributes to the bond angle in the actual water molecule being greater than 90°. The actual state of the molecule, as it normally exists, is that with the lowest total energy, and only detailed calculations can reveal the various contributions. At a qualitative level, as carried out so far in this section, the decision from molecular orbital theory is that the water molecule should be bent.

The $1\pi_u$ orbitals in the $D_{\infty h}$ case (these are the non-bonding $2p_x$ and $2p_y$ orbitals of the oxygen atom) lose their degeneracy in the bent, C_{2v}, molecule. The $2p_x$ orbital retains its non-bonding character, but the 2p orbital makes a very important contribution to the $1b_2$ bonding m.o. of the C_{2v} molecule. It is this factor which is critical in determining the shape of the water molecule. Fig. 4.7 shows how the

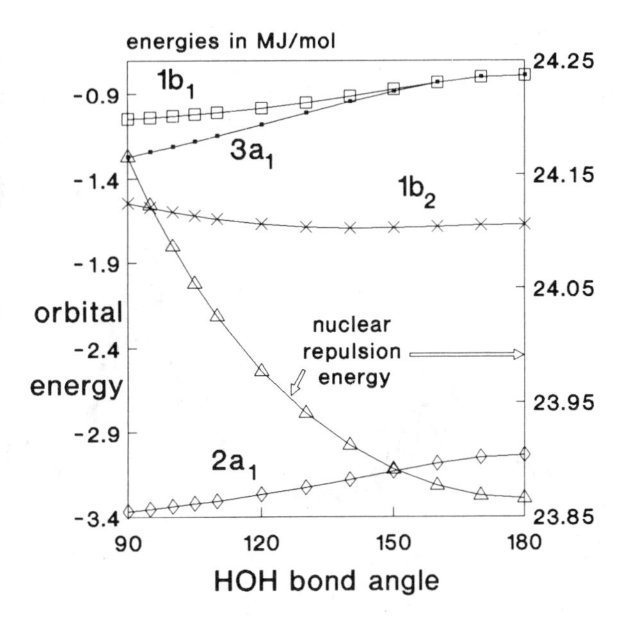

Fig. 4.7 — The variation of the energies of the $1b_1$, $3a_1$, $1b_2$ and $2a_1$ orbitals (left-hand *y* axis) and the nuclear repulsion energy (right-hand axis) of the water molecule with the bond angle.

energy of the filled orbitals, $1\pi_u - 3a_1$ and $1b_1$, $1\sigma_u^+ - 1b_2$, and $2\sigma_g^+ - 2a_1$ orbitals vary with bond angle. It also shows the variation of the internuclear repulsion energy due to the varying interproton distance. The information is derived from quantum mechanical calculations, and it is the balancing between these energies which is largely responsible for the observed bond angle of the ground state water molecule.

Qualitative orbital correlations similar to those of Fig. 4.7 were used by Walsh in 1953 to make predictions of whether AH_2 molecules (A representing any main group element) would be linear or bent. The qualitative Walsh diagrams for various systems are still of use in the predictions of molecular shape and electronic spectra, but the underlying ideas are not fully justifiable. They seem to give generally good results because of the cancelling out of various factors. Most small molecules can now be treated on an *ab initio* (from the beginning) basis in which the appropriate wave equation is solved to give reliable m.o. information. This includes the nuclear repulsion energy as well as the electronic and total energies.

Experimental confirmation of the order of molecular orbital energies for the water molecule is given by its photoelectron spectrum. Fig. 4.8 shows the helium line photoelectron spectrum of the water molecule. There are three ionizations at 1216, 1322 and 1660 kJ mol^{-1}. A fourth ionization at 3474 kJ mol^{-1} may be measured by using suitable X-ray quanta instead of the helium emission. That there are the four

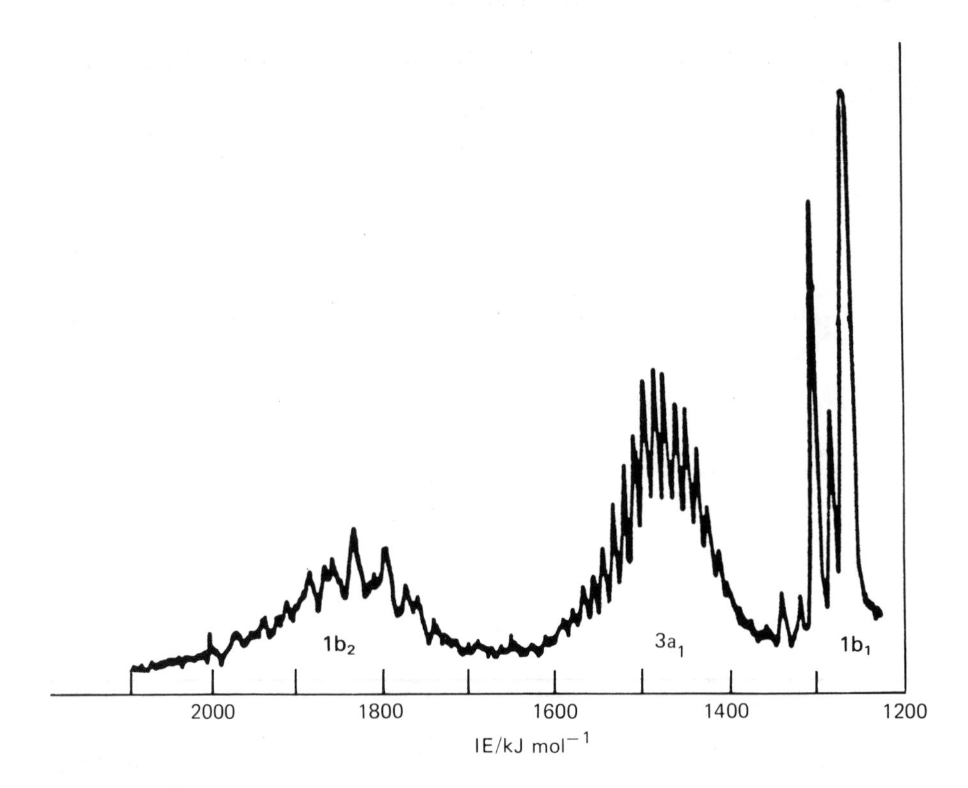

Fig. 4.8 — The photoelectron spectrum of the water molecule. (Reprinted in part from C. R. Brundle, M. B. Brown, N. A. Kuebler and H. Basch, *J. Am. Chem. Soc.*, **94**, 1455 (1972). Copyright 1972 American Chemical Society.)

ionization energies is consistent with expectations from the m.o. levels for a C_{2v} molecule (see Fig. 4.4).

The study of the magnitudes of the various vibrational excitations associated with the second and third ionizations confirms that the second ionization is from the $3a_1$ orbital (both O−H and H−H bonding) and that the third ionization is from the $1b_2$ orbital (O−H bonding but H−H anti-bonding). VSEPR theory suggests that there should only be two ionization energies of the water molecule, one representing the identical bonding pairs of electrons and the other the two identical pairs of non-bonding electrons.

Molecular orbital theory of CO_2 and XeF_2

The linear CO_2 and XeF_2 molecules belong to the $D_{\infty h}$ point group and, for purposes of classifying their atomic orbitals, the principal quantum numbers (2 for C, 5 for Xe) of their valence electrons may be dispensed with.

The s orbital transforms as σ_g^+, the p_x and p_y transform as π_u, and the p_z as σ_u^+. The p–s energy gaps in oxygen and fluorine atoms are very large, and their 2s orbitals must be regarded as being virtually non-bonding. They formally give rise to the orbitals, $1\sigma_g^+$ and $11\sigma_u^+$ (the 1s orbitals of the atoms being ignored). The 2p orbitals of the ligand atoms may be dealt with in two sets — the σ orbitals being the $2p_z$ lying along the molecular axis, and the π orbitals being made up from the $2p_x$ and $2p_y$ orbitals perpendicular to the molecular axis. The σ orbital combinations may be written as:

$$\phi(\sigma_g^+) = \psi_{2p_z}(O,F)_A + \psi_{2p_z}(O,\ F)_B \tag{4.1}$$

the plus sign being used to indicate a bonding interaction (the plus to plus overlap), and

$$\phi(\sigma_u^+) = \psi_{2p_z}(O,F)_A - \psi_{2p_z}(O,\ F)_B \tag{4.2}$$

the minus sign indicating an anti-bonding interaction (the plus to minus overlap).

The π combinations are:

$$\phi(\pi_u) = \psi_{2p_{x,y}}(O,F)_A + \psi_{2p_{x,y}}(O,\ F)_B \tag{4.3}$$

and

$$\phi(\pi_g) = \psi_{2p_{x,y}}(O,F)_A - \psi_{2p_{x,y}}(O,\ F)_B \tag{4.4}$$

with the signs having the same significance as those in equations (4.1) and (4.2). In the above equations, the subscripts A and B are used to distinguish between the two ligand atoms.

A summary of the above classifications follows.

C or Xe	O_2 or F_2
$\sigma_g^+(s)$	σ_g^+(2s bonding), σ_g^+($2p_z$ bonding)
$\sigma_u^+(p_z)$	σ_u^+(2s anti-bonding) σ_u^+($2p_z$ anti-bonding)
$\pi_u(p_{x,y})$	π_u($2p_{x,y}$ bonding), π_g($2p_{x,y}$ anti-bonding)

Taking into account the first ionization energies of the atoms involved in the CO_2 and XeF_2 molecules it is possible to use one qualitative m.o. energy diagram to discuss the disposition of the electrons which they contain. The diagram is shown in Fig. 4.9. The electronic configuration of the CO_2 molecule may be written as $(1\sigma_g^+)^2\ (1\sigma_u^+)^2\ (2\sigma_g^+)^2\ (2\sigma_u^+)^2\ 1\pi_u^4\ 1\pi_g^4$ (the 1s pairs of the three atoms being ignored in the numbering).

The first two pairs of electrons are non-bonding (they are the $2s^2$ pairs of the oxygen atoms) but the next four pairs (two σ and two π) occupy three-centre bonding molecular orbitals and are responsible for the high bond strength of the so-called 'double bonds' of the molecule (743 kJ mol^{-1}). The remainder of the electrons occupy the three-centre non-bonding π_g orbitals.

The photoelectron spectrum of the CO_2 molecule indicates ionization energies of 1330, 1671, 1744 and 1872 kJ mol^{-1}, for the removal of electrons from the $1\pi_g$, $1\pi_u$, $2\sigma_u^+$ and $2\sigma_g^+$ molecular orbitals respectively.

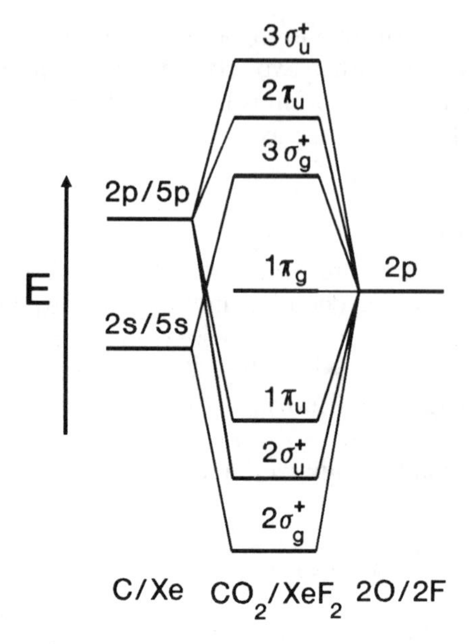

Fig. 4.9 — Molecular orbital diagram for CO_2 and XeF_2.

The electronic configuration of the XeF_2 molecule is written as $(1\sigma_g^+)^2$ $(1\sigma_u^+)^2$ $(2\sigma_g^+)^2$ $(2\sigma_u^+)^2$ $1\pi_u^4$ $1\pi_g^4$ $(3\sigma_g^+)^2$ $2\pi_u^4$ (the 1s pairs of the fluorine atoms and the forty-six electron core of the xenon atom being ignored). It follows the same pattern as that of the CO_2 molecule with the additional occupancy of the $3\sigma_g^+$ and the doubly degenerate $2\pi_u$ orbitals, all three orbitals being anti-bonding. Their occupancy reduces the bonding of the XeF_2 molecule to the single electron pair in the $2\sigma_u^+$ bonding orbital. The bonding of the molecule is equivalent to each Xe$-$F 'bond' possessing a bond order of 0.5. The molecule is only marginally stable with respect to its formation; $\Delta H_f^\ominus(XeF_2,g) = -82$ kJ mol^{-1}, compared with CO_2; $\Delta H_f^\ominus(CO_2,g) = -394$ kJ mol^{-1}. A fairer comparison would be to use the standard heat of formation of CO_2 from carbon in the gaseous state, which is -1109 kJ mol^{-1}.

The molecular orbital treatment of XeF_2 does not depend upon (and so does not over-emphasize) the inclusion of d orbital contributions from the xenon atom. Reference to the $D_{\infty h}$ character table shows that the d_z2 orbital of xenon transforms as σ_g^+, and that the d_{xz} and d_{yz} orbitals transform as π_g. Those orbitals have suitable symmetries to interact with the appropriate orbitals of the fluorine atoms. The other two d orbitals (xy and x^2-y^2) transform as the representation, δ_g, and cannot participate in the bonding of XeF_2. It is doubtful whether any significant d orbital involvement occurs because of their relatively higher energies.

4.3 SOME POLYATOMIC MOLECULES

This section consists of the treatment by VSEPR and m.o. theories for some small polyatomic molecules. The shapes and bonding of the following molecules are discussed:

(i) NH_3,
(ii) CH_4,
(iii) BF_3, NF_3 and ClF_3,
(iv) B_2H_6,
(v) and the HF_2^- ion.

Ammonia, NH_3

The VSEPR treatment takes the nitrogen atom ($2s^22p^3$) and adds the 1s electrons from the three hydrogen atoms producing the $2s^22p^6$ configuration in the valence shell of the nitrogen atom. The four pairs of σ-type electrons adopt a tetrahedral distribution to minimize electron pair repulsions. One of the four tetrahedral positions is occupied by a lone pair, the other three being those of the three bonding pairs. The molecular shape should be trigonally pyramidal with bond angles somewhat smaller than those of a regular tetrahedron. The molecule is a trigonal pyramid with HNH bond angles of 107°.

Molecular orbital treatment of the ammonia molecule

Molecular orbital theory may be used to establish stability preferences between extreme geometries for any molecular system. Those for the ammonia molecule are the trigonally planar, D_{3h}, the trigonally pyramidal, C_{3v}, and the T-shaped, C_{2v}, symmetries.

The 2s and 2p orbitals of the nitrogen atom and the 1s orbital combinations of the three hydrogen atoms transform, with respect to the point groups, D_{3h}, C_{3v} and C_{2v}, as indicated in Table 4.2.

Table 4.2 — Orbitals of the nitrogen and hydrogen atoms of NH_3

Orbital	Point group		
	D_{3h}	C_{3v}	C_{2v}
2s(N)	a_1'	a_1	a_1
$2p_x$(N)	e'	e	b_1
$2p_y$(N)			b_2
$2p_z$(N)	a_2''	a_1	a_1
3 × 1s(H)	$a_1' + e'$	$a_1 + e$	$a_1 + a_1 + b_2$

In the D_{3h} case, interaction between the nitrogen and hydrogen orbitals is possible for those with a_1' and e' symmetry, with the a_2'' orbital being non-bonding. The a_1' bonding combination is only slightly bonding because of the mismatch of energies of the contributing 1s(H) (ionization energy = 1312 kJ mol^{-1}) and 2s(N) (ionization energy = 1400 + 1060 = 2460 kJ mol^{-1}) orbitals. The major interaction occurs with the energetically favourable e′ orbitals giving low-energy bonding and high-energy anti-bonding sets of molecular orbitals. The molecular orbital diagram of the D_{3h} ammonia molecule is shown in the correlation diagram of Fig. 4.10. The

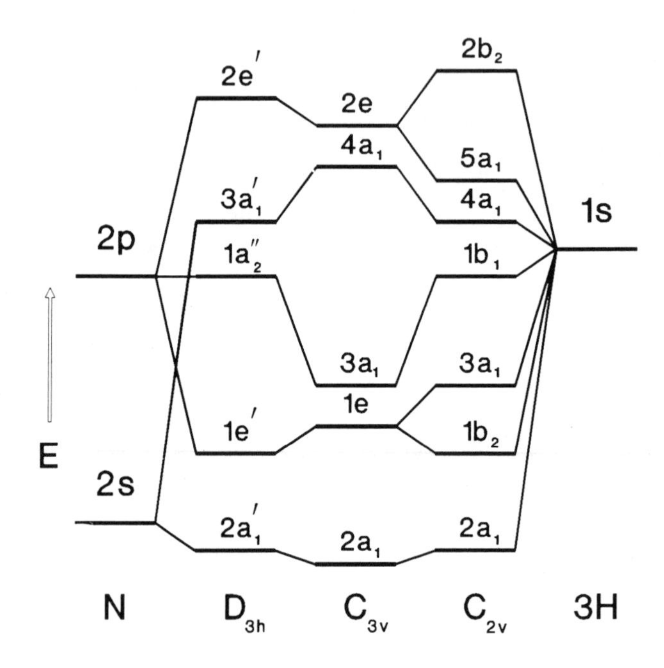

Fig. 4.10 — Correlation diagram for the molecular orbitals of the ammonia molecule with D_{3h}, C_{3v} and C_{2v} symmetries.

orbital numbering takes into account the 1s^2 electrons of the nitrogen atom which would be labelled as $1a_1'$.

In the C_{3v} and C_{2v} symmetries the 2s and $2p_z$ orbitals of the nitrogen atom belong to identical representations (a_1) indicating that mixture is possible. Because the 2p–2s energy gap is large, any mixing is minimal and is best considered after the molecular orbitals are formed.

In the C_{3v} case the main a_1 interaction is between the $2p_z$(N) orbital and the a_1 hydrogen group orbital, shown in Fig. 4.11, the 2s(N) involvement being minimal because of the mismatch of energies mentioned above. As may be seen from Fig. 4.11 the interaction between the hydrogen group orbital (fully H−H bonding) and the nitrogen $2p_z$ orbital (zero in the D_{3h} case) becomes bonding as the group orbital leaves the xy plane. There is less +/− overlap and more +/+ overlap as the bond angle decreases. It is this interaction which is mainly responsible for the C_{3v} symmetry being more stable than the D_{3h}. Fig. 4.12 shows the extent by which the $3a_1$

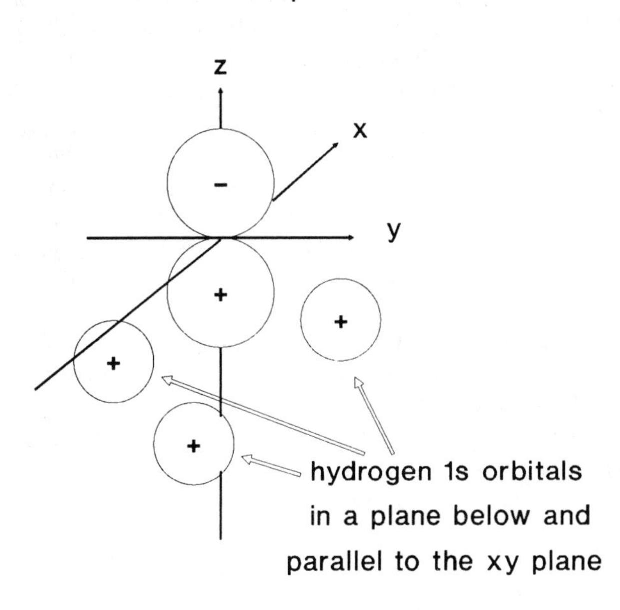

Fig. 4.11 — Diagrammatic representation of the atomic orbital contributions to the $3a_1$ molecular orbital of the ammonia molecule.

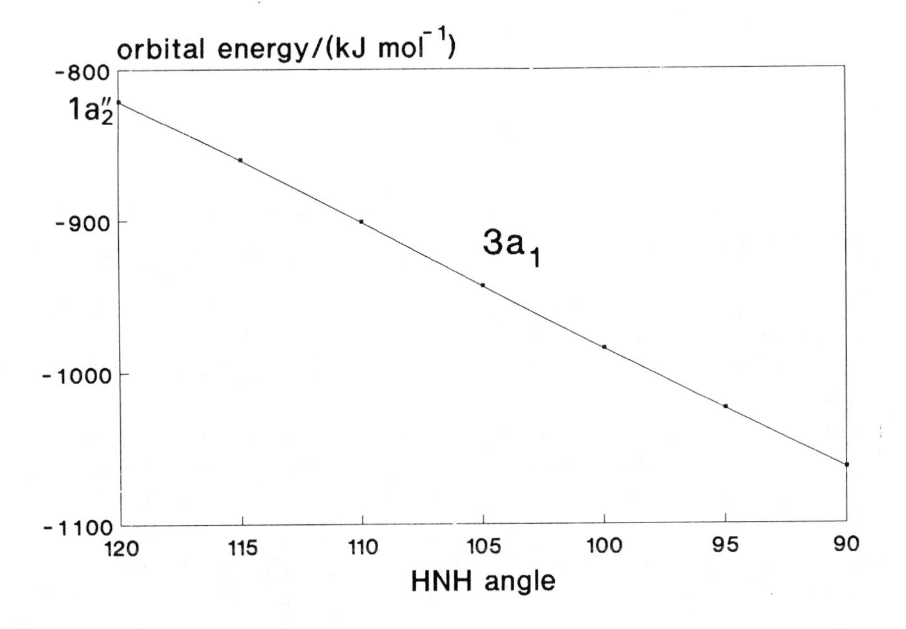

Fig. 4.12 — The variation of the $1a_2''$–$3a_1$ orbital of the ammonia molecule with bond angle.

orbital becomes stabilized as the symmetry deviates from D_{3h} (based on quantum mechanical calculations). The e-type interactions are not as effective as in the D_{3h} case since the hydrogen orbitals are no longer in the *xy* plane. Slight mixing of the $2a_1$ and $3a_1$ orbitals stabilizes the former at the expense of the latter. The m.o. diagram is shown in Fig. 4.10.

The third option for the ammonia molecule is the T-shape which belongs to the C_{2v} point group. The $3a_1$ interaction is mainly between the $2p_z$(N) orbital and the hydrogen orbital positioned along the *z* axis. The interaction between the 2s(N) orbital and the 1s(H) + 1s(H) combination along the *y* axis produces the $2a_1$ (slightly bonding) and 4a (slightly anti-bonding) molecular orbitals. The $2p_x$(N) and $2p_y$(N) orbitals lose their degeneracy in C_{2v} symmetry. The $2p_x$ (N) orbital becomes b_1 and reverts to being a non-bonding orbital (the $1a_2''$ orbital of the D_{3h} case) and the $2p_y$(N) orbital interacts strongly with the appropriate hydrogen group orbital. The m.o. diagram is shown in Fig. 4.10.

The electronic configurations of the ammonia molecule in its three symmetries are shown in Table 4.3.

Table 4.3

NH_3 symmetry	Electronic configuration
D_{3h}	$(2a_1')^2(1e')^2(1a_2'')^2$
C_{3v}	$(2a_1)^2(1e)^4(3a_1)^2$
C_{2v}	$(2a_1)^2(1b_2)^2(3a_1)^2(1b_1)^2$

A plot of the total energies of the D_{3h}, and various C_{3v} and C_{2v}, molecules (taken from accurate calculations) is shown in Fig. 4.13.

The main stabilizing factor in the ammonia molecule is the bonding nature of the $3a_1$ orbital in C_{3v}. symmetry. The two electrons in that orbital cause the molecule to assume C_{3v} symmetry in its ground state. The stabilization is opposed by the increase in interproton repulsion. The alternative D_{3h} and C_{2v} shapes are less stable because the $2p_z$(N) orbital is non-bonding in those cases.

The photoelectron spectrum of the ammonia molecule is consistent with the m.o. conclusions — there are three ionization energies at 1050, 1445 and 2605 kJ mol^{-1}, corresponding to the removal of electrons from the $3a_1$, 1e and $2a_1$ molecular orbitals respectively.

It is obvious from this example that the operation of the molecular orbital theory is far more complicated than the use of the VSEPR method. It is also obvious that the m.o. treatment leads to a greater understanding of the molecule and gives results which are testable by experiment.

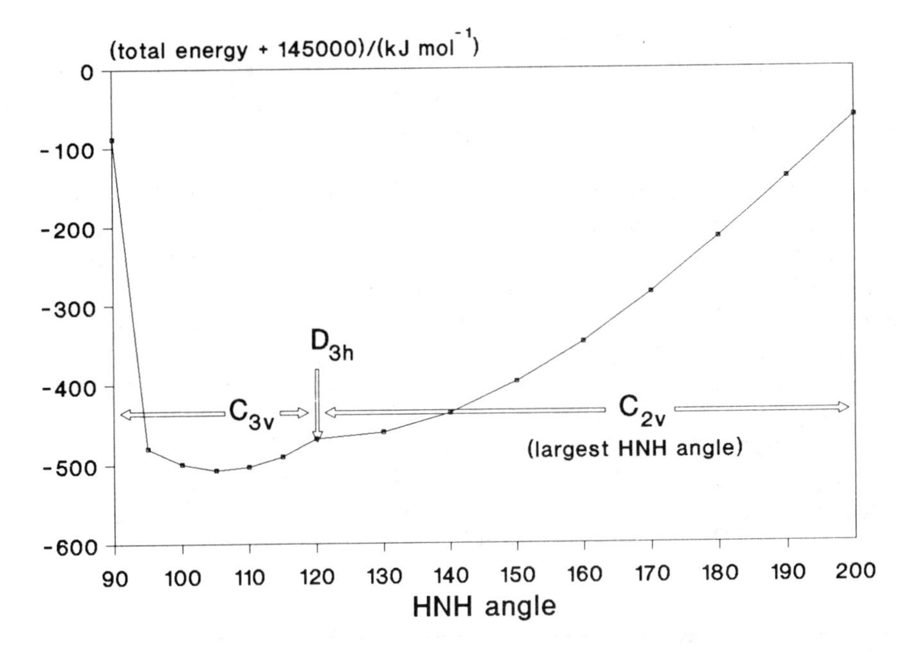

Fig. 4.13 — The variation of the total energies of ammonia molecules of D_{3h}, C_{3v} and C_{2v}, symmetries with the value of the largest bond angle.

Methane, CH_4

The methane molecule is a very important molecule in **organic chemistry**, the geometry around the four-valent carbon atom being basic to the understanding of the structure, isomerism and optical activity of a very large number of organic compounds. It is a tetrahedral molecule belonging to the special tetrahedral point group, T_d. In this section the VSEPR and m.o. theories are compared.

The VSEPR treatment of the methane molecule raises a number of serious problems. The carbon atom has the electronic configuration $2s^2 2p^2$ in its valence shell and, if the four electrons from the ligand hydrogen atoms are added, it becomes $2s^2 2p^6$ — so that there are four pairs of electrons which lead to the correct prediction of the tetrahedral shape of CH_4.

The consequent assertions that are made of there being four identical localized electron pair bonds in the four C–H regions, made up from sp^3 hybrid carbon orbitals and the hydrogen 1s orbitals, are not consistent with the experimentally observed photoelectron spectrum. The photoelectron spectrum is shown in Fig. 4.14 (produced by W. C.Price of King's College London in 1968) and shows that there are two ionization energies at 1230 and 2160 kJ mol^{-1}corresponding to the removal of electrons from the $t_2$and a_1 molecular orbitals respectively. The spectrum in the 1230 kJ mol^{-1} region consists of two main overlapping systems. This is consistent with the expectation that the 'odd' t_2^5configuration (the $^2T_2$ state of the CH_4^+ ion) is subject to the Jahn–Teller effect (dealt with in Chapter 8) and should lose its triple degeneracy and realize two states, $^2B_2$ and ^2E, if the symmetry of the CH_4^+ is reduced to D_{2h} (a

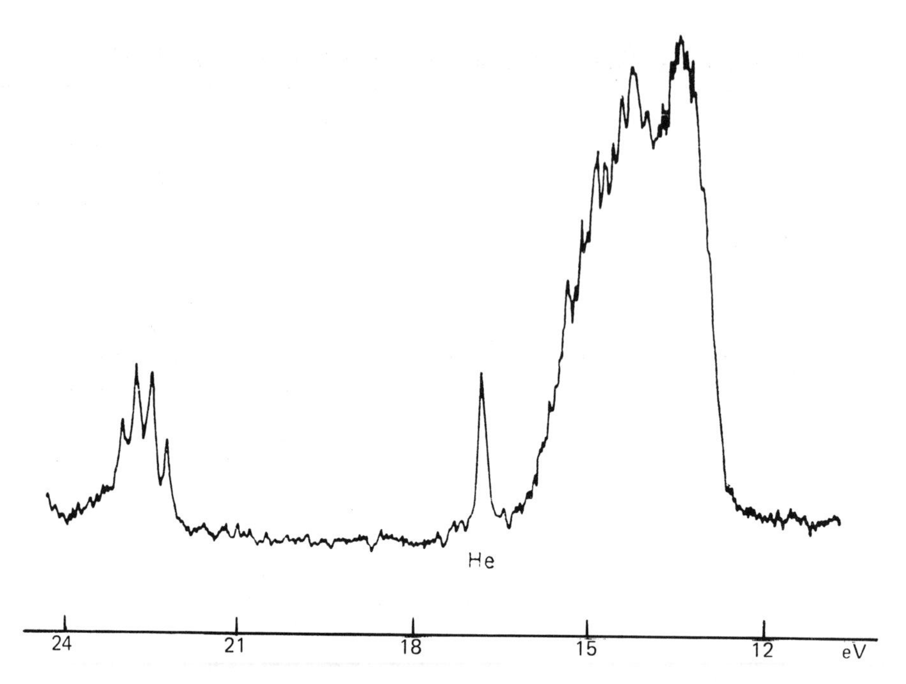

Fig. 4.14—The photoelectron spectrum of the methane molecule. (Reprinted with permission from W. C. Price, Developments in photoelectron spectroscopy, paper presented at Institute of Petroleum Conference on Molecular Spectroscopy, 1968, published by Elsevier, pp. 221–236.)

distorted tetrahedron) by the accompanying distortion. The valence electrons in CH_4 have two levels of energy rather than the single level indicated by the hybridization approach.

Molecular orbital theory of tetrahedral and square planar methane

It is instructive to compare the stabilities of the tetrahedral and square planar forms of the methane molecule. The transformation properties of the orbitals of the carbon atom and the group orbitals of the four hydrogen atoms, with respect to the T_d and D_{4h} point groups, are given in Table 4.4.

The molecular orbital diagrams for the two symmetries of the methane molecule are shown in Fig. 4.15. In T_d symmetry there is a total match between the two a_1 and two t_2 sets of orbitals. In D_{4h} symmetry the a_{1g} combinations are similar in energy to the a_1 in T_d. There is a smaller amount of H–H bonding character in the D_{4h} case, which is the reason for placing the $1a_{1g}$ orbital slightly higher in energy than the $1a_{1g}$ orbital. The $1t_2$ orbitals lose their degeneracy in D_{4h} symmetry where they become the e_u bonding pair of orbitals and the non-bonding $1b_{1g}$ orbital. The latter orbital is placed at a fairly high level since it is considerably H–H anti-bonding. This character makes its energy higher than that of the carbon $2p_z$–$1a_{2u}$ orbital. The anti-bonding $2t_2$ orbitals lose their degeneracy in D_{4h} symmetry and become the anti-bonding $2e_u$ and non-bonding $1a_{2u}$ orbitals.

Table 4.4 — Orbitals of the carbon and hydrogen atoms of CH_4

Orbital	Point group	
	T_d	D_{4h}
2s(C)	a_1	a_{1g}
$2p_x$(C)	t_2	e_u
$2p_y$(C)		
$2p_z$(C)		a_{2u}
4 × 1s(H)	$a_1 + t_2$	$a_{1g} + e_u + b_{1g}$

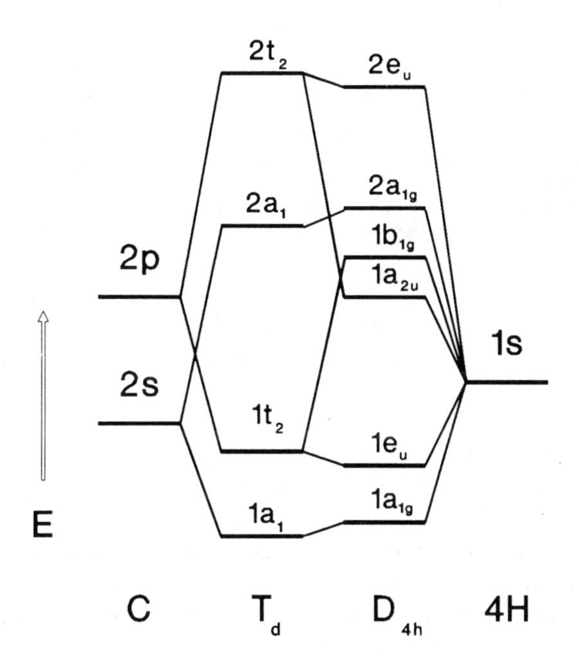

Fig. 4.15 — Correlation diagram for the molecular orbitals of methane with T_d and D_{4h} symmetries.

The electronic configurations of the methane molecule with T_d and D_{4h} symmetries are $(1a_1)^2$ $(1t_2)^4$ and $(1a_{1g})^2(1e_u)^4(1a_{2u})^2$ respectively. In the T_d case, all eight electrons occupy bonding orbitals, but in D_{4h} symmetry, two of the electrons occupy the non-bonding $1a_{2u}$ orbital, which is the reason for methane being tetrahedral in its electronic ground state. As the symmetry deviates from D_{4h}, the $2p_z$ orbital of the carbon atom is able to participate in the formation of a bonding orbital.

The m.o. theory is consistent with the two ionizations shown in the photoelectron spectrum of CH_4 and implies that the bonding consists of four electron pairs which occupy the $1a_1$ and $1t_2$ five-centre molecular orbitals.

Boron, nitrogen and chlorine trifluorides

Most of the principles by which such AB_3 molecules as BF_3, NF_3, and ClF_3, are to be treated have been dealt with in previous sections. For this reason their treatments are presented in a concise fashion with only additional points being stressed.

As was the case with the ammonia molecule, the molecules of this section may adopt one of the three symmetries, D_{3h}, C_{3v} or C_{2v}. The VSEPR and molecular orbital approaches are used and are compared and criticized, and factors influencing structure preference are discussed.

Boron trifluoride

The VSEPR treatment of the shape of the BF_3 molecule depends upon the three electron pairs repelling each other to a position of minimum repulsion, giving the molecule its trigonally planar configuration. The molecule belongs to the D_{3h} point group, but experimental observation of the B–F internuclear distance (130 pm) shows it to be smaller than that in the BF_4^- ion (153 pm). Between the same two atoms a shorter distance implies a higher bond order. The VSEPR treatment ignores any possible contribution to the bonding, which may be made by the other electrons of the ligand fluorine atoms. It assumes that the fluorine contributions are just the three σ electrons.

In the molecular orbital treatment of the D_{3h} BF_3 molecule the orbitals of the boron atom are classified as follows:

2s(B):	a_1'
$2p_{x,y}$(B):	e'
$2p_z$(B):	a_2''

The three 'sigma'-type 2p orbitals of the fluorine atoms may be positioned along the lines joining them to the boron atom, respectively. The arrangement is shown in Fig. 4.16. They transform as the sum $a_1' + e'$, and so may be involved in the production of six molecular orbitals'of those symmetries, three bonding and three anti-bonding.

It is convenient to consider the $2p_z$ orbitals of the three fluorine atoms as a separate set for purposes of simplicity. They are shown, together with the $2p_z$ orbital of the central boron atom, in Fig. 4.17. Inspection of that figure shows that there are bonding possibilities between the boron and fluorine orbitals of a 'pi' type. The three $2p_z$ orbitals of the fluorine atoms have the character:

	E	C^3	C_2	σ_h	S_3	σ_v
$3 \times 2p_z$	3	0	−1	−3	0	1

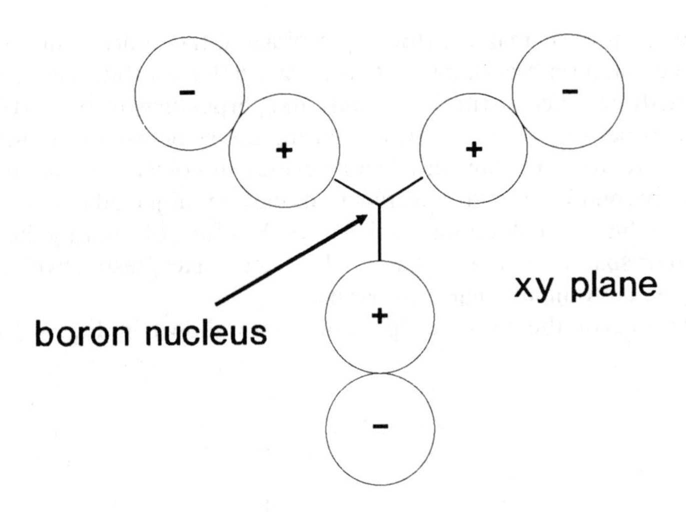

Fig. 4.16 — The 'sigma'-type 2p orbitals of the three fluorine orbitals in BF_3.

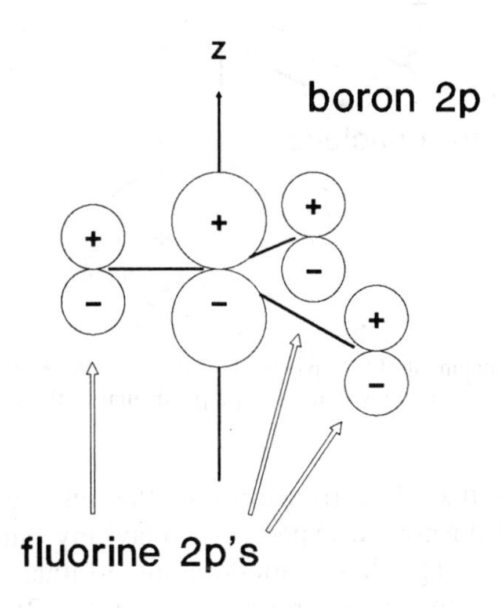

Fig. 4.17 — The atomic orbital contributions to the 'pi' system of BF_3

and this reduces to the sum $a_2' + e''$.

Although there is a formal method of carrying out such a reduction, it is not necessary to use it for most purposes. Reduction of the above character may be achieved by the realization of the bonding potential between the $2p_z$ orbitals of the boron and fluorine atom, reference to which has already been made. In order for this to be so, one of the group orbitals made from the $2p_z$ orbitals of the fluorine atoms must belong to the a_2'' representation. If the characters of a_2'' are subtracted from those of the $3 \times 2p_z$ orbitals, as derived above, the remaining characters correspond to those of the representation, e''.

For completeness the remaining three 2p orbitals of the fluorine atoms may be classified. Notice that no subscript is used to distinguish the 2p orbitals used in the 'σ' bonding and the three which are in the xy plane and perpendicular to the B−F bond directions. This is because of the particular setting up of the σ-type orbitals along directions which are in the xy plane but do not necessarily coincide with either of the x or y axes. The $2p_x$ and $2p_y$ orbital wavefunctions may be subjected to any two of an infinite number of linear combinations which have the effect of rotating the original orbitals from their spatial orientations along the x and y axes, respectively, to ones which are at some convenient angle to those axes.

The orientations of the fluorine 2p orbitals are shown in Fig. 4.18. Their

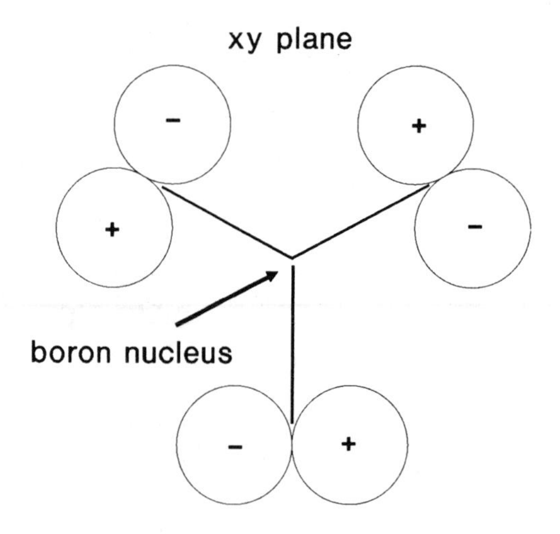

Fig. 4.18 — An arrangement of the 2p orbitals of the three fluorine atoms in BF_3 which are in the horizontal plane and perpendicular to the C_2 axes.

characters are such that they transform as the sum: $a_2' + e'$. Because of their orientation they would not be expected to play any part in the bonding of the molecule and are to be regarded as non-bonding orbitals.

A summary of the atomic and group orbitals of the BF_3 molecule follows:

Boron atoms	Fluorine atoms
$a_1'(2s)$	a_1' and e'(σ-type 2p)
$e'(2p_{x,y})$	a_2'' and e''(π-type perpendicular to xy)
$a_2''(2p_z)$	a_2' and e'(in xy perpendicular to B−F)

The m.o. diagram for BF_3 is shown in Fig. 4.19. There are three virtually non-bonding orbitals (derived from the three 2s(F) orbitals of the fluorine atoms) which transform as a_1' and e' molecular orbitals and are at very low energies (the 2p–2s energy gap in fluorine is 2026 kJ mol^{-1}). They are important for orbital numbering and electron counting purposes only, and are omitted from the diagram.

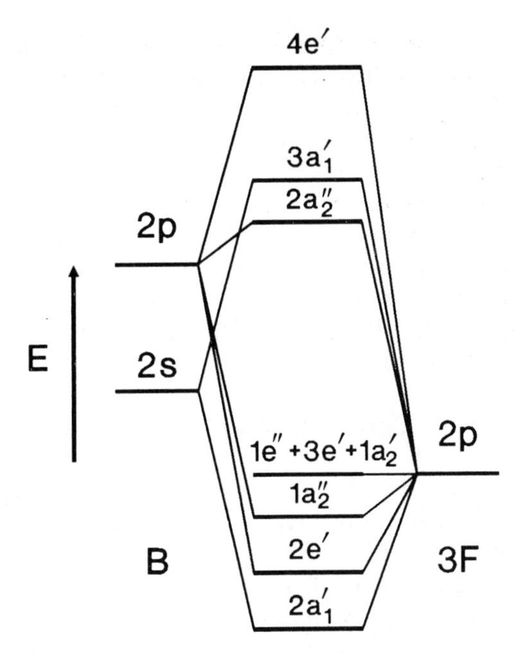

Fig. 4.19 — Molecular orbital diagram of BF_3 (with D_{3h} symmetry).

There are four bonding orbitals: the $2a_1'$, the doubly degenerate $2e'$, and the π-type $1a_1''$. Then all with very similar (but not identical) energies come the non-bonding $3e'$, $1e''$ and $1a_2'$ orbitals, and higher still in energy are the anti-bonding $2a_2''$, $3a_1'$ and $4e'$ orbitals. The electronic configuration of the trigonally planar BF_3 molecule is thus:

$$BF_3(D_{3h}):\ (2a_1')^2(2e')^4(1a_2'')^2(3e')^4(1e'')^4(1a_2')^2$$

The first two levels contain the three σ-type bonding pairs, with the fourth bonding pair occupying the π-type $1a_2''$ orbital. The bond order of each B–F linkage is therefore $1\frac{1}{3}$, and is in accordance with the bond length being smaller than in BF_4^- which has a bond order of 1.0. The photoelectron spectrum of BF_3 is consistent with ionizations from the $1a_2'$, $1e''$, $3e'$, $1a_2''$ and $2e'$ orbitals with energies of 1502, 1573, 1633, 1824 and 1924 kJ mol^{-1} respectively.

Nitrogen trifluoride, NF_3

The VSEPR theory of NF_3 takes the valence shell configuration of the nitrogen atom: $2s^22p^3$ and adds the three σ-type electrons from the fluorine atoms, giving: $2s^22p^6$ — four pairs of σ-type electrons which distribute themselves tetrahedrally around the nitrogen atom. The three identical bonding pairs are squeezed together somewhat by the lone pair to give a trigonally pyramidal molecule with bond angles slightly less than the regular tetrahedral angle. The observed angle is 104°.

In the molecular orbital treatment of the C_{3v} molecule the orbitals of the nitrogen atom are classified with respect to the C_{3v} character table:

2s(N):	a_1
$2p_{x,y}$(N):	e
$2p_z$(N):	a_1

The 2s and $2p_z$ orbitals (both are a_1) may mix to an extent which depends on their relative energies (a similar situation to the NH_3 case). Although the 2p–2s energy gap is large (1060 kJ mol^{-1}) the ionization energy of the fluorine atom (1680 kJ mol^{-1}) makes possible the interaction of its 2p orbitals with the 2s and 2p orbitals of the nitrogen atom almost equally likely.

The σ-type fluorine orbitals (2p) may be placed so that they lie along the N–F directions, and may be classified as the sum $a_1 + e$. The other six 2p(F) orbitals may be dealt with in two sets. One set is formed by the three 2p(F) orbitals which are no longer in the correct position to form π-type overlap with the $2p_z$ orbital of the nitrogen atom. They transform as the sum $a_1 + e$. The other set, made up from the other 2p orbitals which are perpendicular to the N–F bond directions, behaves as the sum a_2^{+e}. The two sets of orbitals are virtually non-bonding. The m.o. diagram for NF_3 is shown in Fig. 4.20. The three fluorine 2s orbitals which form the $1a_1$ and 1e

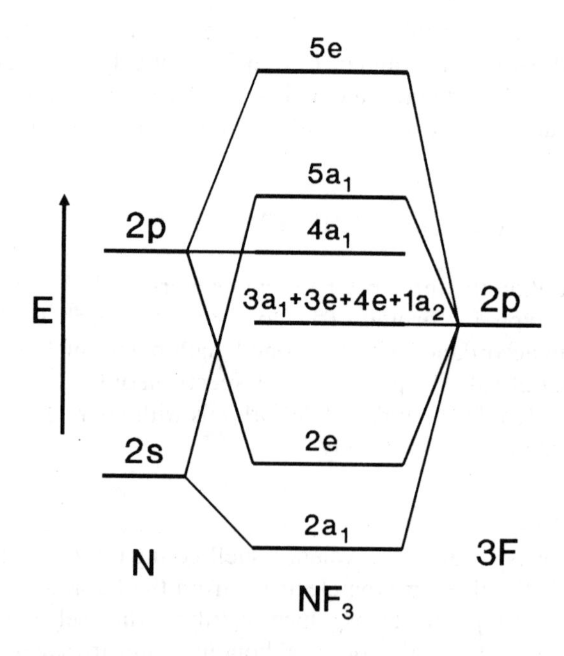

Fig. 4.20 — Molecular orbital diagram for NF_3 (with C_{3v} symmetry).

orbitals are omitted from the diagram. They are non-bonding because of their very low energy compared to any of the nitrogen orbitals. The $2a_1$ bonding orbital is constructed from the nitrogen a_1 (mainly 2s) and the fluorine σ-type a_1 group orbital. The next molecular orbitals of higher energy are the 2e doubly degenerate set. The non-bonding 2p orbital combinations of the fluorine atoms are next highest in energy and are labelled as $3a_1$, 3e, 4e and $1a_2$ molecular orbitals. In fact these have slightly different energies and are not degenerate, as indicated by the simplified diagram of Fig. 4.20.

The next highest orbital is the non-bonding a_1, mainly $2p_z$(N) orbital, now labelled as $4a_1$ and is the HOMO*. The higher orbitals are the $5a_1$ and 5e which are normally vacant and anti-bonding.

The electronic configuration of the NF_3 molecule may be written as:

$$NF_3(C_{3vb}): 2a_1^2 2e^4 3a_1^2 1a_2^2 3e^4 4e^4 4a_1^2$$

there being three bonding pairs of electrons in the $2a_1$ and 2e four-centre molecular orbitals.

If the NF_3 molecule possessed a trigonally planar structure its electronic configuration would be that of the $BF_3(D_{3h})$ molecule, with the extra pair of electrons occupying the anti-bonding $2a_2''$ orbital. That would cancel out any bonding effect of the two electrons in the bonding $1a_2''$ orbital and it would not have the extra stability of the π-type bonding possessed by the BF_3 molecule. For $NF_3(D_{3h})$ the HOMO would be the anti-bonding $2a_2''$ orbital, whereas with the actual C_{3v} symmetry the HOMO is non-bonding.

Chlorine trifluoride, ClF_3

The VSEPR treatment of the ClF_3 molecule is of some interest in that it uses a chlorine 3d orbital to contain one of the electron pairs. The chlorine atom is normally $3s^2 3p^5$, and as such is expected to be univalent. To arrange for it to be trivalent, one of the paired-up 3p electrons must be excited to a 3d level. Then the three electrons from the ligand fluorine atoms may be added to give the configuration $3s^2 3p^6 3d^2$, making five electron pairs, three of which are bonding and two are non-bonding. The basic electron pair distribution is expected to be trigonally bipyramidal. There are three possible shapes for ClF_3 which may result from permuting the three fluorine atoms between the two different positions in the trigonal bipyramid, in the trigonal plane or in a position along the C_3 axis. There is a general rule which is based upon a detailed consideration of the repulsions between bonding–bonding and bonding–non-bonding, and between two non-bonding pairs of electrons. This is that (for the case of five electron pairs) non-bonding pairs are more stable in the trigonal plane. Applied to ClF_3 this rule implies that the molecule should be T-shaped. The relatively greater repelling effects of the two in-plane non-bonding pairs cause the angle F−Cl−F to be slightly less than 90° . The observed angle is 87°, as is shown in the diagram of Fig. 4.21.

The molecule belongs to the C_{2v}. point group, and for molecular orbital purposes the C_2 axis is setup along the z axis.

The orbitals of the chlorine atom transform as:

*highest occupied molecular orbital.

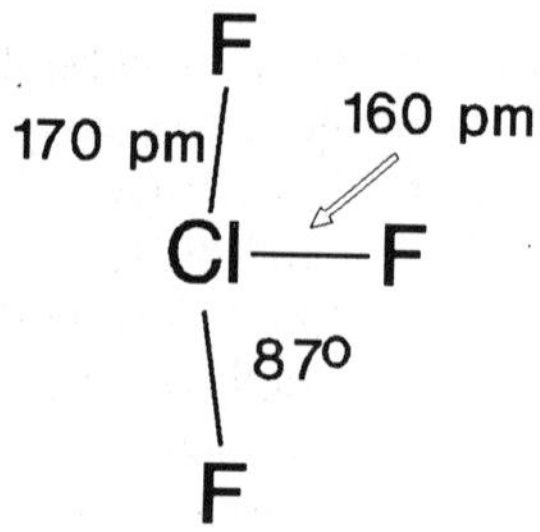

F
170 pm 160 pm
Cl—F
87°
F

Fig. 4.21 — The structure of the ClF_3 molecule.

3s(Cl):	a_1
$3p_x$(Cl):	b_1
$3p_y$(Cl):	b_2
$3p_z$(Cl):	a_1

The 3s and $3p_z$ orbitals may mix, the appropriate atomic orbital combinations being given by the equations:

$$c_1(a_1) = \psi_{3s} + \lambda\psi_{3p_z} \tag{4.5}$$

and

$$c_2(a_1) = \lambda\psi_{3s} - \psi_{3p_z} \tag{4.6}$$

the factor, λ, allowing for incomplete mixing.

The fluorine 2p orbitals used in the σ-type bonds, and which are placed along the Cl−F bond directions, are of two kinds. The $2p_z$(F) orbital placed along the C_2 axis is unique and transforms as an a_1 representation. Two $2p_y$(F) orbitals are used for σ-type bonding and transform, in linear combinations, as a_1 and b_2 group orbitals:

$$f_1(a_1) = \psi_{2p_y}(F)_A + \psi_{2p_y}(F)_B \tag{4.7}$$

and

$$f_2(b_2) = \psi_{2p_y}(F)_A - \psi_{2p_y}(F)_B \tag{4.8}$$

The three $2p_x$(F) orbitals of the fluorine atoms transform as the sum $a_2 + 2b_1$ (i.e. there are two b_1 combinations), and may take part in π-type interactions. The other three 2p(F) orbitals not connected with the σ-type bonding transform as:

$2p_y$(F on z axis):	b_2
$2p_z$(F's along y axis):	$a_1 + b_2$

A summary of the symmetry properties of the atomic and group orbitals of the $ClF_3(C_{2v})$ molecule follows:

Chlorine orbitals		Fluorine orbitals
2s(Cl):	a_1	$2p_z$(F on z axis): a_1
$2p_x$(Cl):	b_1	$f_1(a_1)$, $f_2(b_2)$, (both along y axis)
$2p_y$(Cl):	b_2	$3 \times 2p_x$(F): $a_2 + 2b_1$
$2p_z$(Cl):	a_1	$2p_y$(F), (b_2) (on z axis)
		$2 \times 2p_z$(F), ($a_1 + b_2$) (on y axis)

The molecular orbital diagram is shown in Fig. 4.22. There are two σ-type

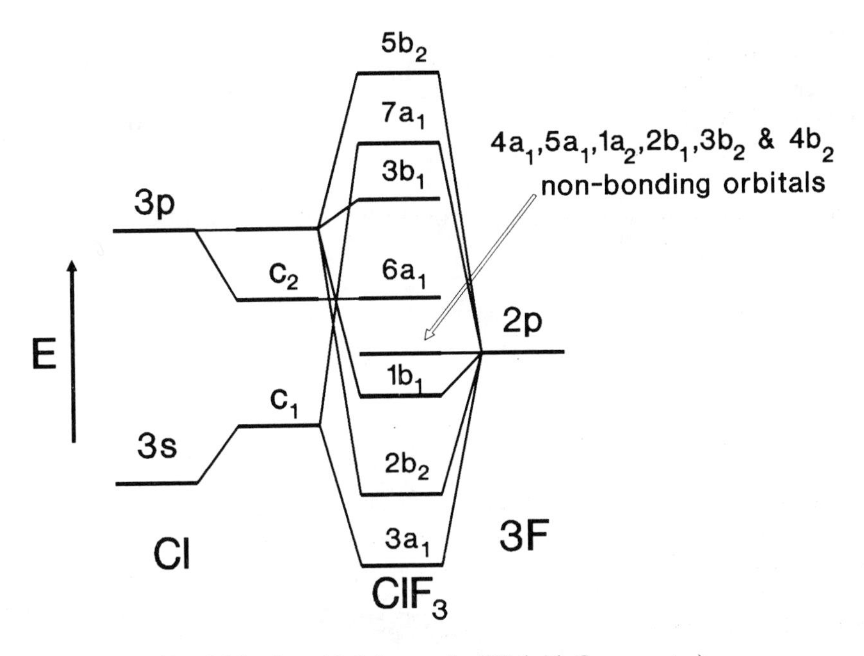

Fig. 4.22 — Molecular orbital diagram for ClF_3(with C_{2v} symmetry).

bonding orbitals, labelled $3a_1$ (with the c_1 orbital of the chlorine atom), and $2b_2$. The only other significant bonding is a π-type combination of the b_1 orbitals perpendicular to the molecular plane, yz. The other fluorine group orbitals (two a_1, one a_2, one b_1 and two b_2) are non-bonding and are labelled as $4a_1$, $5a_1$, $1a_2$, $2b_1$, $3b_2$ and $4b_2$, respectively. The $6a_1(c_2)$ orbital is non-bonding. There are three anti-bonding orbitals: $3b_1$, $7a_1$ and $5b_2$.

The electronic configuration of the ClF_3 molecule is written as:

$$ClF_3(C_{2v}){:}\ 3a_1^2 2b_2^2 1b_1^2 4a_1^2 1a_2^2 2b_1^2 3b_2^2 4b_2^2 5a_1^2 6a_1^2 3b_1^2$$

The bonding in the molecule is due to the pairs of electrons occupying the $3a_1$ and $2b_2$ orbitals. The bonding effects of the pair of electrons occupying the $1b_1$ orbital are cancelled out by the anti-bonding pair in the $3b_1$ orbital. With an excess of only two bonding pairs of electrons it is not surprising that the ClF_3 molecule is not very stable. The electron pair in the $2b_2$ orbital accounts for the weak bonding which is observed in the two 'bonds' along the y axis — their individual bond orders are only 0.5. The bond order of the Cl–F bond along the z axis is 1.0. There is no need to invoke the use of the very-high-energy 3d level of chlorine to explain the bonding in ClF_3, but m.o. theory should give a justification for the C_{2v} symmetry of the molecule. The possibility of ClF_3 being D_{3h} is remote since the four electrons extra to those possessed by BF_3 would occupy the $2a_2''$ and $3a_1'$ orbitals, both of which are anti-bonding. With the C_{3v} symmetry, the σ-type anti-bonding $5a_1$ orbital would be occupied which would rule out that shape for ClF_3. Molecular orbital theory predicts the symmetry to be C_{2v}.

The results of quantum mechanical calculations for the total energies of the three extreme symmetries of the three molecules discussed in this section are shown in Fig. 4.23. The conclusions from the calculations are in agreement with those arrived at

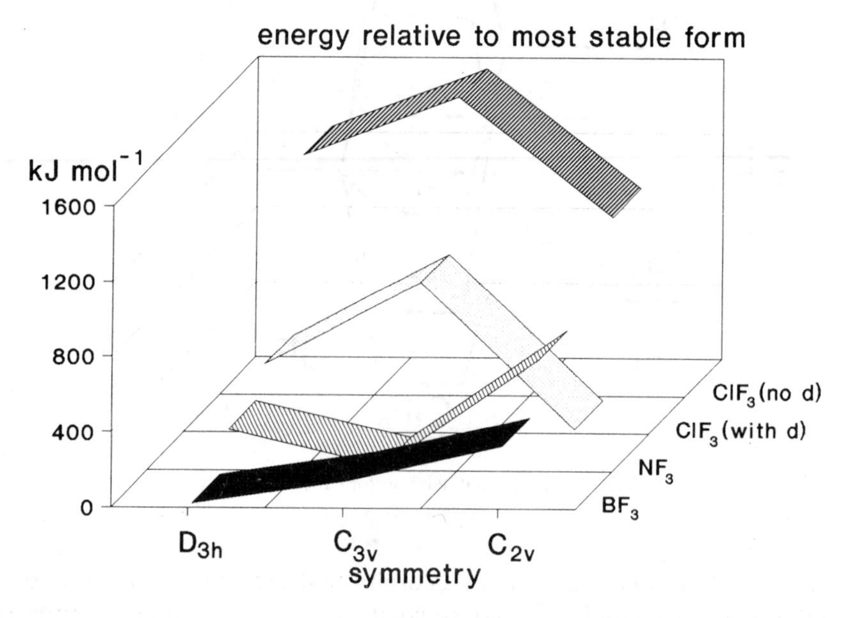

Fig. 4.23 — A comparison of the total energies of the molecules BF_3, NF_3 and ClF_3 (with and without 3d orbital contributions) with the symmetries D_{3h}, C_{3v} and C_{2v}, relative to the form with the lowest energy.

qualitatively. It is obvious from the results shown in Fig. 4.23 for ClF_3 that d orbital participation is of some significance. This would be expected from their symmetries, the $d_{x^2-y^2}$ and d_{z^2} orbitals transforming as a_1, the d_{xz} as b_1 and the d_{yz} as b_2, in the C_{2v} point group. The d_{xy} orbital transforms as a_2 and cannot participate in the C_{2v}, ClF_3 molecule. The omission of the consideration of the d orbitals of the chlorine atom from the qualitative m.o. description of ClF_3 does not alter the conclusions about its shape, and to that extent the d orbital participation is not an important factor.

Diborane, B_2H_6

The diborane molecule, B_2H_6, is the simplest of the boron hydrides (BH_3 has a transient existence) and, in addition to being a molecule with two 'central' atoms in the VSEPR sense, it poses an unusual problem in chemistry. The molecule is 'electron deficient' in the sense that there are only twelve valency electrons available for the bonding of the eight atoms. The geometry of the molecule is shown in Fig. 4.24. The boron atoms are surrounded by four hydrogen atoms, two of which (the

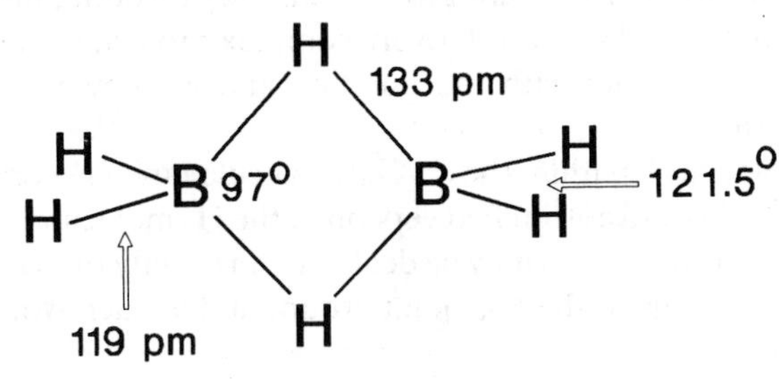

Fig. 4.24 — The structure of the diborane molecule.

bridging atoms) are shared between them. The coordination of the boron atoms is irregularly tetrahedral, with the HBH angles being 97° for the bridging hydrogen atoms, and 121.5° for the terminal angles. The molecule belongs to the D_{2h} point group.

The VSEPR theory can be applied to the geometry around either of the boron atoms. The boron atom has the configuration $2s^2 2p^1$ and could be made 'trivalent' by the excitation of one of the 2s electrons to an otherwise vacant 2p orbital: $2s^1 2p^2$. Bearing in mind that the boron is involved in some form of bonding to four hydrogen atoms, the only sensible distribution is that in which the three electrons from the hydrogen atoms (being half the number of 1s(H) electrons available) are placed in the valence shell of the boron atom to give the configuration: $2s^2 2p^2 2p^1 2p^1$ with two singly occupied 2p orbitals. The normal practice of making electron pairs has to be abandoned. The VSEPR argument continues by considering the minimization of the repulsions between the two pairs of electrons and the two unpaired electrons. This gives a distorted tetrahedral distribution with the electron pairs (those involved with the terminal B–H bonding) being further apart than the 'bonds' associated with the single electrons (which are involved with the bridging B–H bonding). The two irregular tetrahedra, so formed, join up by sharing the bridging hydrogen atoms to give the D_{2h} symmetry of the molecule in full agreement with observation. Localized bonding could be inferred from the treatment, and this leads to the concept of there being four localized one-electron bonds in the bridging between the two boron atoms. The one-electron bond in the H_2^+ molecule-ion is reasonably strong (section 3.2). It is a general rule that whenever **delocalization** can occur, it does. This is because the larger orbitals involved allow for a minimization of interelectronic repulsion.

The molecular orbital theory as applied so far would classify the orbitals of the boron atoms, and then those of the hydrogen atoms, with respect to their transformation properties in the D_{2h} point group. This may be done, but is a very lengthy and complex procedure. With a relatively complex molecule, such as B_2H_6, there is a simpler way of dealing with the bonding and is known as the **molecules within molecule** method.

The 'molecules' within the B_2H_6 molecule are chosen to be the two BH_2 (terminal) groups and a stretched version of the H_2 molecule. As a separate exercise, the bonding in a BH_2 group may be dealt with in exactly the same way as was the H_2O molecule. It belongs to the C_{2v} point group, and as such would have the electronic configuration:

$$BH_2(C_{2v}):\quad 1b_2^2 2a_1^2 3a_1^1$$

with the non-bonding $3a_1$ orbital (it is the $2p_x$(B) orbital) being singly occupied. There are two important differences from the H_2O case. The 2p–2s energy gap in the boron atom is relatively small (350 kJ mol^{-1}) and this allows the 2s(B) and $2p_z$(B) orbitals (both belonging to a_1) to mix. This has the consequence of making the higher-energy $3a_1$ orbital non-bonding. The bonding $2a_1$ is of higher energy than the bonding $1b_2$ orbital because of the very large HBH bond angle which favours the latter orbital.

The two BH_2 groups may be set up so that their z axes are collinear and coincident with the Cartesian z axis as in Fig. 4.24. Their molecular planes are contained within the Cartesian yz plane. This ensures that the $1b_1(2p_x(B))$ orbitals of the two BH_2 groups are aligned in the x direction and in the xz plane. The two hydrogen atoms responsible for the bridging are situated along the x axis. The two hydrogen group orbitals are used in the construction of those molecular orbitals responsible for the cohesion of the three molecules in their formation of diborane. This arrangement of orbitals is shown in Fig. 4.25.

In order to construct the m.o. diagram for the bridging between the two BH_2 groups it is essential to rename the participating orbitals within the framework of the D_{2h} point group to which diborane belongs. This may be done by inspecting the D_{2h} character table, and gives the following results:

The orbitals of BH_2, $3a_1$ and $1b_1$ form the following linear combinations:

$$\phi(a_g) = \phi(3a_1)_A + \phi(3a_1)_B \tag{4.9}$$
$$\phi(b_{1u}) = \phi(3a_1)_A - \phi(3a_1)_B \tag{4.10}$$
$$\phi(b_{3u}) = \phi(1b_1)_A + \phi(1b_1)_B \tag{4.11}$$

and

$$\phi(b_{2g}) = \phi(1b_1)_A - \phi(1b_1)_B \tag{4.12}$$

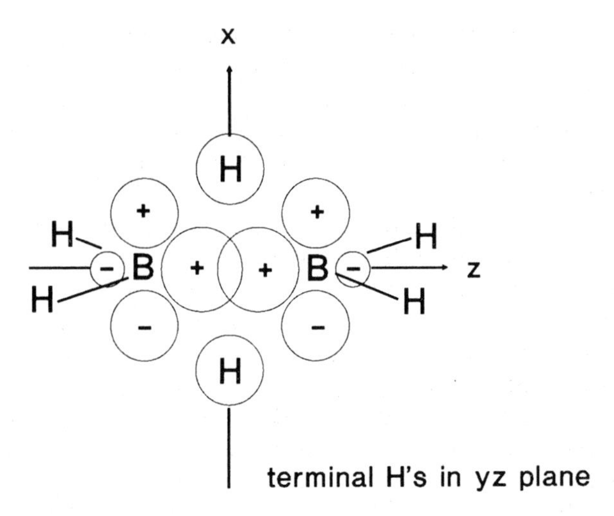

Fig. 4.25 — The atomic orbitals contributing to the bridge bonding in the diborane molecule.

with their new D_{2h} names indicated.

The hydrogen 1s orbitals may be regarded in their usual group forms:

$$\phi(a_g) = \psi_{1s}(A) + \psi_{1s}(B) \tag{4.13}$$

and

$$\phi(b_{3u}) = \psi_{1s}(A) - \psi_{1s}(B) \tag{4.14}$$

also with their new D_{2h} names indicated.

The m.o. diagram for the bridging in diborane may be constructed by the formation of bonding and anti-bonding combinations of the contributing orbitals of the same symmetries.

The m.o. diagram is shown in Fig. 4.26. There are two bonding orbitals, of a_g and b_{3u} symmetries respectively, which contain the four available electrons. The two orbitals are four- centre molecular orbitals which allow the minimization of interelectronic repulsion and give the molecule some stabilization compared to the formulation with four localized one-electron bonds.

The hydrogen difluoride ion, HF_2^-

The hydrogen difluoride ion has a central hydrogen atom and belongs to the $D_{\infty h}$ point group. It is important in explaining the weakness of hydrogen fluoride as an acid in aqueous solution, and as an example of **hydrogen bonding**. Hydrogen bonding occurs usually **between** molecules which contain a suitable hydrogen atom and one or more of the very electronegative atoms: N, O or F. It is to be considered, in general, as a particularly strong **intermolecular force**.

In the case of HF_2^- it may be considered that the ion is a combination of an HF molecule and a fluoride ion, F^-. Both of those chemical entities have considerable thermodynamic stability independently of each other. Their combination in HF_2^-

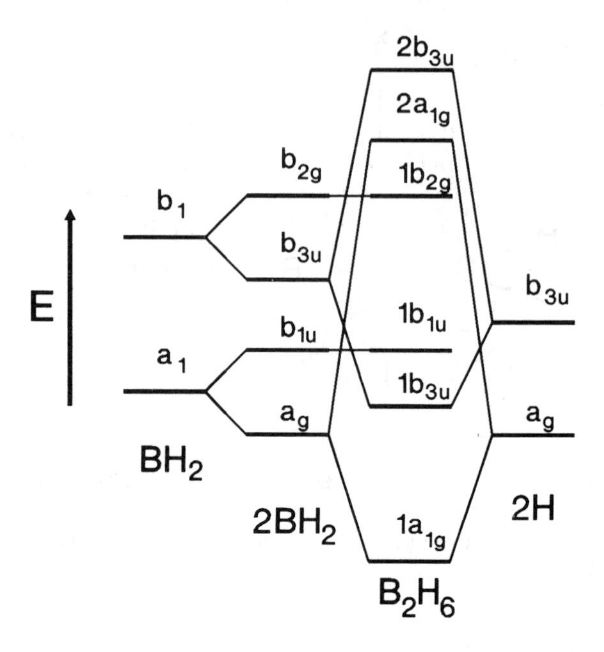

Fig. 4.26 — Molecular orbital diagram for the bridge bonding in the diborane molecule.

indicates the operation of a force of attraction which is significantly greater than the normal intermolecular forces which operate in all systems of molecules. The latter are responsible for the stability of the liquid and solid phases of chemical substances. They counteract the tendency towards chaos.

The majority of hydrogen bonding occurs in systems where the hydrogen atom, responsible for the bonding, is asymmetrically placed between the two electronegative atoms (one from each molecule). In those cases the bonding may be regarded as an extraordinary interaction between two dipolar molecules, the extraordinary nature of the effect originating in the high effectiveness of the almost unshielded hydrogen nucleus in attracting electrons from where they are concentrated — around the very electronegative atoms of neighbouring molecules. The hydrogen bond in the HF_2^- ion is capable of treatment by bonding theories.

The VSEPR treatment is best approached by considering the ion as being made up from three ions: F^-–H^+–F^-. The central proton possesses no electrons until the ligand fluoride ions supply two each. The two pairs of electrons repel each other to give the observed linear configuration of the three atoms. The two pairs of electrons would occupy the 1s and 2s orbitals of the hydrogen atom and, what with a considerable amount of interelectronic repulsion, would not lead to stability.

The molecular orbital theory is straightforward. The hydrogen 1s orbital transforms as σ_g^+ within the $D_{\infty h}$ point group. The $2p_z$ orbitals of the fluorine atoms (since the ion is set up with the molecular axis coincident with the Cartesian z axis) may be arranged in the linear combinations:

$$\phi(\sigma_g^+) = \psi_{2p_z}(A) + \psi_{2p_z}(B) \tag{4.15}$$

and

$$\phi(\sigma_u^+) = \psi_{2p_z}(A) - \psi_{2pz}(B) \tag{4.16}$$

Molecular orbitals may be constructed from the two orbitals of σ_g^+ symmetry to give bonding and anti-bonding combinations. The orbital, localized on the fluorine atoms, of σ_u^+ symmetry remains as a non-bonding orbital. The m.o. diagram is shown in Fig. 4.27. The four electrons, which may be regarded to have been supplied

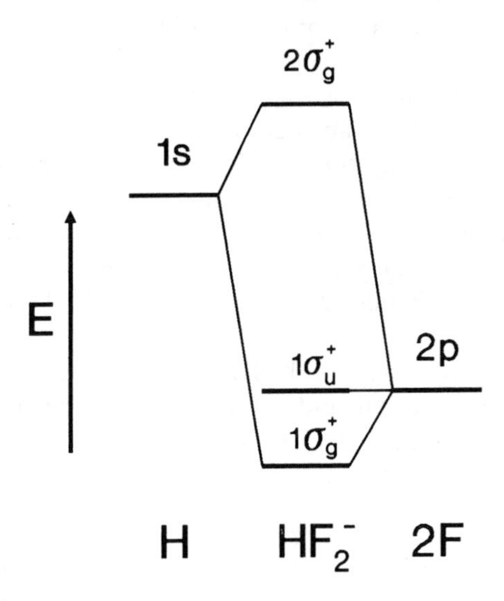

Fig. 4.27 — Molecular orbital diagram for the hydrogen difluoride anion.

by two fluoride ions, occupy the bonding and non-bonding orbitals, and the F–H bond order is 0.5. This is consistent with the observed weakness of the hydrogen bond of 126 kJ mol^{-1} (the $\Delta H^{\ominus}$ of the reaction $HF_2^- \rightarrow H^+ + HF$) although that figure is a very large amount relative to 'normal' hydrogen bond strengths of 10-30 kJ mol^{-1}.

The HF_2^- ion is isoelectronic with the unknown compound, helium difluoride, HeF_2. The latter compound would have a very similar electronic configuration to that of HF_2^-. The reason for its non-existence is indicated by the values of the ionization energies of H^- and He (which are isoelectronic) of 73 and 2372 kJ mol^{-1} respectively. The attraction for electrons represented by the large ionization energy of the helium atom is greater than that exerted by the two fluorine atoms, making the formation of a stable compound impossible. It is of interest to note that the molecule XeF_2 has a stable existence (even though the ionization energy of Xe is 1170 kJ mol^{-1}) although it is very weakly bonded.

4.4 CONCLUDING SUMMARY

In this chapter the VSEPR and molecular orbital theories are extended to cover a variety of problems associated with the bonding and shapes of polyatomic molecules.

The protocol for carrying out the derivation of a qualitative m.o. diagram for a molecule is dealt with in some detail and applied to several examples. It is summarized as:

(1) Determine the point group to which the molecule belongs.
(2) Classify the valency orbitals of the central atom.
(3) Classify the ligand group orbitals.
(4) Draw the m.o. diagram, allowing interaction between the orbitals of the central atom and the ligand group orbitals of the same symmetry.

Molecular orbital correlation diagrams are introduced, and their success in explaining bonding and molecular shape demonstrated. Molecular orbital theory is used to demonstrate structure preference factors for the molecules of water, ammonia and methane. The correlations, although qualitative, are backed up by some *ab initio* calculations. It is now possible for students to carry out such calculations, and it is strongly recommended that they should do some as part of their course. 'Practical' theoretical chemistry is an excellent method for gaining a deeper understanding of the subject.

A detailed treatment of the shapes of the BF_3, NF_3 and ClF_3 molecules is carried out. Hydrogen bonding is introduced, and its relationship to the instability of HeF_2 is discussed.

5

Structures of elements, metallic bonding ionic compounds

5.1 INTRODUCTION

In this chapter the factors which are responsible for determining the forms of elements (metallic or non-metallic) are discussed, together with an introductory treatment of metallic bonding. Ionic compounds have structures which are related to those of metals, and ionic crystal lattices are discussed from that standpoint. The energetics of ionic bond formation are considered.

5.2 ELEMENTARY FORMS

The majority of the elements are solid metals in their standard states. Of the metallic elements, only mercury is a liquid at 298 K (gallium melts at 302.9 K). The non-metallic elements exist as either discrete small molecules, in the solid (S_8), liquid (Br_2) or gaseous (H_2) states, or as extended atomic arrays in the solid state (C as graphite or diamond), the elements of Group 18 being monatomic gases at 298 K.

Metallic character decreases across any period and increases down any group, the non-metals being situated towards the top right-hand region of the periodic table. The trends in ionization energies and electronegativity coefficients are of an opposite nature to that of metallic character. A large ionization energy is associated with a large effective nuclear charge, and the atoms in this class are those which can participate in the formation of strong covalent bonds. Metal atoms, on the contrary, can only form weak covalent bonds with each other, and usually exist in the solid state with lattices in which the coordination number of each atom is relatively high (between eight and fourteen). Diagrams of the three most common lattices are shown in Figs 5.1, 5.2, and 5.3. In the body-centred cubic lattice (Fig. 5.1) the coordination number of each atom is eight. There are six next-nearest neighbours in the centres of the adjacent cubes so that the coordination number may be regarded as

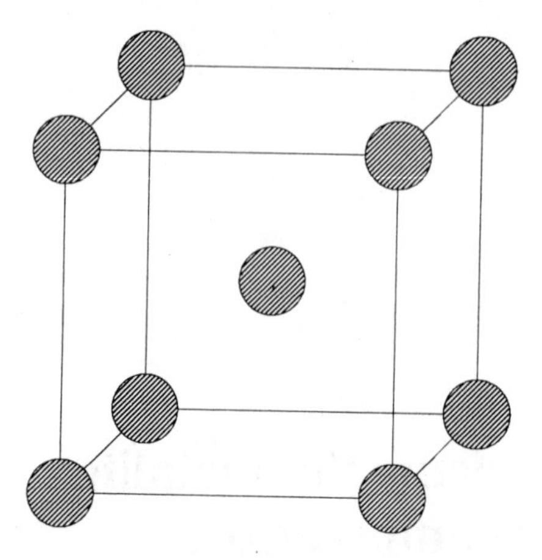

Fig. 5.1 — A representation of atoms arranged in a body-centred cubic structure.

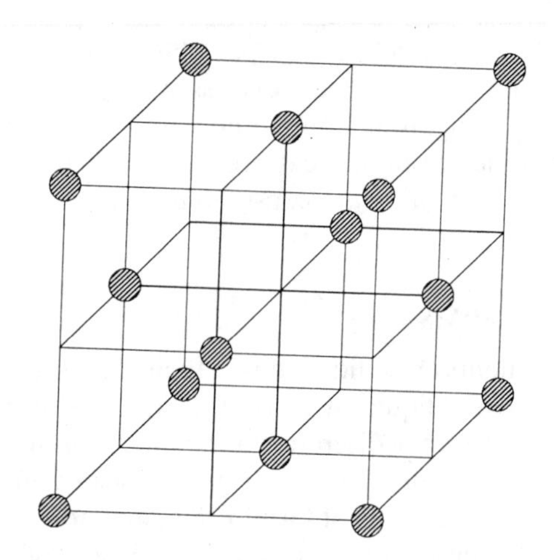

Fig. 5.2 — A representation of atoms arranged in a cubic closest-packed structure.

being fourteen. The cubic closest- packed lattice (Fig. 5.2) may be regarded as a face-centred cubic arrangement in which there is an atom at the centre of each of the six faces of the basic cubic of eight atoms. The coordination number of any particular atom is twelve, consisting of (considering an atom in one of the face centres of the diagram in Fig. 5.2) four atoms at the corners of that face and the eight atoms in the centres of adjacent faces of the two cubes sharing the atom in question. In the hexagonally closest-packed arrangement (Fig. 5.3) the coordination consists of six

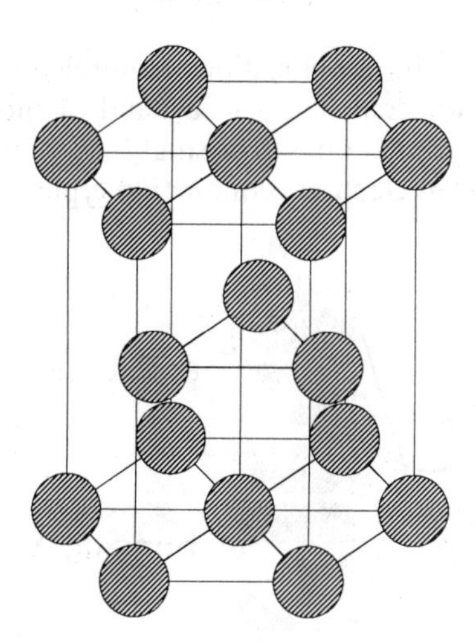

Fig. 5.3 — A representation of atoms arranged in a hexagonal closest-packed structure.

atoms in the same plane as the atom under consideration plus three atoms from both of the adjacent planes, making a total of twelve.

A discussion of the elementary forms of lithium and hydrogen serves to illustrate the factors which determine their metallic or non-metallic nature.

Elementary lithium

The atom of lithium has the outer electronic configuration, $2s^1$, and might be expected to form a diatomic molecule, Li_2. The molecule, Li_2, does exist in lithium vapour and has a dissociation energy of only 108 kJ mol^{-1}. The enthalpy of atomization of the element is 161 kJ mol^{-1}, which is a measurement of the strength of the bonding in the solid state. To produce two moles of lithium atoms (as would be produced by the dissociation of one mole of dilithium molecules) would require $2\times161=322$ kJ of energy. From such figures it can be seen that solid metallic lithium is more stable than the dilithium molecular form by $(322-108)/2=107$ kJ mol^{-1}. The weakness of the covalent bond in dilithium is understandable in terms of the low effective nuclear charge which allows the 2s orbital to be very diffuse. If the effectiveness of the nuclear charge is low, a 2s electron would not be expected to show much interaction with an electron from another atom. The diffuseness of the 2s orbital of lithium is indicated by the large bond length (267 pm) in the dilithium molecule. The metal exists in the form of a body-centred cubic lattice in which the radius of the lithium atoms is 152 pm — again a very high value indicative of the low cohesiveness of the structure.

The high electrical conductivity of lithium (and metals in general) indicates considerable electron mobility. This is consistent with the molecular orbital treatment of a three-dimensional array of atoms in which the 2s orbitals are completely delocalized over the system, with the formation of $n/2$ bonding orbitals and $n/2$ antibonding orbitals for the n atoms concerned. Fig. 5.4 shows a diagrammatic representation of the 2s band of lithium metal. The formation of a band of delocalized molecular orbitals allows for the minimization of interelectronic repulsion. The small gaps in energy between adjacent levels in a band of molecular orbitals allow a considerable number of the higher ones to be singly occupied. Such single occupation is important in explaining the electrical conduction typical of the metallic state.

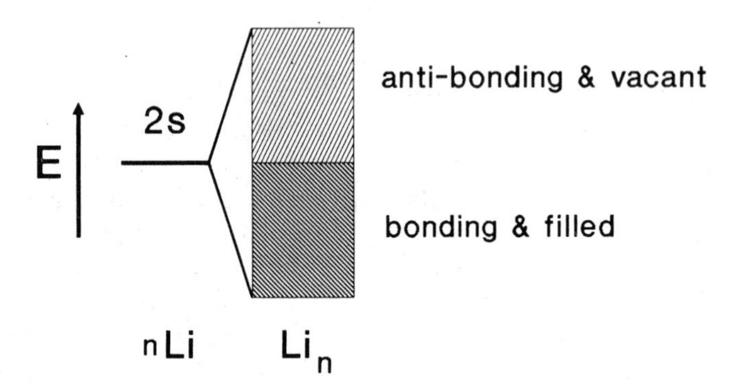

Fig. 5.4 — The 2s band of orbitals in lithium metal.

Elementary hydrogen

The simplest element exists normally as the dihydrogen molecule, H_2. The high value of the ionization energy of the hydrogen atom is indicative of the electron-attracting power of the proton which contributes to the large single bond energy in the dihydrogen molecule. It is instructive to consider the energetics of formation of larger molecules such as H_4.

The hypothetical H_4 molecule could have the possible structures, (i) tetrahedral, (ii) square planar or (iii) linear. The results of some molecular orbital calculations on these molecules are shown in Table 5.1, the nearest-neighbour distances being assumed to be 100 pm.

Table 5.1 — The calculated energies of some H_4 molecules

Symmetry	Energies/kJ mol^{-1}		
	Electronic	Nuclear	Total
T_d	−12138	8344	−3794
D_{4h}	−12333	7518	−4815
$D_{\infty h}$	−11524	6017	−5507

The most stable symmetry for H_4 is the linear $D_{\infty h}$ form, the main factor being the low value of the internuclear repulsion energy. The calculated values for the internuclear repulsion, electronic and total energies for the H_2 molecule (r_{eq}=74 pm) are 1876, −4807 and −2930 kJ mol respectively. The reaction:

$$H_4(\text{linear}, 100\,\text{pm}) \rightarrow 2H_2(74\,\text{pm}) \qquad (5.1)$$

would have a ΔU value of $(2\times-2930)-(-5507)=-353\,\text{kJ mol}^{-1}$. There would be a positive change in the entropy of the system which would also contribute to the instability of the H_4 molecule. These and other calculations show that any form of hydrogen which is more than diatomic is less stable than H_2 and that the main reason for this is the nuclear repulsion term. For hydrogen to exist in the metallic form would require tremendous pressure to overcome the internuclear repulsion. It is supposed that such conditions exist in the core of Jupiter, and the presence of metallic hydrogen could explain the magnetic properties of the planet.

The non-metallic elements whose cohesiveness depends upon participation in strong covalent bond formation exist in the following forms.

(i) Diatomic molecules which may be
  (a) singly bonded (e.g. H_2 and F_2)
  (b) doubly bonded (e.g. O_2), or
  (c) triply bonded (e.g. N_2).
(ii) Polyatomic small molecules involving single covalent bonds between adjacent atoms (e.g. P_4 and S_8)
(iii) Three-dimensional arrays (graphite, diamond, boron and silicon). A capacity to be at least three-valent is necessary for a three-dimensional array to be formed.

The individual molecules form crystals in which the cohesive forces are intermolecular ones and the form of a particular crystal is usually determined by the economy of packing the units together.

Of the elements which participate in catenation (chain formation) in their elementary states, only carbon retains the property in its compounds to any great extent. The relatively high strength of the single covalent bond between two carbon atoms gives rise to such a large number, and wide variety, of compounds which form the basis of the branch of the subject known as organic chemistry. Boron and silicon chemistry contains some examples of compounds in which catenation is exhibited.

5.3 THE METALLIC BOND

The bonding in lithium metal has been described in the previous section. The band of molecular orbitals formed by the 2s orbitals of the lithium atoms is half filled by the available electrons. Metallic beryllium, with twice the number of electrons, might be expected to have a full '2s band'. If that were so, the material would not exist since the 'anti-bonding' half of the band would be fully occupied. Metallic beryllium does exist and does so because the band of molecular orbitals produced from the 2p atomic orbitals overlaps (in terms of energy) the 2s band. This makes possible the partial filling of both the 2s and the 2p bands, giving metallic beryllium a greater cohesiveness and a larger electrical conductivity than lithium. The overlapping of the 2s and 2p bands of beryllium, and their partial occupancies, are shown diagrammatically in Fig. 5.5.

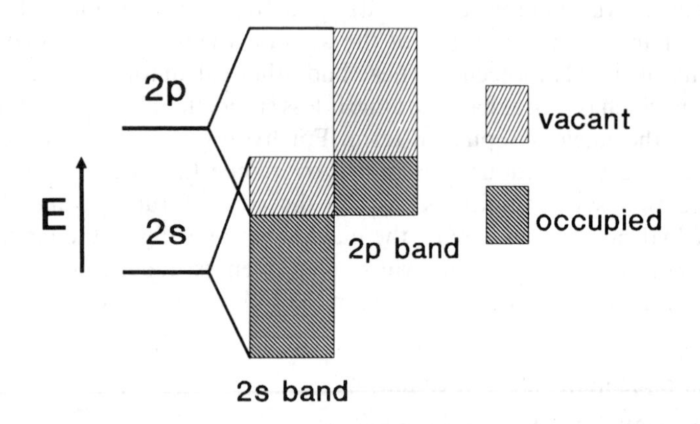

Fig. 5.5 — The 2s and 2p bands of beryllium metal.

If overlap between bands does not occur, the size of the band gap (between the lowest level of the vacant band and the highest level of the filled band) determines whether the element exhibits insulator properties (large band gap), or is a semiconductor (small band gap). The band gaps in the diamond form of carbon and in silicon (which has the diamond, four-coordinate structure) are $521\,kJ\,mol^{-1}$ and $106\,kJ\,mol^{-1}$ respectively. Diamond is an insulator, but silicon is a semiconductor.

Electrical conduction in a semiconductor may occur because of (i) thermal excitation of electrons from the highest filled band (HFB) to the lowest vacant band (LVB), (ii) photo-excitation from the HFB to the LVB, or by the presence of an impurity element. If such an element is introduced deliberately this is known as doping. Ultra-pure silicon is an *intrinsic* semiconductor, but would be of no practical use (the band gap being rather high), most commercial semiconductors being based on doped silicon (extrinsic semiconductors). The difference between an insulator and various semiconductors is shown in Fig. 5.6.

There are two main types of extrinsic semiconductor which are based upon whether a suitable band of the impurity element overlaps with (or is close enough to allow thermal or photo-excitation to occur) the HFB or the LVB of the silicon. If an otherwise *vacant* impurity band overlaps with the HFB of the silicon its partial occupancy by electrons from the silicon allows for p-type semiconduction, where the conduction may be thought of in terms of the movement of positive holes in the main conduction band. If a filled impurity band overlaps with the LVB of the silicon and

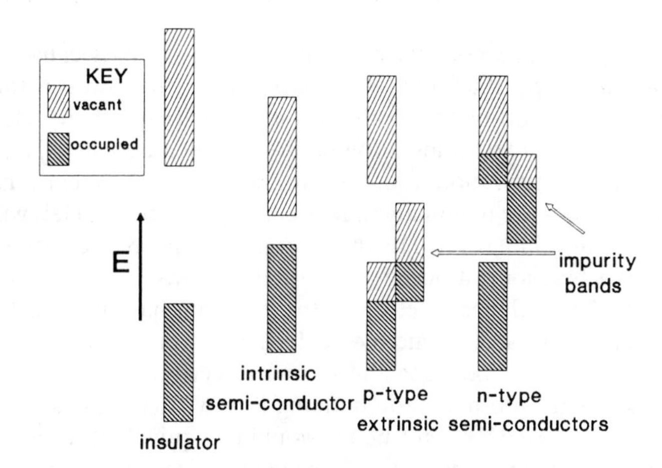

Fig. 5.6 — A diagrammatic representation of band gaps and band occupation in insulators, intrinsic semiconductors and p-type and n-type extrinsic semiconductors.

causes the latter to be partially populated, this gives rise to n-type (normal) semiconduction by electrons in the previously vacant upper band of the silicon. Even if there is no overlap in either of the cases, semiconduction may occur by either thermal or photo-excitation. Thermal semiconduction depends upon the extent to which a conduction band is populated, and that is governed by a Boltzmann factor, $e^{-(\text{bandgap})/kT}$ (see Chapter 7). Such conductivity increases as the temperature increases.

Metallic conduction shows a reduction in conductivity with an increase in temperature, in spite of the population of more of the band with unpaired electrons (there being a very small difference between adjacent levels in any one band). This inverse relationship is thought to be due to higher temperatures causing greater amplitudes of atomic vibrations or even discontinuities in the metal lattice (less extensive delocalization of the electrons) which would lead to an increase in resistance at their boundaries. At very low temperatures (such as that of liquid helium, 4 K) some metals, alloys and compounds become superconductive. This means that their resistance to the passage of an electrical current becomes zero.

Recently, interest has developed into the superconductive properties of a range of mixed oxides. The latest developments have resulted in the production of a compound of the composition, $Ca_{0.8}Y_{0.2}Sr_2Tl_{0.5}Pb_{0.5}Cu_2O_7$, which is superconductive at temperatures up to 107 K.

Further progress towards the production of materials which show superconductivity at even higher temperatures will depend upon further experimentation, and the development of a theory with predictive value. It is possible that under suitable circumstances (interionic distances and temperature) population of an otherwise vacant '3s' band of the oxide ions allows superconductivity to occur.

5.4 THE IONIC BOND

The interaction between any two atomic orbitals is at a maximum if the energy difference between them is zero and that they have suitable symmetry properties. A difference in energy between the two participating orbitals causes the bonding orbital to have a greater than 50% contribution from the lower of the two atomic orbitals. Likewise the anti-bonding orbital contains a greater than 50% contribution from the higher of the two atomic orbitals. In a molecule such as HF, the bonding orbital has a major contribution from the fluorine 2p orbital, and in consequence the two bonding electrons are unequally shared between the two nuclei, which results in the molecule being dipolar. In HF the energies of the participating orbitals are indicated by the first ionization energies of the atoms H (1312 kJ mol^{-1}) and F (1681 kJ mol^{-1}). The valence electron of the Na atom has an ionization energy of only 494 kJ mol^{-1}, and a diatomic molecule, NaF, would involve a difference of $1681-494=1187$ kJ mol^{-1} between the valence electrons 3s(Na) and 2p(F). That difference is so great as to make the bonding level virtually the same as the 2p(F) orbital so that if the bond were formed it would be equivalent to the transfer of an electron from the sodium atom to the fluorine atom, causing the production of the two ions, Na^+ and F^-.

The standard enthalpy change for the reaction:

$$\mathrm{Na(s)}+\tfrac{1}{2}F_2\mathrm{(g)} \rightarrow \mathrm{Na^+(g)+F^-(g)} \tag{5.2}$$

is composed of

(i) $\Delta H_a^{\ominus}$(Na) , 107 kJ mol^{-1}

(ii) $\frac{1}{2}\times D(F_2,g)$, 79 kJ mol^{-1}

where $D(F_2,g)$ is the dissociation energy of the fluorine molecule

(iii) I_1(Na) , 494 kJ mol^{-1}

where I_1 is the first ionization energy of the sodium atom, and

(iv) $-E$(F) , -322 kJ mol^{-1}

where E(F) is the electron affinity of the fluorine atom, and is $\Delta H^{\ominus}$ (5.2)=358 kJ mol^{-1}—considerably endothermic and would not be conducive for the feasibility of the reaction.

The above calculation does not take into account the electrostatic attraction between the oppositely charged ions. Assuming that the ions approach each other to within their ionic radii (r_i (Na^+), 98 pm, r_i(F^-), 133 pm), 231 pm, Coulomb's law may be used to calculate the energy of stabilization due to electrostatic attraction:

$$E(\mathrm{Na^+/F^-})=-Ne^2/4\pi\varepsilon_0\, r \tag{5.3}$$

where N is the Avogadro number, e the electronic charge, ε_0 the vacuum permittivity and r the interionic distance. If we put the values for the terms in equation (5.3) we obtain:

$$E(Na^+/F^-) = -601\ kJ\ mol^{-1} \quad (5.4)$$

The standard enthalpy of formation of the substance Na^+F^- with discrete ion pairs (not interacting with each other) is then calculated to be:

$$\Delta H_f^\ominus(Na^+F^-, g) = 358 - 601 = -243\ kJ\ mol^{-1} \quad (5.5)$$

The actual substance formed when sodium metal reacts with difluorine is solid sodium fluoride, and the observed standard enthalpy of its formation is $-569\ kJ\ mol^{-1}$. The actual substance is $326\ kJ\ mol^{-1}$ more stable than the hypothetical ion pair, $Na^+F^-(g)$ described above. The added stability of the observed compound arises from the long-range interactions of all the positive Na^+ ions and negative F^- ions in the solid **lattice** which forms the structure of sodium fluoride. The ionic arrangement is shown in Fig. 5.7.

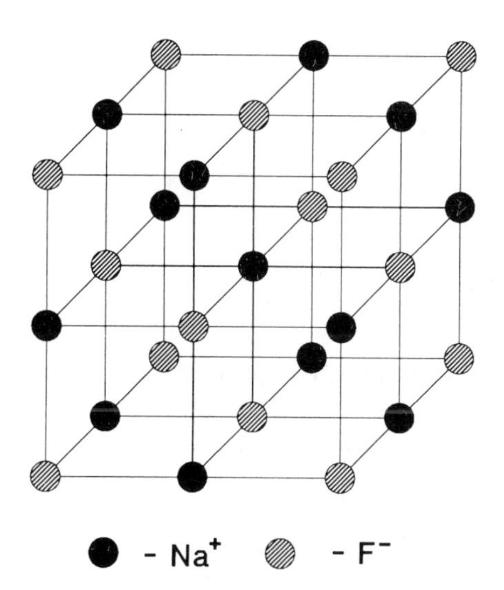

Fig. 5.7 — A representation of the arrangement of sodium and fluoride ions in the sodium fluoride lattice.

Each Na^+ ion is octahedrally surrounded (coordinated) by six fluoride ions, and the fluoride ions are similarly coordinated by six sodium ions.

The overall stability of the NaF lattice is represented by the resultant of the many stabilizing attractions (Na^+–F^-) and destabilizing repulsions (Na^+–Na^+ and F^-–F^-)

which amount to a stabilization which is 1.747 times that of the interaction between the individual Na^+–F^- ion pairs. The factor 1.747 is the so-called Madelung constant for the lattice and arises from the forces experienced by each ion. These are composed of:

six repulsions at a distance r
twelve attractions at a distance $r\sqrt{2}$
eight repulsions at a distance $r\sqrt{3}$
six attractions at a distance $r\sqrt{4}$
twenty-four repulsions at a distance $r\sqrt{5}$
and so on to infinity.

The series:

$$6-12/\sqrt{2}+8/\sqrt{3}-6/\sqrt{4}+24/\sqrt{5}-\ldots$$

eventually becomes convergent and gives the value for the Madelung constant for the sodium chloride type of lattice (adopted by sodium fluoride). The values of Madelung constants for all types of lattice are available from data books.

The electrostatic contribution to the **lattice energy**, L, for the sodium fluoride arrangement — the negative value of the standard enthalpy change for the reaction:

$$Na^+(g)+F^-(g) \rightarrow Na^+F^-(s) \tag{5.6}$$

is given by the equation:

$$L(Na^+F^-)=MNe^2/4\pi\varepsilon_0 r \tag{5.7}$$

where M is the Madelung constant. In addition to the electrostatic forces there is a repulsive force which operates at short distances between ions as a result of the overlapping of filled orbitals. This repulsive force may be represented by the equation:

$$E_{rep}=B/r^n \tag{5.8}$$

B being a proportionality factor. The full expression for the lattice energy becomes:

$$L=MNe^2/4\pi\varepsilon_0 r-B/r^n \tag{5.9}$$

When r is the equilibrium interionic distance, the differential dL/dr has the value zero, so that:

$$dL/dr = 0 = -MNe^2/4\pi\varepsilon_0 r^2 + nB/r^{n+1} \tag{5.10}$$

which gives for B:

$$B = MNe^2 r^{n-1}/4\pi\varepsilon_0 n \tag{5.11}$$

and substituting this value into equation (5.9) gives:

$$L = MNe^2/4\pi\varepsilon_0 r[1 - 1/n] \tag{5.12}$$

which is known as the Born–Landé equation. The equation may be applied to any lattice, provided that the appropriate value of the Madelung constant is used. The absolute value of the product of the ionic charges, $|z_A z_B|$, should be included in the equation to take into account any charges which differ from ± 1.

The value of the exponent, n, is estimated from compressibility experiments to be 9 for most ionic systems. Using that value for sodium fluoride produces the value of $L = 933\,\text{kJ mol}^{-1}$.

It is now possible to produce a theoretical value for the standard enthalpy of formation of sodium fluoride based upon the ionic model described above. The necessary equation is produced by adding together equations (5.2) and (5.6) to give:

$$Na(s) + \tfrac{1}{2}F_2(g) \rightarrow Na^+F^-(s) \tag{5.13}$$

and the calculated value for $\Delta H_f^\ominus(Na^+F^-)$ is 358–933 = -575 kJ mol which is almost identical to the observed value of $-569\,\text{kJ mol}^{-1}$. Such good agreement between theory and observation is rare in chemistry and gives considerable respectability to the theory. Other evidence for the ionic nature of sodium fluoride is that the molten substance is a good conductor of electricity, as are aqueous solutions of the compound. Those ionic substances, which are soluble in water, dissociate to give their component hydrated ions which cause the solution to exhibit good electrical conductance.

In general terms the feasibility of ionic bond production depends upon the $\Delta H_f^\ominus$ value for the compound, MX_n, the entropy of formation being of minor importance. The value of $\Delta H_f^\ominus$ may be estimated from known thermodynamic quantities and the calculated value for the lattice energy:

$$\Delta H_f^\ominus(\text{ionic}) = \Delta H_a^\ominus(M) + (n/2)\Delta H_a^\ominus(X_2) + \Sigma I_M - nE_X - L \tag{5.14}$$

the enthalpy of atomization of the metal, $\Delta H_a^\ominus(M)$, sometimes being called the **sublimation energy**, and that of the non-metal, $\Delta H_a^\ominus(X_2)$, being a combination of the energy required to produce the gas phase molecule plus the dissociation energy of the molecule. The ionization energy term is the appropriate sum of the first n ionization energies to produce the required n-positive ion. The n electrons are used to produce either n singly negative ions or $n/2$ doubly negative ions' and the E_X term has to be modified accordingly. The L value is calculated for the appropriate ionic arrangement (if known, otherwise guesses have to be made).

The first three terms of equation (5.14) are positive and can only contribute to the feasibility of ionic bond production by being minimized. The last two terms are negative, and for ionic bond feasibility should be maximized. From such considerations it is clear that ionic bond formation will be satisfactory for:

(1) very electropositive and easily atomized metals
(2) very electronegative and easily atomized non-metals.

The lattice energy term may be increased, for ions with charges, z_+ and z_-, by the product of the charges, z_+z_-. To increase the cation charge involves the expense of an increase in ΣI_M and, although there is some payback in the production of more anions (a larger E_X term), there are severe limits of such activity. This is because the successive ionization energies of an atom increase more rapidly than does the lattice energy with increasing cation charge. The stoichiometry of the ionic compound produced from two elements is determined by the values of the five terms of equation (5.14) which minimize the value of the enthalpy of formation of the compound.

Calculations may be made for hypothetical compounds such as NaF_2. Assuming that the compound contains Na^{2+} ions, and that the lattice produced is the fluorite (CaF_2) type (see Fig. 5.11), the standard enthalpy of formation is given by:

$$\Delta H_f^\ominus(NaF_2,\text{ionic}) = \Delta H_a^\ominus(Na) + D(F_2) + I_1(Na) + I_2(Na) - 2E_F - L(NaF_2) \quad (5.15)$$

The numerical values of the terms in equation (5.15) are respectively 107, 158, 494, 4562(!), 2×322 and 1543 kJ mol^{-1}, which gives a value of +3134 kJ mol^{-1} for the standard enthalpy of formation of ionic NaF_2 — a compound which would not be expected to exist. The reason for that is the very large value of the second ionization energy of the sodium atom, the electron being removed from one of the very stable 2p orbitals.

The relation of ionic structures to the packing of anion/cation lattices

One very instructive way of thinking about any ionic structure is to consider that the anions (usually the larger of the two participating ions in a binary compound) are arranged in either cubic or hexagonal closest packing. Such arrangements of spherical objects may be studied at the orange seller's stall! With a hard sphere model in mind the **interstices** (or holes) in, for instance, the cubic closest-packed lattice would have a size dependent upon the size of the spheres from which it was composed. Fig. 5.8 contains a diagram of two layers of closest-packed spheres, which shows that there are two kinds of hole. One is bounded by four spheres and is termed

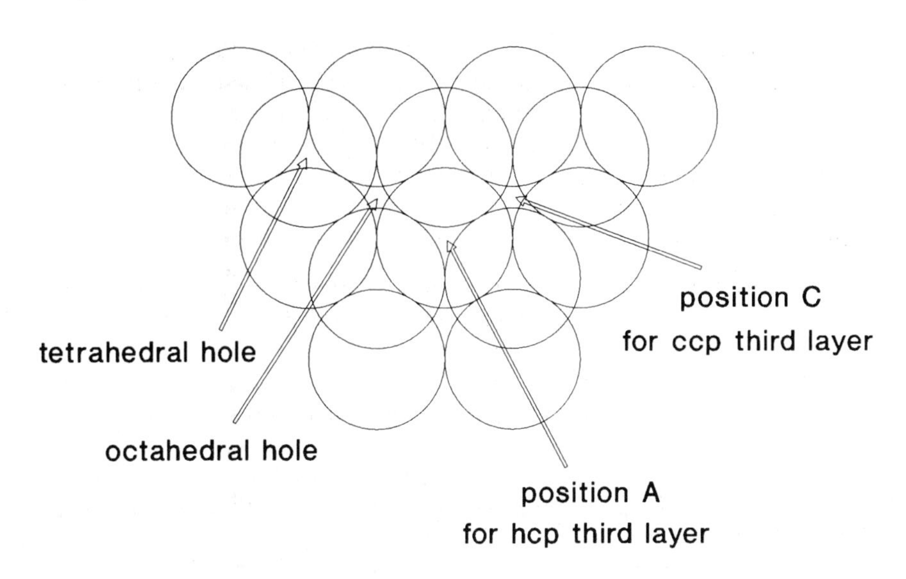

Fig. 5.8 — Two layers of closest-packed spheres.

a tetrahedral hole, the other being bounded by six spheres and is termed an octahedral hole. The radius, r_+, of an entering sphere which is to fit exactly into either a tetrahedral or an octahedral hole is dependent upon the radius, r_-, of the cubic closest-packed spheres.

The calculations for the two types of hole are illustrated by the diagrams in Fig. 5.9. Both diagrams show the small (cation) sphere in contact with two large (anion)

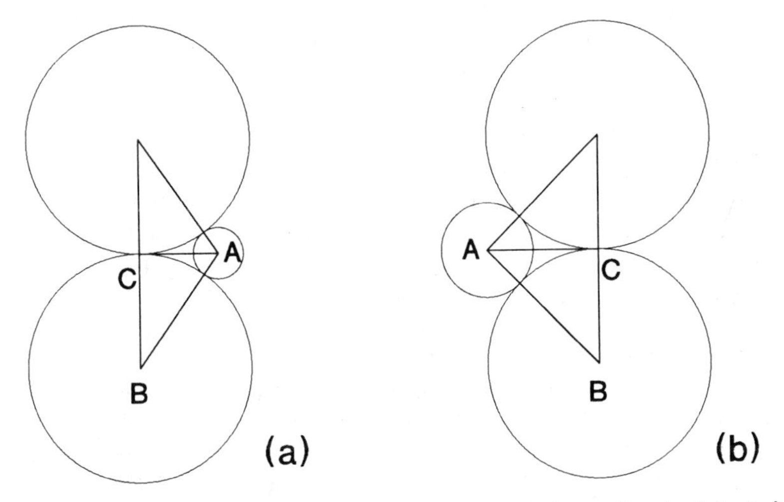

Fig. 5.9 — Geometrical constructions for tetrahedral and octahedral holes in a closest-packed lattice.

spheres, with the two large spheres in contact with each other. In diagram (a) the anion–cation–anion angle is the tetrahedral angle of 109°28′, and in diagram (b) the angle is 90°, appropriate to tetrahedral and octahedral holes respectively. The right-angled triangles, ABC, in both cases, are constructed so that the side, BC, is equal to the anion radius, r_-, and the side, AB, is equal to the sum of the two radii, $r_+ + r_-$. The respective angles, BAC, in cases (a) and (b), are 109°28′/2=54°44′ and 45°. The sines of these angles are given by the equation:

$$\sin \mathrm{BAC} = r_-/(r_- + r_+) \tag{5.16}$$

Since the triangle, ABC, in the case of the octahedral hole contains a right angle the theorem of Pythagoras may be used as an alternative to equation (5.16) to achieve the desired result. In that case the equation to be solved is $2r_-^2 = (r_+ + r_-)^2$.

The solutions of the equation (5.16) for the two angles show that for a tetrahedral hole the ratio, r_+/r_-, is equal to 0.213, and for an octahedral hole has the value, 0.414. The calculations indicate that an octahedral hole could accommodate an ion which was around 94% larger in radius than a tetrahedral hole.

Ions behave only very roughly like hard spheres, but the calculations above do indicate that only relatively very small ions would be expected to be able to occupy tetrahedral holes. The ratio of the sodium fluoride ionic radii is 95/136=0.698, but the sodium fluoride lattice is that which would be produced if the octahedral holes in a cubic close-packed fluoride lattice were filled by sodium ions.

Fig. 5.8 offers an insight into the difference between the cubic and hexagonal closest-packing arrangements which are adopted by many metals. The first layer of closest-packed spheres in the diagram is overlain by the second layer in the manner shown. The third layer of spheres may be arranged so that position A is occupied (so excluding position C from use). The atoms in the third layer, in this case, are directly above corresponding atoms in the first layer. This ABABAB . . . arrangement is that of the hexagonally closest-packed lattice. If, alternatively, the third layer makes use of position C in Fig. 5.8 the arrangement is that of the cubic closest-packed lattice, ABCABC . . .

The relationship between the two views of the cubic closest packing, as described by Figs 5.2 and 5.8, is made clear by the view in Fig. 5.10 which consists of a C_8 rotation (45°) of the diagram of Fig. 5.2 around an axis perpendicular to the paper. It is more obvious from the diagram of Fig. 5.10 that the atoms belong to four layers of the type exhibited by Fig. 5.8 of the ABCA . . . type. It should be realized that, atom for atom, the second and third layers are not directly above each other.

Consideration of an extended version of Fig. 5.8, leads to the conclusion that, for a lattice of N atoms arranged in either cubic or hexagonal closest packing, there are N octahedral holes and $2N$ tetrahedral holes.

Some other examples of the use of the above view of ionic structures are given below.

(1) In the zinc blende form of zinc sulphide, ZnS, half of the tetrahedral holes in a sulphide cubic close-packed lattice are occupied by zinc ions (Fig. 5.11).

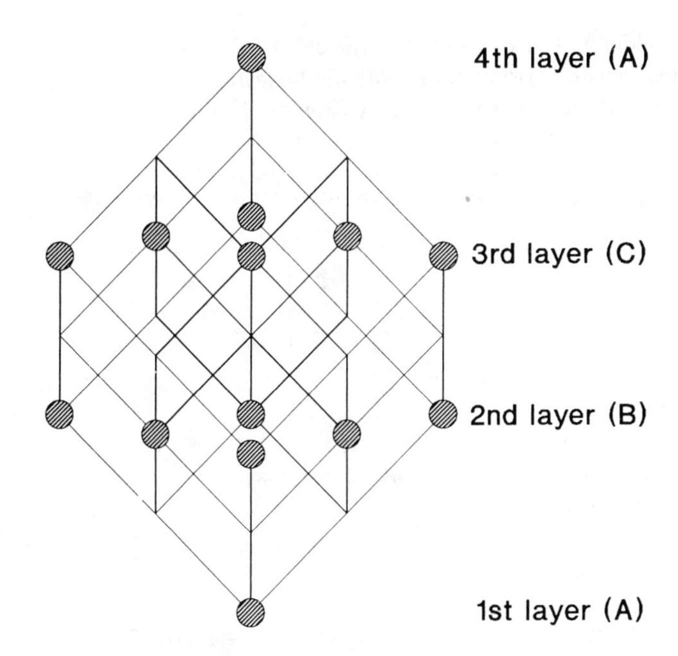

Fig. 5.10 — An alternative view of the cubic closest-packed (face-centred cubic) arrangement of atoms. The atom in the fourth layer is directly above the atom in the first layer. The atoms in the second and third layers are not directly above each other.

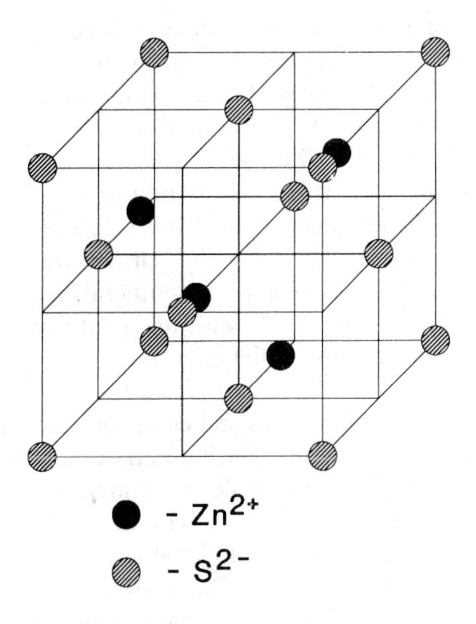

Fig. 5.11 — The zinc blende structure.

(2) In calcium fluoride (fluorite), the calcium ions may be considered to have a cubic close-packed arrangement with all the tetrahedral holes filled by the fluoride ions to give the CaF_2 stoichiometry (Fig. 5.12).

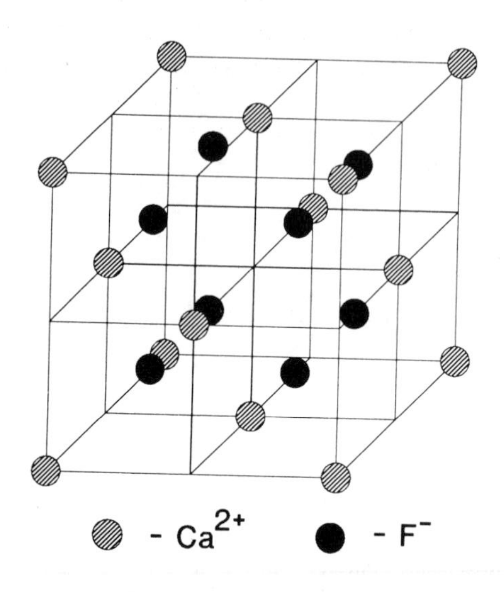

Fig. 5.12 — The fluorite structure.

The above approach can be widened to include those compounds which may be thought of in terms of the filling of the tetrahedral and octahedral holes in hexagonally close-packed host lattices of one or other of the ions. In addition it is very useful in the discussion of the structures of so-called interstitial compounds. For example, the majority of the transition metal hydrides may be thought of as slightly expanded metallic lattices with some or all of the tetrahedral and octahedral interstices occupied by hydrogen atoms. Thus all stoichiometries from the metal, M, to MH_3, are understandable. In general such hydrides are 'non-stoichiometric' in that the hydrogen/metal ratios are usually non-integral, their formulae depending upon a balance between the number of holes filled with hydrogen atoms and the higher entropies associated with non-stoichiometry.

The relationship between lattice diagrams and compound stoichiometry

There are simple rules which allow the stoichiometry of a compound to be derived from its lattice diagram. Fig. 5.7 serves to demonstrate the application of the rules. The diagram of the NaF structure in Fig. 5.7 contains representations of thirteen sodium ions and fourteen fluoride ions. Only one ion — a sodium ion — is wholly contained within the cubic arrangement. There are twelve sodium ions half way along each cube edge. In the actual crystalline material the ions along such edges will each be shared by **four** cubes so that each ion on an edge counts only as a quarter for purposes of determining the stoichiometry of the compound. Therefore, the twelve edge-type sodium ions of Fig. 5.7 count as 12/4 = three whole sodium ions. There is a total of four whole sodium ions represented by the diagram of Fig. 5.7. There are six fluoride ions in the cube faces, which count as three whole ions, since each face is shared by two cubes. The eight fluoride ions at the cube corners count as a single whole ion since any ion at a corner is shared by eight cubes. There is a total of four whole fluoride ions represented by the Fig. 5.7 diagram. Thus the stoichiometry of NaF as given by applying these rules to the lattice diagram is 4:4=1:1, so all is well! Similar rules apply to non-cubic systems.

5.5 CONCLUDING SUMMARY

This chapter contains an introductory treatment of the following topics.

(1) The factors which cause an element to be metallic.
(2) The forms of non-metallic elements.
(3) The energetics of ionic bond formation.
(4) The factors favouring the formation of ionic compounds.
(5) The metallic bond and the electrical conductivities of metals, semiconductors and insulators.
(6) Some simple ionic lattices.

6

Vibrational and electronic spectroscopy

6.1 INTRODUCTION

This chapter is concerned with two areas of investigation which are very helpful in the understanding of molecular structure: vibrational and electronic spectroscopy.

Vibrational spectroscopy is based upon the understanding of the normal modes of vibration which molecules may undergo, and the quantum rules which govern their excitation.

There are two branches of vibrational spectroscopy which provide information about molecular vibrations. These are (i) infrared absorption spectroscopy and (ii) Raman spectroscopy.

The two types of vibrational spectroscopy are differentiated by their mechanisms of excitation of the vibrational energy of a molecule. The complete absorption of a quantum (photon) in the infrared region of the spectrum (frequency range, 4×10^{14}–6×10^{11} Hz, wavelength range, 700 nm–500 μm) causes a molecule to be vibrationally excited. In the Raman effect, a higher-energy quantum (usually in the visible or ultraviolet regions) interacts with a molecule, causing it to gain or lose vibrational energy. The two mechanisms are subject to different selection rules, and are important in the determination of molecular symmetry.

Electronic spectroscopy is concerned with electronic transitions in atoms or molecules. An electronic transition may be described in terms of a change in electronic configuration brought about by either absorption or emission of a photon of suitable energy. It is essential to appreciate that a full description of an electronic transition must be given in terms of the two electronic *states* concerned, since any electronic configuration usually gives rise to more than one state. The energies of electronic transitions normally correspond to those of photons in the visible (wavelength range, 400–700 nm) or the ultraviolet (wavelength range, 200–400 nm) regions of the electromagnetic spectrum. Some are observed in the far ultraviolet region at wavelengths below 200 nm and some in the near infrared region at wavelengths above 700 nm.

6.2 VIBRATIONAL SPECTROSCOPY

It is essential to consider the number of independent vibrational modes of a molecule and their symmetries before an understanding of vibrational spectroscopy may be achieved.

Number of vibrational modes possessed by a molecule

The number of vibrational modes possessed by a molecule depends upon (i) the number of atoms in the molecule, and (ii) whether the molecule is linear or non-linear.

It may be considered that each atom in a molecule may move such that its movement may be resolved along the three Cartesian axes. For a molecule possessing n atoms this amounts to $3n$ possible movements (or degrees of freedom) of the atoms in the molecule. Three of the combinations of the $3n$ degrees of freedom are **translational** in that all the atoms move in the same direction, and which are resolvable along the x, y and z axes. These are the three degrees of translational freedom of the molecule. Some other combinations of the $3n$ degrees of freedom will amount to **rotations** about the Cartesian axes. For a linear molecule there will be only two such modes of rotation — the rotation about the molecular axis is discounted since it does not correspond to any movement of atoms with respect to their Cartesian coordinates. For any non-linear molecule there are three rotational modes.

From the above consideration the conclusions are that for linear molecules there are $3n-5$ vibrational degrees of freedom (or modes), and that for non-linear molecules there are $3n-6$. For linear molecules the non-vibrational modes are the three translational modes and the two rotations. Non-linear molecules have an extra rotational mode.

The vibrational energy of a molecule is quantized, and any particular vibration is governed by a vibrational quantum number, υ, which has integral values including zero.

Molecules at normal temperatures are usually in their zero-point vibrational states (υ=0). The residual, or zero-point, vibration of molecules (which persists even at absolute zero) is a consequence of the uncertainty principle. The extent of the vibration represents the uncertainty in the positions of the nuclei.

The most common vibrational transition is from the zero-point level (vibrational quantum number, υ=0) to the next higher level (υ=1). The transition, υ=0, to υ=2, is known as the first overtone of the $0 \rightarrow 1$ transition. Other transitions that are observable are combinations of transitions of the same symmetry.

Two examples are dealt with in this chapter - the non-linear water molecule and the linear carbon dioxide molecule.

The vibrational modes of the water molecule

The water molecule is triatomic ($n=3$) so that it has $3\times3-6=3$ vibrational modes. It is important to decide the nature of these modes — whether they are stretching or bending — and what their symmetry properties are.

The water molecule belongs to the C_{2v} point group. The motions of the three atoms along their individual x, y and z directions may be characterized by writing down the number of such movements which are unaffected by each of the four symmetry operations of the C_{2v} point group. This produces the character:

	E	C_2	$\sigma_v(xz)$	$\sigma_v'(yx)$
$9\times(x, y \text{ or } z)$	9	−1	1	3

The C_2 rotation moves the six motions of the two hydrogen atoms, leaves the z motion of the oxygen atom alone, but reverses the motions of the oxygen atom in the x and y directions, giving an overall character of −1 for that operation. Reflexion in the xz plane again moves all six of the hydrogen motions but leaves the oxygen atom's motions in the x and z directions alone. The oxygen motion in the y direction is reversed so the character has a resultant value of unity. Reflexion in the molecular plane (yz) leaves the six y and z motions alone but reverses the three x motions so the character is 3 for that operation.

Inspection of the character table for the C_{2v} point group reveals that the translations (movements along the x,y and z directions) transform as the irreducible representations b_1, b_2 and a_1 respectively. The rotations about the x, y and z axes transform as b_2, b_1 and a_2 respectively. The characters of these six representations must be subtracted from the total character of the nine degrees of freedom to give the character of the three vibrational modes:

	E	C_2	$\sigma_v(xz)$	$\sigma_v'(yx)$
9 d.f.:	9	−1	1	3
3 trans d.f.	3	−1	1	1
	6	0	0	2
3 rot. d.f.:	3	−1	−1	−1
3 vib. d.f.:	3	1	1	3

The character of the three vibrational modes reduces to the sum:

$$\text{3 vib. d.f.} = 2\times \mathrm{a}_1+\mathrm{b}_2 \tag{6.1}$$

In a simple case, as is the one under discussion, it is a very straightforward matter to identify the vibrational modes appropriate to their symmetries. The three vibrational modes of the water molecule are shown in Fig. 6.1. The symmetrical

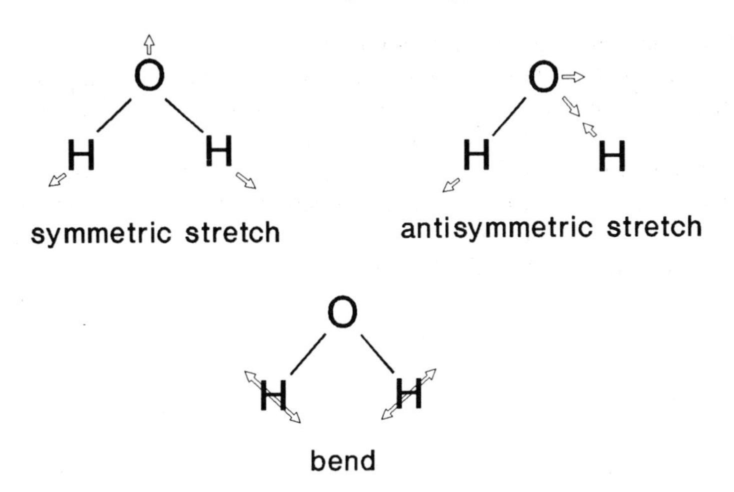

Fig. 6.1 — The vibrational modes of the water molecule.

stretch is of a_1 symmetry, as is the bending mode, with the antisymmetric stretch being b_2. In a more formal way it is possible to derive the character of the stretching modes by writing down the number of the O–H 'bonds' which are unaffected by the four symmetry operations. This gives:

	E	C_2	$\sigma_v(xz)$	$\sigma_v'(yx)$
2×O–H	2	0	0	2

which reduces to the sum:

$$2\times\text{O–H} = a_1 + b_2 \tag{6.2}$$

The bond angle is unaffected by any of the four symmetry operations and therefore the bending mode is of a_1 symmetry.

Absorptions arising from the 0→1 transitions of the three vibrational modes may be observed in the infrared spectrum of water vapour as bands at 3756 cm^{-1} (b_2), 3652 cm^{-1} (a_1, stretch) and 1595 cm^{-1} (a_1, bend). In general, stretching frequencies are greater than bending frequencies, the former being more involved with the forces responsible for the strength of the bonds. The frequencies of vibrational transitions are in general higher for stronger bonds and are lower for more massive atoms.

The units for vibrational bands that are usually quoted are reciprocal centimetres, cm^{-1}. If the absorption frequency, ν (in Hz, with units of s^{-1}), is divided by the velocity of light, c ($cm\,s^{-1}$), the resulting number has units of cm^{-1}, and is known as a **wavenumber**, symbolized as $\bar{\nu}$ ($=\nu/c$). This juggling with units is carried out to make the quoted numbers of manageable proportions. To convert a wavenumber into more meaningful energy units of $kJ\,mol^{-1}$. it is necessary to multiply it by 0.01196.

The above factor arises from the relationship between energy, E, and wavenumber (worked out for the Rydberg constant in Chapter 1, equation (1.3), the Rydberg constant being expressed as a wavenumber in reciprocal metres) given by the equation:

$$E = Nhc\bar{\nu} \qquad (6.3)$$

For example, the $1595\,\text{cm}^{-1}$ bending mode of the water molecule would be excited by a photon of energy: $1595 \times 0.01196 = 19.08\,\text{kJ mol}^{-1}$. The fraction of water molecules which would be so excited at equilibrium at 298 K is given by:

$$n/N = e^{-h\nu/kT} = 0.00046 \qquad (6.4)$$

The value of kT at 298 K is only $2.48\,\text{kJ mol}^{-1}$ and only if $h\nu < kT$ does the above ratio exceed unity. Only very low energy vibrations are excited to an appreciable extent at the standard temperature.

The basis of the above calculation is given in in Chapter 7.

The vibrational modes of the carbon dioxide molecule

Since the carbon dioxide molecule is linear it possesses four vibrational modes ($3 \times 3 - 5 = 4$). The identification procedure is similar to that used for the water case, but use is made of the $D_{\infty h}$ character table. The nine x, y and z motions of the three atoms have the character given in Table 6.1. The table shows the characters of the three translational and two rotational modes (only rotations around the x and y axes are involved) and shows the reduction of the vibrational modes into irreducible representations.

Table 6.1 — Irreducible representations of CO_2 vibrations

	E	C_∞	σ_v	i	S_∞	C_2
9 d.f.:	9	$3+6\cos\phi$	3	-3	$2\cos\phi-1$	-1
2 rot. d.f.:	2	$2\cos\phi$	0	2	$-2\cos\phi$	
3 trans. d.f.:	3	$1+2\cos\phi$	1	-3	$2\cos\phi-1$	-1
Σ rot.+trans.:	5	$1+4\phi$	1	-1	-1	-1
4 vib. d.f.:	4	$2+2\cos\phi$	2	-2	$2\cos\phi$	0
σ_g^+:	1	1	1	1	1	1
4 vib.$-\sigma_g^+$:	3	$1+2\cos\phi$	1	-3	$2\cos\phi-1$	-1
π_u:	2	$2\cos\phi$	0	-2	$2\cos\phi$	0
σ_u^+:	1	1	1	-1	-1	-1

The operations carried out in Table 6.1 are:

(i) the numbers of the nine x, y and z motions of the three atoms which are unaffected by the symmetry operations are written down in the first row, under the appropriate operations,
(ii) the characters of the two rotational and three translational degrees of freedom are written down next and added together,
(iii) the sum of the rotational and translational characters is subtracted from the total to give the character of the vibrational degrees of freedom,
(iv) first the characters of the σ_g^+ representation are subtracted from the vibrational characters, then those of the π_u representation are subtracted from the remainder, resulting in the characters corresponding to the σ_u^+ representation.

Thus the four vibrational modes of the carbon dioxide molecule are:

$$4 \text{ vib. d.f.:} = \sigma_g^+ + \pi_u + \sigma_u^+ \tag{6.5}$$

The σ_g^+ mode corresponds to the symmetric stretch, and the σ_u^+ to the antisymmetric stretch, the doubly degenerate π_u modes being the bending vibrations (which can occur in the xz and yz planes), as shown in Fig. 6.2.

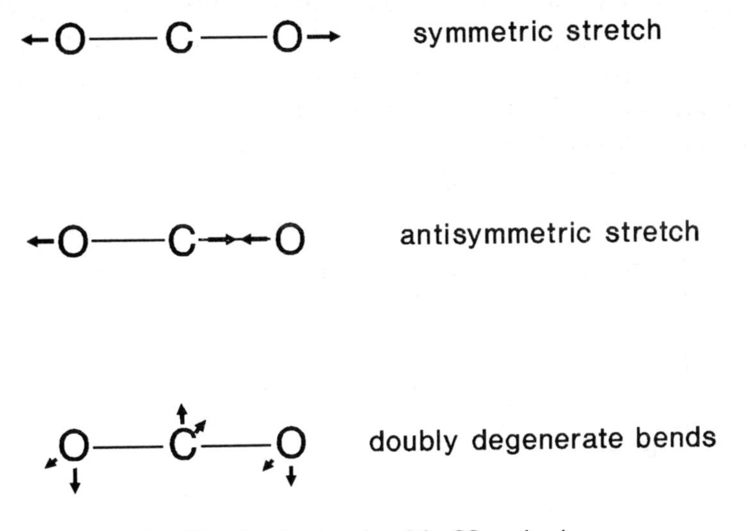

Fig. 6.2 — The vibrational modes of the CO_2 molecule.

There are only two bands observed in the infrared absorption spectrum of the CO_2 molecule. They correspond to the antisymmetric stretch (2349 cm^{-1}) and the degenerate bending mode (667 cm^{-1}). The symmetric stretch is not observed because of a **selection rule**.

Selection rules for infra-red and Raman spectroscopy

Selection rules are derived by considering the values of integrals which determine the probability of any given transition. The rules are different for changes in vibrational energy brought about by (i) absorption of quanta in the infrared region of the spectrum and (ii) the Raman effect. A Raman spectrum is obtained by using an incident beam of radiation (conventionally of either ultraviolet or visible radiation) which alters the vibrational state of the target molecule. The effect is usually observed at 90° to the incident beam to avoid the detection of the incident radiation. As well as a large intensity of scattered incident radiation, there are bands corresponding to frequencies which are given by:

$$\nu_i - \nu_1$$
$$\nu_i - \nu_2$$
$$\nu_i - \nu_3 \,, \qquad \text{etc.}$$

where ν_i is the frequency of the incident radiation and ν_1, ν_2, ν_3, etc., are the Raman frequencies of the molecule. Such bands are known as the Stokes bands and correspond to an increase in the vibrational energy of the target molecule. In a minority of cases there are anti-Stokes bands observed which are at frequencies $\nu_i+\nu_1$, $\nu_i+\nu_2$, $\nu_i+\nu_3$, etc. Such bands are produced when the vibrational state of the target molecule is decreased. Since the majority of molecules, at room temperature, are in their zero-point vibrational states, the anti-Stokes bands are not commonly observed.

1. Infrared absorption

The quantum theory of the absorption of radiation by a molecule indicates that the value of at least one of the integrals:

$$\int \psi_1 x \psi_0 \, d\tau$$
$$\int \psi_1 y \psi_0 \, d\tau$$
$$\int \psi_1 z \psi_0 \, d\tau$$

appropriate to the vibrational transition from the zero-point level ($\upsilon=0$) to the first excited state ($\upsilon=1$), should be non-zero. The x, y and z terms represent the dipole moments, in those directions, possessed by the molecule. The nature of electromagnetic radiation is dipolar in that there is oscillation of the electric component, and transfer of energy to a molecule can only take place when that molecule is also dipolar. The above integrals do not need to be solved completely — the necessary information required is whether they are zero or finite. If they are zero then infrared absorption will not occur. If at least one is finite then absorption of infrared radiation is permitted.

Symmetry theory may be used to decide whether any integral is of zero value or otherwise. It is essential to realize a very important property of the vibrational wavefunctions, ψ_0 and ψ_1, (the subscripts being the values of υ), for any vibrational mode of a given molecule. The ψ values are dependent upon two terms: one is a function of the **square** of the displacement of the molecule from its equilibrium position (and is therefore always completely symmetric), and the other is a polynomial term, which is **unity** for the zero-point state and x (or y, or z) for the first excited state. This has the consequences that:

(i) the symmetry of the wavefunction for the zero-point vibration of any mode of any molecule is always that corresponding to symmetric behaviour with respect to all the contained symmetry operations of the point group.

(ii) the symmetry of the wavefunction for the associated first vibrationally excited state is that of the particular vibration. This is because the displacements from the equilibrium positions of the atoms of the molecule have the same symmetries as the changes in the x, y and z coordinates.

The various integrals for the $0 \rightarrow 1$ transitions of the water molecule are evaluated as in Table 6.2.

Table 6.2 — Infrared activity of the vibrational modes of H_2O

Integral	a_1 (str. and bend)	b_2 (str.)
$\int \psi_1 x \psi_0 \, d\tau$	$a_1.b_1.a_1 = b_1$	$b_2.b_1.a_1 = a_2$
$\int \psi_1 y \psi_0 \, d\tau$	$a_1.b_2.a_1 = b_2$	$b_2.b_2.a_1 = a_1$
$\int \psi_1 z \psi_0 \, d\tau$	$a_1.a_1.a_1 = a_1$	$b_2.a_1.a_1 = b_2$

The entries in Table 6.2, under each vibrational mode, are the symmetries, respectively, of the zero-point vibrational wavefunction, the appropriate dipole and the wavefunction of the first vibrationally excited state. The triple products are shown, and the operation of the selection rule is that if one or more are completely symmetric the transition is allowed as an absorption process. The two a_1 vibrations and the b_2 vibration are all active in the infra-red region.

The situation with the CO_2 molecule is outlined in Table 6.3.

Table 6.3 — Infrared activity of the vibrational modes of CO_2

Integral	σ_g^+ (sym. str.)	σ_u^+ (antisym. str.)	π_u (bends)
$\int \psi_1(x,y)\psi_0\,d\tau$	$\sigma_g^+.\pi_u.\sigma_g^+ = \pi_u$	$\sigma_u^+.\pi_u.\sigma_g^+ = \pi_g$	$\pi_u.\pi_u.\sigma_g^+ = \sigma_g^+ + \sigma_u^+ + \delta_g$
$\int \psi_1 z\psi_0\,d\tau$	$\sigma_g^+.\sigma_u^+.\sigma_g^+ = \sigma_u^+$	$\sigma_u^+.\sigma_u^+.\sigma_g^+ = \sigma_g^+$	$\pi_u.\sigma_u^+.\sigma_g^+ = \pi_g$

In this case the x and y dipoles form a doubly degenerate, π_u, set and must be considered together. The triple product for the symmetric stretching vibration, σ_g^+, does not contain σ_g^+ and is therefore inactive in the infrared region. The other two vibrations are active since their triple products do contain the completely symmetric representation, σ_g^+.

There is a general rule concerning the activity (or otherwise) of any vibrational transition excited by the absorption of an infrared frequency. This is that there should be a change in the dipole moment of the molecule brought about by any vibration which may be excited by absorption. Those vibrations which do not cause the dipole moment to change are inactive in the infrared. The three vibrations of the water molecule all cause the dipole moment of the molecule to change and are infrared active. The symmetric stretch of the centrosymmetric CO_2 molecule does not cause the dipole moment of the molecule to change and is, therefore, inactive.

The above rule stems from the symmetry properties of the vibrations. In the case of a centrosymmetric molecule, such as CO_2, the dipole moment operators are 'u', and since the ground ($\upsilon=0$) and first excited ($\upsilon=1$) states of a symmetric stretching vibration are both 'g', the triple products must be 'u', causing the transition to be forbidden. Only if the first excited state is 'u' is the transition allowed. The 'u'-type vibrations (σ_u^+ and π_u modes of CO_2) of a molecule are such that they cause a change in the dipole moment of the molecule.

2. *Raman Spectra of H_2O and CO_2*

The criteria which determine whether a particular vibrational mode is active in the Raman spectrum are the values of integrals of the type:

$$\int \psi_1 P \psi_0\,d\tau$$

where ψ_0 and ψ_1 have their previous meanings and where P represents any quadratic combination of x, y or z. Thus x^2, y^2, z^2, xy, xz or yz are appropriate functions, and combinations of these such as x^2-y^2 are also in order. They are part of what is known as the polarizability tensor and have the symmetries indicated in each point group character table. For a vibrational mode to be Raman active the appropriate integrals should contain the completely symmetric representation for the point group to which the molecule belongs. Raman activity occurs if the vibration produces a change in the polarizability of the molecule. Sometimes this may be self-evident and for some cases the symmetries of the appropriate integrals must be determined.

For the water molecule; the possibilities are tabulated in Table 6.4.

Table 6.4 — Raman activity of the vibrational modes of H_2O

P (symmetry)	$a_1.P.a_1$	$b_2.P.a_1$
x^2,y^2 or $z^2(a_1)$	a_1	b_2
$xy(a_2)$	a_2	b_1
$xz(b_1)$	b_1	a_2
$yz(b_2)$	b_2	a_1

In Table 6.4 the various symmetries of the P factor have been used to determine the triple product symmetries for the excitation of the a_1 (symmetric stretching and bending modes) and the b_2 antisymmetric stretch. Both sets of integrals contain the completely symmetric representation, and so all three vibrational modes of the water molecule are active in its Raman spectrum.

The results for the triple products for the CO_2 molecule are tabulated in Table 6.5. In some instances it is has not been necessary to determine the triple product, as it is clear that all its components must be u type.

From Table 6.5 it is clear that only the symmetric stretching vibrational mode of the CO_2 molecule is Raman active. The other vibrational modes are inactive. There is a clear distinction between the vibrational spectra (infrared and Raman) of the two molecules, H_2O and CO_2. The water molecule (C_{2v}) has no inversion centre and its

Table 6.5 — Raman activity of the vibrational modes of CO_2

P (symmetry)	$\sigma_g^+.P.\sigma_g^+$	$\sigma_u^+.P.\sigma_g^+$	$\pi_u.P.\sigma_g^+$
$z^2(\sigma_g^+)$	σ_g^+	σ_u^+	π_u
$xz,yz(\pi_g)$	π_g	π_u	'u'
$x^2-y^2,xy(\delta_g)$	δ_g	δ_u	'u'

three vibrational modes are all infrared and Raman active. The CO_2 molecule ($D_{\infty h}$) does possess an inversion centre and there is mutual exclusion between the infrared and Raman activities of its vibrational modes. The symmetric stretch is not observed in the infrared spectrum but is present in the Raman spectrum. The opposite is true for the other modes; the antisymmetric stretching and the degenerate bending absorptions which are present in the infrared spectrum but are absent from the Raman spectrum. The mutual exclusion rule is useful in the experimental determination of whether a molecule possesses an inversion centre or not.

The Raman spectrum of the CO_2 molecule is complicated by its containing two bands corresponding to vibrations with wavenumbers of 1285 cm^{-1} and 1388 cm^{-1}. This is because of the interaction of the actual symmetric stretching vibrational mode with the first overtone of the bending mode. The wavenumber of the bending mode is 667 cm^{-1}. and that of its first overtone would be at approximately $2 \times 667 =$ 1334 cm^{-1}. The transition from the zero-point bending mode to its first overtone is allowed in the Raman effect since both ground and excited states have σ_g^+ symmetry. Since the *P* factor also contains the completely symmetric representation, the appropriate integrals are finite. Because the symmetric stretching mode has identical symmetry properties to those of the first overtone of the bending mode, and their frequencies are very similar, they interact to give two new frequencies. The Raman band observed at 1388 cm^{-1} is the first overtone of the bending mode with an admixture of the fundamental symmetric stretching frequency; that at 1285 cm^{-1}. is the 'negative' combination of the two modes. Such a phenomenon is known as **Fermi resonance**. This interaction between excited states in vibrational spectroscopy is common and can lead to the misinterpretation of infrared and Raman spectra if ignored.

Compared to the water molecule, the frequencies of the symmetric and antisymmetric stretching modes of the CO_2 molecule are very different at 2349 and 1300 cm^{-1} respectively. This may be understood in general terms by reference to Fig. 6.2 from which it may be seen that the symmetric stretch does not involve the movement of the carbon atom. The frequency of the symmetric stretch is determined by the CO bond strength and the mass of the moving oxygen atoms. The antisymmetric stretch may be regarded as the movement of the central carbon atom, with the ligand oxygen atoms remaining relatively in their equilibrium positions. The main movement of the mass six atom, rather than that of two mass sixteen atoms, is the reason for the antisymmetric stretch having its relatively low frequency.

6.3 ELECTRONIC SPECTRA OF ATOMS AND MOLECULES

The understanding of the electronic spectra of atoms is important in two respects: that of extending atomic orbital theory to the excited states of atoms, and that of contributing to the elucidation of the spectra of complexes. Likewise the treatment of the electronic spectra of molecules is important in that it leads to an extension of molecular orbital theory, and is essential to the full understanding of the spectra of complexes.

Electronic spectra of atoms

To understand the electronic transitions which can occur when an atom absorbs radiation of appropriate frequencies (or when an electronically excited atom emits its characteristic frequencies) it is essential to make the distinction between electronic **configurations** and electronic states. The electronic configurations of atoms are dealt with in Chapter 1, and represent the occupancy of the available atomic orbitals in any particular case. Hund's rules are followed so that the number of unpaired electrons is determined. It must be appreciated that any one electronic configuration may represent one, **or more than one**, electronic state. As is shown in Chapter 1, the electronic states ^3P, ^1D and ^1S arise from the $2s^2\ 2p^2$ configuration of the carbon atom.

The outer electronic configuration of the mercury atom is normally $6s^2$, but is more accurately described as the 6^1S_0 electronic state. There is only one microstate corresponding to the s^2 configuration which produces L and S values which are both zero. If one electron is excited to one of the 6p orbitals the configuration becomes s^1p^1, and there are twelve possible microstates which are represented in Table 6.6.

Table 6.6 — Microstates of the s^1p^1 configuration

m_l(s)	m_l(p)			Σm_l	Σm_s
0	1	0	−1		
↑	↑			1	1
↑		↑		0	1
↑			↑	−1	1
↓	↓			1	−1
↓		↓		0	−1
↓			↓	−1	−1
↑	↓			1	0
↓	↑			1	0
↑		↓		0	0
↓		↑		0	0
↑			↓	−1	0
↓			↑	−1	0

The twelve microstates correspond to the nine required for the full definition of the ^3P state, leaving three (all with Σm_s values of zero) which constitute the ^1P state. The full term symbols of the electronic states of the mercury atom *described* above are:

Ground state: 6^1S_0
Excited states: 6^1P_1, 6^3P_2, 6^3P_1 and 6^3P_0.

The 6 prefix is the value of the principal quantum number, and is usually omitted from the description. The J values follow the rule outlined in Chapter 1.

The relative energies of the electronic states of the mercury atom arising from the s^2 and s^1p^1 configurations are shown in Fig. 6.3. The ^3P state has a lower energy than that of the ^1P state, which is to be expected from a consideration of interelectronic repulsions.

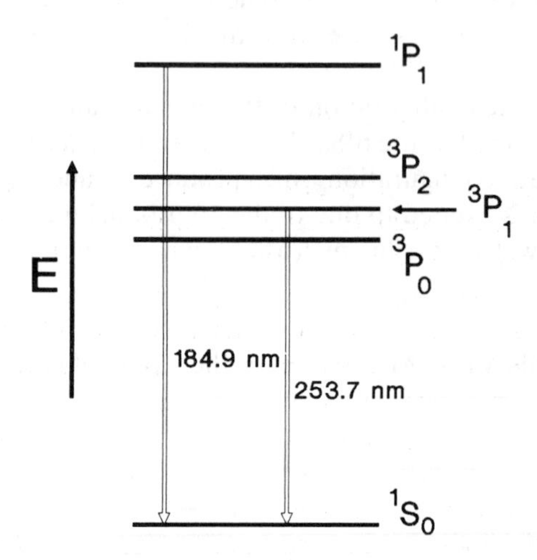

Fig. 6.3 — Some electronic states of the mercury atom and two emissive transitions.

Selection rules for atomic spectra

There are three selection rules which apply to transitions between electronic states of atoms. These are that allowed changes in the values of the quantum numbers l (for the electron transferred) and J are ± 1, and that there must not be a change in the multiplicity between the two states. The consequences of these selection rules are that the $^1P_1 \rightarrow {}^1S_0$ transition is allowed ($\Delta l = -1$, $\Delta J = -1$, and no change multiplicty), the $^3P \rightarrow {}^1S$ transitions being forbidden by the in multiplicity rule. Some breakdown of the quantum rules occurs with heavy atoms, and the transition, $^3P_1 \rightarrow {}^1S_0$, is observed. The other possible transitions, $^3P_{2,0}$, is observed. The other possible transitions, $^3P_{2,0} \rightarrow {}^1S_0$, are not observed. They are forbidden by both the multiplicity and ΔJ rules. The transitions which are observed in the emission spectrum of the mercury atom are indicated in Fig. 6.3 and are two of the 'lines' which are used extensively to initiate photochemical reactions.

The basis of the Δl and multiplicity rules may be dealt with as the problem of determining the intensity of an electronic transition. The transition probability depends upon the square of the integral:

$$\int \psi_f M \psi_i \, d\tau$$

where ψ_f and ψ_i are the wavefunctions of the final and initial states, and M is the dipole moment operator. The ψ functions may expressed as the product of ϕ (the **orbital** wavefunction) and σ (the **spin** wavefunction): $\phi \times \sigma$. The above integral then becomes:

$$\int \phi_f M \phi_i \, d\tau . \int \sigma_f \sigma_i \, d\sigma$$

where the subscripts have their previous meanings. The first integral is concerned with orbital changes, the second being concerned with changes of spin. The two integrals may be dealt with separately.

The integral concerned with orbital changes can be seen to be either zero or finite by the application of symmetry theory. If the discussion is limited to the one electron which is transferred in the electronic transition, the only symmetry property which is relevant is the inversion centre. Atomic orbitals are either symmetric to the inversion operation (s and d, for example) or antisymmetric (p and f, for example). The dipole moment operators (in the x, y and z directions) are always antisymmetric to inversion (like arrows). In order that the integral should be non-zero it should have g-type symmetry. Since the dipole moment operators are always u-type it must be that for a non-zero value to result, the symmetry of the product $\phi_f . \phi_i$ must be u-type. The only way that can be achieved is by ϕ_f and ϕ_i having different symmetries. This means that the transitions, p→s, d→p, etc., are allowed and that s→s, p→p, etc., are forbidden. This is consistent with the Laporte rule, which restricts electronic transitions in atoms to changes in l of ± 1. The above symmetry considerations do not rule out transitions such as s→f (g→g), but the Laporte rule does. The necessary basis for the understanding of the forbiddenness of atomic transitions is the consideration that a photon has a spin of unity. Angular momentum should be conserved so that the following transition (caused by absorption of a photon) is allowed:

$$S(l=0, s=\tfrac{1}{2}) + h\nu(s=1) \rightarrow p(l=1, s=\tfrac{1}{2}) \qquad (6.6)$$

there being no change in total angular momentum. A transition from a p orbital to a d orbital, accompanied by the emission of a photon, is allowed by the Laporte rule. The conservation of angular momentum indicates that the photon spin, in this case, would be -1.

The spin integral is easily seen to be either zero or finite. There are only two values of the spin quantum number ($\pm\tfrac{1}{2}$), and these may be represented by α and β. The spin integral for an electronic transition in which there is conservation of spin is $\int \alpha^2 d\sigma$ (or $\int \beta^2 d\sigma$), which has a non-zero value (think of the area under a plot of the equation $y=x^2$). The spin integral for a transition in which there is a spin change is $\int \alpha\beta d\sigma$, which has a zero value (similar to the area under a plot of $y=x$). This is the basis of the spin rule, which implies that no spin changes take place in a spin-allowed transition.

Relevance of atomic spectra to those of complexes of transition elements

The basis of atomic spectra has been dealt with in sufficient detail for the understanding of term symbols of electronic states and the rules which permit transitions between any two states. The electronic spectra of complexes containing central atoms (or ions) which possess d electrons may be understood (partially) in terms of what are called d–d transitions. The five-fold degeneracy of the d orbitals of such atoms is lost in non-spherical environments and leads to the possibility of transitions between the resulting levels. Although the d electrons are involved in bonding and anti-bonding combinations with ligand orbitals, they may be considered as having largely retained their d atomic character.

It is important to consider the atomic states which are associated with a given d electronic configuration.

The d^1 case is straightforward. If there are no other unpaired electrons to consider then the L value is 2 and $S=\frac{1}{2}$, which produces the ^2D state. There are two possible J values of $2\frac{1}{2}$ and $1\frac{1}{2}$.

A d^2 configuration gives rise to 45 microstates — there are ten ways of placing one electron in the five d orbitals (taking spin into account) and nine ways of placing the second electron. The indistinguishability of electrons causes there to be only 45 distinguishable microstates. The exercise of writing out these microstates should be carried out. Using the criteria outlined in Chapter 1 it is possible to decide that the d^2 configuration gives rise to the following electronic states:

$$d^2 \rightarrow {}^3F,\ {}^1D,\ {}^3P,\ {}^1G \text{ and } {}^1S \tag{6.7}$$

The application of Hund's rules indicates that the ground state is the ^3F state. A useful 'rule-of-thumb' is available to give the ground state for any d configuration. Taking the values of m_l in the order 2, 1, 0, −1 and −2, and considering electrons to have values of m_l in that order as they are fed into the five orbitals (with maximum single occupancy as per Hund's rules), the value of L (for the electronic ground state) is given. For example, the ground term for a d^4 configuration has an L value of:

$$L(d^4)=2+1+0-1=2 \tag{6.8}$$

so that it corresponds to a ^5D state. The multiplicity of 5 is given by $2S+1$, where S (total spin) is $4\times\frac{1}{2}$. In general it is necessary to define all the microstates of a given configuration in order to decide which states exist. Tabulations are available, and the relevant states are included in the appropriate Orgel and Tanabe–Sugano diagrams in this (see Chapter 8 and Appendix II) and more advanced texts.

Electronic spectra of molecules

The discussion of electronic spectroscopy of molecules is restricted to a few examples of small molecules and ions (which are relevant to the spectra of complexes when they act as ligands), and to a very general treatment of polyatomic molecules. As with atomic spectroscopy it is essential to bear in mind the distinction between configurations and states.

Small molecules and ions

The examples used in this section are (i) the chloride ion, (ii) dioxygen, (iii) water, (iv) ammonia, (v) trioxygen (ozone) and (vi) carbon monoxide

(i) Aqueous chloride ion

The chloride ion has a $3s^2 3p^6$ configuration, the lowest vacant level being the 4s orbital. Theoretically a 3p to 4s transition would be permitted by the rules, but has not been observed. It would be in the far ultraviolet region because of the very large energy difference between the two orbitals. The aqueous chloride ion exhibits a very broad double-peaked absorption in the far ultraviolet region, which may be interpreted as a **charge transfer to solvent** (c.t.t.s.) transition. The transition involves one of the 3p electrons being transferred to the water molecules which form the primary solvation sphere of the chloride ion:

$$Cl^{-}.H_2O \xrightarrow{h\nu} Cl.H_2O^{-} \longrightarrow Cl + H_2O^{-} \quad (6.9)$$

and this may be followed by the decomposition of the negative water molecule:

$$H_2O^{-} \longrightarrow H + OH^{-} \quad (6.10)$$

The dual peaks in the chloride ion spectrum may be interpreted in terms of the production of the two electronic states of the resulting chlorine atom, $^2P_{3/2}$ and $^2P_{1/2}$.

If the chloride ion is acting as a ligand then there is the possibility of a ligand-to-metal **charge transfer** transition. Charge transfer transitions are usually of high intensity. The intensity of an electronic transition is governed by the magnitude of the **molar absorption coefficient**, ε.

When a beam of monochromatic radiation, of intensity I_0, is allowed to be incident upon a cell, containing a solution of the absorbing substance (c=molarity), of optical path length, l cm, the intensity of the transmitted radiation, I_T, is given by the equation:

$$I_T = I_0 . 10^{-\varepsilon c l} \quad (6.11)$$

Equation (6.11) is an expression of the Beer–Lambert Law of light absorption. It may be transformed into a more practical form by rearrangement to:

$$I_0 / I_T = 10^{\varepsilon c l} \quad (6.12)$$

and then taking the logarithm (to base ten) of both sides gives:

$$A = \log (I_0 / I_T) = \varepsilon c l \quad (6.13)$$

where A is termed the **absorbance** of the particular solution.

Equation (6.13) is useful analytically since it shows that the absorbance is directly proportional to the concentration of the absorbing substance (at a constant optical path length, it being usual to use 1 cm cells in the spectrophotometer), which is an expression of Beer's Law. In addition it indicates that, for a given concentration of the absorbing substance, the absorbance is directly proportional to the cell's solution thickness (Lambert's Law) which is also of practical importance. The measurement of absorbance at a given wavelength may be carried out with a spectrophotometer, and with the knowledge of the cell's solution thickness and the concentration of the absorbing substance the ε value can be determined. It is usual to obtain an ε value from a Beer's Law plot of A versus c, a straight line being essential for any confidence to be placed on the result.

The ε for the aqueous chloride ion has a value of $\sim 10^4\,\mathrm{l\,mol^{-1}cm^{-1}}$ at 181 nm and is typical of what is observed for fully allowed transitions.

(ii) Dioxygen

Gaseous dioxygen has a normal electronic configuration which includes two unpaired electrons occupying the doubly degenerate $1\pi_g$ (anti-bonding) level. There are three electronic states associated with such a configuration. Although the derivation of these states is not given in this text, it is very similar to that applied to the p^2 configuration of the carbon atom in Chapter 1. The states of dioxygen resulting from the $1\pi_g^2$ configuration are ${}^3\Sigma_g^-$, ${}^1\Delta_g$ and ${}^1\Sigma_g^+$, in order of increasing energy. The first excited configuration, $1\pi_u^3 1\pi_g^3$, gives rise to the excited states, ${}^3\Sigma_u^+$, ${}^3\Sigma_u^-$ and ${}^1\Delta_u$.

If the selection rules for atomic transitions are applied (conservation of spin and $g \rightarrow u$ being essential for an allowed transition) it might be expected that the transitions, ${}^3\Sigma_g^- \rightarrow {}^3\Sigma_u^+$, and ${}^3\Sigma_u^-$, would be allowed and, in consequence, be observed in the dioxygen spectrum. The ${}^3\Sigma_g^- \rightarrow {}^3\Sigma_u^+$ transition is of very low intensity and is actually forbidden, as a detailed symmetry calculation indicates. The dipole moment operators for a $D_{\infty h}$ molecule transform as Σ_u^+ and Π_u, and the triple products which govern the values of the integrals responsible for the intensity of the transition are derived in Table 6.7.

Table 6.7 — Symmetries of triple products for ${}^3\Sigma_g^- \rightarrow {}^3\Sigma_g^+$ of O^2

Representation	E	C_∞^ϕ	σ_v	i	S_∞^ϕ	C_2
Σ_g^-	1	1	−1	1	1	−1
Σ_u^+	1	1	1	−1	−1	−1
Σ_u^+	1	1	1	−1	−1	−1
$\Sigma_g^- \times \Sigma_u^+ \times \Sigma_u^+$	1	1	−1	1	1	−1
Σ_g^-	1	1	−1	1	1	−1
Π_u	2	$2\cos\phi$	0	−2	$2\cos\phi$	0
Σ_u^+	1	1	1	−1	−1	−1
$\Sigma_g^- \times \Pi_u \times \Sigma_u^+$	2	$2\cos\phi$	0	2	$-2\cos\phi$	0

The first part of Table 6.7 shows that the triple product, $\Sigma_g^- \times \Sigma_u^+ \times \Sigma_u^+$, is equal to Σ_g^-, since the square of any representation must be equal to, or contain, the fully symmetric representation. The second part of Table 6.7 shows that the triple product, $\Sigma_g^- \times \Pi_u \times \Sigma_u^+$, is equal to the Π_g representation. Since neither of the triple products is equal to the fully symmetric representation, Σ_g^+, the transition is forbidden.

The above calculations may be taken to indicate that a $- \rightarrow +$ (or a $+ \rightarrow -$) transition is forbidden. Although the calculations are not given here it is a simple matter to show that the $+ \rightarrow +$, and $- \rightarrow -$ transitions are allowed. This is an extra selection rule which applies to homonuclear diatomic molecules.

The $^3\Sigma_g^- \rightarrow {}^3\Sigma_u^-$ transition is fully allowed and is observed as a very intense absorption starting at a wavelength of 201 nm and extending into the far ultraviolet region as a series of bands. Each band of increasing energy corresponds to the electronic excitation plus an increasing amount of vibrational excitation of the dioxygen molecule. Bands rather than lines are observed, each band being composed of a merging set of lines due to a great number of changes in the rotational states of the ground and excited electronic states of the molecule.

Both the transitions, $^3\Sigma_g^- \rightarrow {}^3\Sigma_u^+, {}^3\Sigma_u^-$, are of importance in the chemistry of the atmosphere, and are largely responsible for screening the Earth's surface from radiation of wavelengths lower than 240 nm. The forbidden transition occurs at low intensity, and absorbs in the region, 240–200 nm, below which wavelengths the fully allowed transition takes over. The two transitions of the dioxygen molecule, together with the ionization of the dinitrogen molecule (and absorptions due to N_2^+), render the lower atmosphere free of radiation of wavelengths lower than 240 nm.

Another aspect of the electronic spectroscopy of dioxygen is that absorption by the $^3\Sigma_g^- \rightarrow {}^3\Sigma_u^-$ transition can cause the photodissociation of the molecule:

$$O_2(^3\Sigma_g^-) \xrightarrow{h\nu} O_2(^3\Sigma_u^-) \longrightarrow O(^3P) + O(^1D) \qquad (6.14)$$

Both electronic states of the oxygen atoms are important in the production and destruction of trioxygen molecules (see section (v)).

(iii) Water

Liquid water absorbs radiation in the region below 200 nm, with maximum absorption occurring at 170 nm, and reference to the m.o. diagram for the molecule (Fig. 4.6) shows that the lowest energy electronic transition is (in terms of C_{2v} nomenclature) $1b_1 \rightarrow 4a_1$. The highest occupied orbital is the non-bonding $1b_1$, and the lowest unoccupied orbital is the anti-bonding $4a_1$.

Written as a transition between electronic states, rather than between molecular orbitals, this becomes (dropping the Mulliken prefixes): $^1A_1 \rightarrow {}^1B_1$. The ground state is totally symmetric since the occupied orbitals are all doubly occupied and since the square of any representation is also totally symmetric. The first excited state is B_1 since the product, $b_1 \times a_1$ is b_1. The dipole moment operators in C_{2v} transform as $a_1(z)$, $b_1(x)$ and $b_2(y)$. The involvement of the b_1 operator in the triple product:

$$b_1 \times b_1 \times a_1 = a_1 \tag{6.15}$$

ensures that the transition is allowed and occurs with considerable intensity.

The transition causes a proportion of the absorbing molecules to undergo photodissociation according to the equation:

$$H_2O(^1A_1) \xrightarrow{h\nu} H_2O(^1B_1) \longrightarrow H(^2S) + OH(^2\Pi) \tag{6.16}$$

to produce a hydrogen atom and a hydroxyl radical, both in their electronic ground states.

Further transitions in the water molecule occur in the far ultraviolet region, but none has been fully characterized.

(iv) Ammonia

The ammonia molecule has C_{3v} symmetry in its electronic ground state ($^1A_1$) and its HOMO is the $3a_1$ orbital, as indicated in Fig. 4.10. The LUMO* is the anti-bonding $4a_1$ level. The lowest energy transition is:

$$^1A_1 \longrightarrow {}^1A_1 \tag{6.17}$$

The dipole moment operators transform as a_1 and e, so that the transition is forbidden ($a_1 \times e \times a_1$) cannot contain a_1). The weak absorption of radiation takes place in the ultraviolet region from 217 to 17 nm. The second electronic transition involves the excitation of an electron from the $3a_1$ orbital to the doubly degenerate 2e level. The transition is allowed ($e \times e \times a_1$ contains a_1) and appears as a high-intensity absorption in the 169–140 nm region.

There are theoretical grounds for suspecting that both the excited states of the molecule are planar. The occupation of the $3a_1$ orbital is the main factor which determines the C_{3v} symmetry of the ground state. The transfer of one of the $3a_1$ electrons to the $4a_1$ orbital is sufficient to cause the electronically excited state to adopt D_{3h} symmetry. Fig. 4.10 shows that the $3a_1'$ orbital in D_{3h} symmetry offers a greater stability for an electron than does the $4a_1$ orbital in C_{3v} symmetry.

(v) Trioxygen (ozone)

The trioxygen molecule, O_3, belongs to the C_{2v} point group, its bond angle being 116.8° and the O–O bond lengths between the central and terminal oxygen atoms being 127.8 pm. The m.o. diagram for O_3 may be derived in the usual manner.

The orbitals of the central oxygen atom transform as:

$2s(O)_c$: a_1
$2p_x(O)_c$: b_1
$2p_y(O)_c$: b_2
$2p_z(O)_c$: a_1

*lowest unoccupied molecular orbital.

the 2s and $2p_x$ orbitals have the same symmetry and have the potential to combine. In practice the energy gap between them is too large to permit significant mixing, and the 2s orbital will be considered to be non-bonding in character.

It is possible to arrange one 2p orbital from each of the ligand oxygen atoms to lie along the O–O bond directions so they may take part in σ-type bonding. The two orbitals taken together form two group orbitals: one (O_1+O_1, bonding) has a_1 symmetry, and the other (O_1-O_1, anti-bonding) has b_2 symmetry. Because the bond angle is larger than 90° the b_2 orbital has a better overlap with the central atom than the a_1 orbital. This affects the order of their energies and that of the anti-bonding combinations.

The 2p orbitals of the ligand atoms which are in the molecular plane but at right angles to the bond directions form two group orbitals which transform as a_1 and b_2 representations and are to be considered as non-bonding orbitals.

The $2p_x$ orbitals of the ligand atoms form two group orbitals which transform as b_1 (O_1+O_1, bonding) and a_2 (O_1-O_1, anti-bonding) representations. The b_1 orbital participates in a π-type interaction with the $2p_x$ (b_1) orbital of the central oxygen atom.

The m.o. diagram is shown in Fig. 6.4, and is formed by allowing orbitals of the central atom to interact with those of the ligand atoms of identical symmetries. The five highest filled levels are consistent with the observation of ionization energies of 1187, 1208, 1305, 1582 and 1856 kJ mol^{-1} from the photoelectron spectrum.

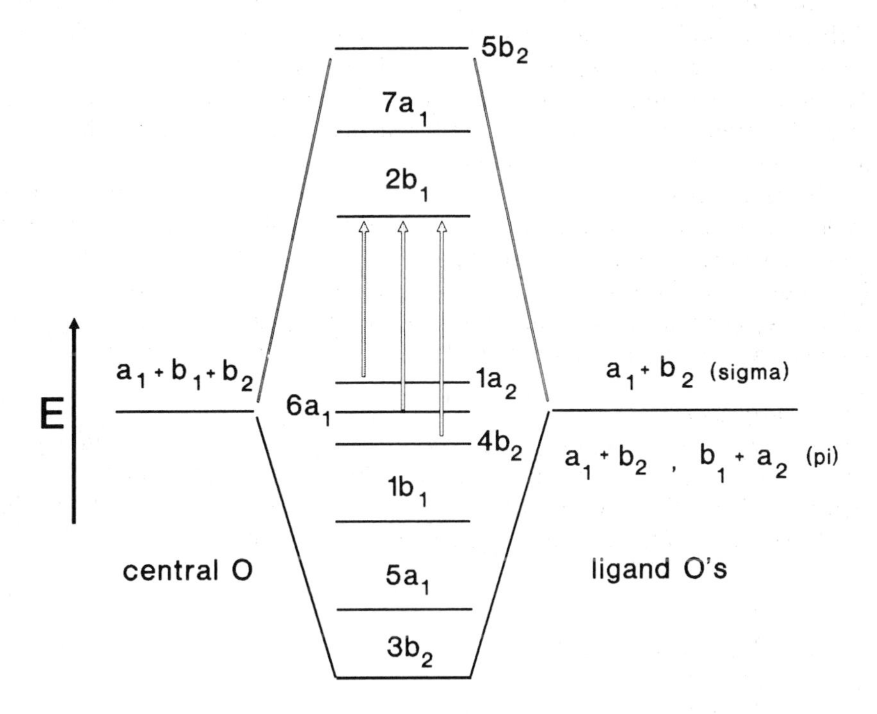

Fig. 6.4 — Molecule orbital diagram for trioxygen.

The non-bonding orbitals, $4b_2$, $6a_1$ and $1a_2$, are occupied, and transitions from these orbitals to the lowest vacant orbital, $2b_1$, explain the observed spectrum of the trioxygen molecule. The transitions may be written as:

$$^1A_1 \rightarrow {}^1B_2;\ (1a_2 \rightarrow 2b_1) \tag{6.18}$$

$$\rightarrow {}^1B_1;\ (6a_1 \rightarrow 2b_1) \tag{6.19}$$

$$\rightarrow {}^1A_2;\ (4b_2 \rightarrow 2b_1) \tag{6.20}$$

The first and second transitions are allowed, the third being forbidden. The first transition is responsible for a broad absorption between 500 and 700 nm and explains the slight blue colour of trioxygen. The second allowed transition is responsible for a strong absorption in the ultraviolet region between 220 and 290 nm. It is responsible for protecting the surface of the Earth from such radiation.

If O_3 was a linear molecule it would have the electronic configuration; $(2\sigma_g^+)^2(2\sigma_u^+)^2 1\pi_u^4 1\pi_g^4 (3\sigma_g^+)^2$ (see Fig. 4.09 for CO_2). The $3\sigma_g^+$ orbital is anti-bonding and would lead to bond weakening. In the bent, C_{2v}, molecule, trioxygen has no anti-bonding electrons.

(vi) Carbon monoxide

The ground state electronic configuration of the CO molecule is $^1\Sigma^+$, since there are no unpaired electrons. The first excited state is derived from the 5σ to 2π orbital transition (see Fig. 3.11) and is $^1\Pi$. The $^1\Sigma^+ \rightarrow {}^1\Pi$ transition occurs at 154 nm in the vacuum ultraviolet region. It is an allowed transition since the appropriate triple products contain the fully symmetric representation, Σ^+.

Polyatomic molecules

In very general terms, polyatomic molecules may possess σ- and π-type bonding orbitals, non-bonding orbitals and σ- and π-type anti-bonding orbitals. These are generally in the order of increasing energy: σ(bonding), π(bonding), n(non-bonding), π^*(anti-bonding) and σ^*(anti-bonding). Those molecules which are **saturated** (totally σ-bonded with no non-bonding electrons) may exhibit only $\sigma \rightarrow \sigma^*$ transitions, which are observed in the far ultraviolet region. **Unsaturated** molecules (with some π bonding) may have $\sigma \rightarrow \pi^*$, $\pi \rightarrow \pi^*$ and $\pi \rightarrow \sigma^*$ transitions in addition to the $\sigma \rightarrow \sigma^*$. The transitions of lowest energy ($\pi \rightarrow \pi^*$) are observed in the ultraviolet region. The **conjugation** of formal double bonds (any two formal double bonds separated by a formal single bond are conjugated) can cause the lowest energy transition to move to longer wavelengths, and some transitions are then observed in the visible region. If the molecule possesses non-bonding electrons then $n \rightarrow \pi^*$ transitions are observable. Their position in the spectrum depends upon the presence or absence of conjugation. They are generally to be observed at longer wavelengths than the corresponding $\pi \rightarrow \pi^*$ transitions. The general types of electronic transition possible for polyatomic molecules are shown diagrammatically in Fig. 6.5.

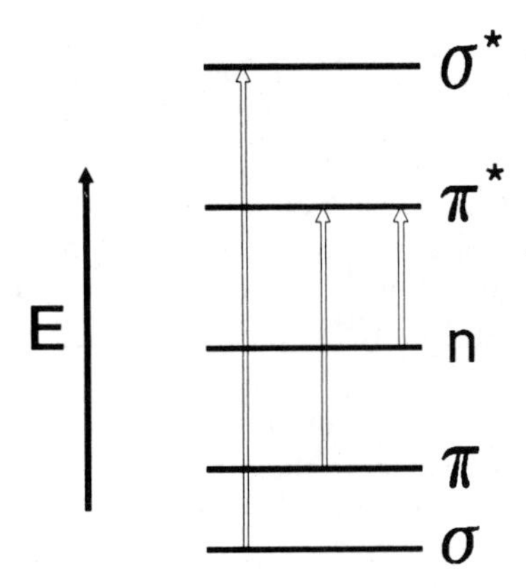

Fig. 6.5 — Generalized orbitals and transitions for polyatomic molecules.

The examples chosen to exemplify the spectra of polyatomic molecules are (i) ethene, C_2H_4, (ii) methanal, CH_2O, (iii) benzene, C_6H_6, and (iv) 1,10-phenanthroline (phen), $C_{12}H_8N_2$.

(i) Ethene, C_2H_4

The lowest energy transition of the ethene molecule is that concerned with the $\pi \rightarrow \pi^*$ transition. The molecule is planar and belongs to the D_{2h} point group, a diagram of the molecule being shown in Fig. 6.6. If the C–C axis is made coincident with the z axis, and the molecular plane is yz, the π bonding orbital $(2p_x+2p_x)$ has b_{3u} symmetry. The anti-bonding π-type orbital $(2p_x-2p_x)$ possesses b_{2g} symmetry. The

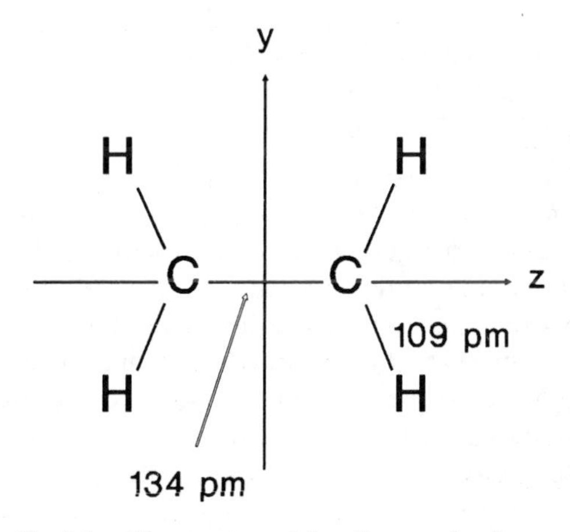

Fig. 6.6 — The structure of the ethane molecule.

'$\pi \rightarrow \pi^*$' transition in orbital terms is $b_{3u} \rightarrow b_{2g}$, and in terms of electronic states this becomes:

$$^1A_g \rightarrow {}^1B_{1u} \tag{6.21}$$

since the product of $b_{3u} \times b_{2g}$ is b_{1u}. The dipole moment operators (x, y and z) transform as b_{3u}, b_{2u} and b_{1u} respectively. The last representation ensures that the transition is allowed, since the triple product of $a_g \times b_{1u} \times b_{1u}$ is a_g. It is observed at 180 nm in the far ultraviolet region, with an ε value of around $10^4\,l\,mol^{-1}cm^{-1}$. A value of such a magnitude is typical of a fully allowed transition.

The excitation of a π bonding electron to the anti-bonding orbital effectively destroys the π-type bonding between the two carbon atoms. In the excited state the bond order between the two CH_2 groups is unity, with the σ-type bonding orbital being cylindrically symmetric with respect to the carbon–carbon axis. This allows the rotation of one CH_2 group with respect to the other with the formation of an excited molecule belonging to the D_{2d} point group (one CH_2 plane being perpendicular to the second CH_2 plane).

(ii) Methanal, CH_2O

The methanal molecule is planar and belongs to the C_{2v} point group. The $\pi \rightarrow \pi^*$ transition, in terms of orbitals, is from b_1 to b_1. In state terminology this becomes ${}^1A_1 \rightarrow {}^1A_1$, and the transition is allowed because the dipole moment operator in the z direction also transforms as a_1. The transition is observed as an absorption at 190 nm, with an ε value of about $10^4\,l\,mol^{-1}cm^{-1}$.

The $n \rightarrow \pi^*$ transition is from an orbital which is essentially the $2p_y$ orbital of the oxygen atom (which transforms as b_2) to the anti-bonding b_1 orbital. Using state terminology this is written as:

$$ {}^1A_1 \longrightarrow {}^1A_2 \qquad (6.22) $$

(since $b_1 \times b_2 = a_2$), the transition being forbidden since none of the dipole moment operators transforms as a_2. The 'forbidden' transition is actually observed at around 250 nm, but with the low value of ε of $20\,l\,mol^{-1}cm^{-1}$.

(iii) Benzene, C_6H_6

The benzene molecule has a regular hexagonal structure and belongs to the D_{6h} point group. There has been a great deal of controversy surrounding the bonding of the molecule, but only the molecular orbital theory can fully satisfactorily describe the electronic arrangement. One of the earlier formulations of the bonding is the molecule cyclo-1,3,5-hexatriene, which possesses three **localized** π-type bonds. The six carbon atoms in real benzene each contribute a $2p_z$ orbital (if the molecular plane is coincident with the xy plane) which transform, with respect to the D_{6h} point group, as:

	E	C_6	C_3	C_2	C_2'	C_2''	i	S_3	S_ϕ	σ_h	σ_d	σ_v
$6\times 2\pi_z$:	6	0	0	0	−2	0	0	0	0	−6	0	2

which reduces to:

$$6\times 2p_z = b_{2g}+e_{1g}+a_{2u}+e_{2u} \tag{6.23}$$

The a_{2u} combination is the fully bonding orbital, and the b_{2g} is the fully anti-bonding orbital. The doubly degenerate e_{1g} and e_{2u} sets of orbitals are overall bonding and overall, anti-bonding respectively. Their relative energies are shown in Fig. 6.7. The a_{2u} and e_{1g} orbitals are fully occupied and the six c-type bonding

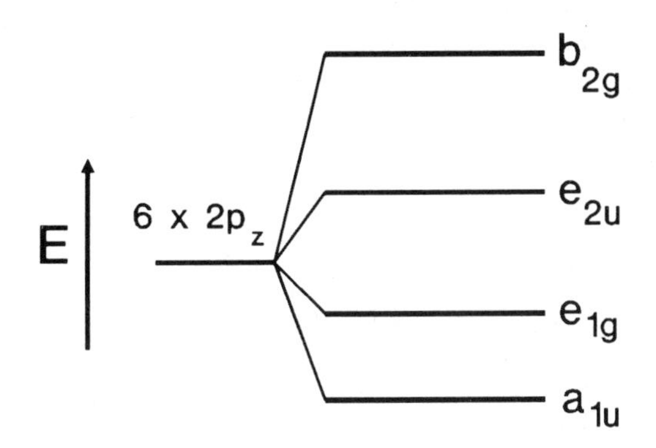

Fig. 6.7 — Molecular orbital diagram for the 'pi' orbitals of the benzene molecule.

electrons are **delocalized** around the six carbon centres. Delocalization occurs when it is permitted by symmetry and because it leads to a general stabilization of the molecule, the interelectronic repulsion energy being minimized in the larger, more extensive, orbitals.

The ground state of the benzene molecule is completely symmetric and is written as $^1A_{1g}$. If an orbital transition from the HOMO, e_{1g}, to the LUMO, e_{2u}, occurs, the excited states are given by the product: $e_{1g}\times e_{2u}$. This produces the states $^1B_{1u}$, $^1B_{2u}$ and $^1E_{1u}$ (for spin-allowed transitions). The dipole moment operators transform as $A_{2u}(z)$ and $E_{1u}(xy)$. Only the $^1A_{1g}\rightarrow{}^1E_{1u}$ transition is fully allowed, the triple product containing the A_{1g} representation. The other two transitions are forbidden. All three transitions are observed in the absorption spectrum of benzene. Fig. 6.8 shows the spectrum of benzene vapour. The lowest energy transition is $^1A_{1g}\rightarrow{}^1B_{2u}$ and is centred at 255 nm with an ε value of $220\,l\,mol^{-1}cm^{-1}$. The band associated with the $^1A_{1g}\rightarrow{}^1B_{1u}$ transition has an absorption maximum at 200 nm with an ε value of $7000\,l\,mol^{-1}cm^{-1}$. The fully allowed $^1A_{1g}\rightarrow{}^1E_{1u}$ transition occurs at 180 nm with an ε value of $47\,000\,l\,mol^{-1}cm^{-1}$. The high value of ε for the $^1A_{1g}\rightarrow{}^1B_{1u}$ transition does not signify a breakdown in the selection rules but is due to the close proximity of the fully allowed band. The forbidden band at 200 nm appears as an addition to the long wavelength end of the fully allowed band. The two lower frequency bands exhibit some vibrational fine structure.

(iv) 1,10-Phenanthroline, $C_{12}H_8N_2$

The 1,10-phenanthroline molecule, which belongs to the C_{2v} point group, is shown in Fig. 6.9. Its absorption spectrum (hexane solution) is shown in Fig. 6.10. There are

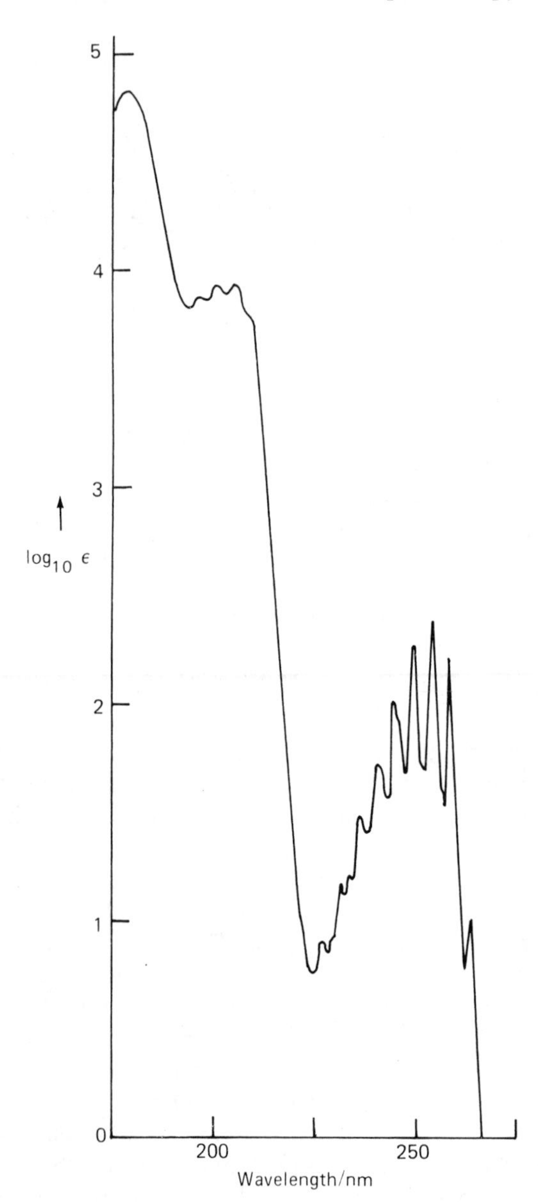

Fig. 6.8 — The electronic spectrum (ultraviolet) of benzene vapour.

three aromatic rings in the molecule, with the two nitrogen atoms suitably situated so as to allow it to act as a bidentate ligand. The π-type orbitals of the carbon and nitrogen atoms (composed from the 2p atomic orbitals perpendicular to the molecular plane) are delocalized over the whole molecule. There are seven b_1 and seven a_2 π-type orbitals.

Each of the nitrogen atoms possesses a non-bonding pair of electrons which are used in *e*-type donation to metal ions when the molecule acts as a bidentate ligand.

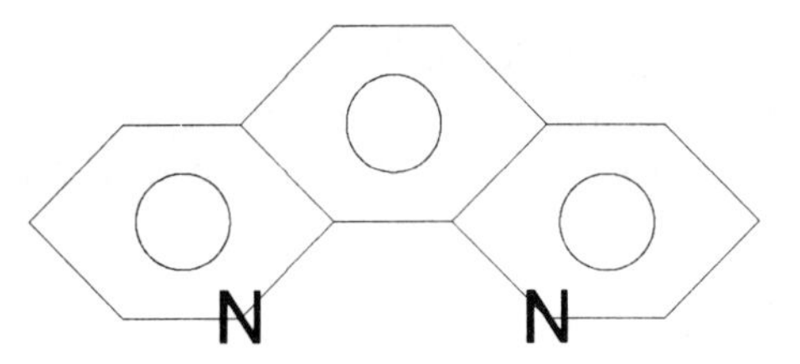

Fig. 6.9 — The structure of the 1,10-phenanthroline molecule.

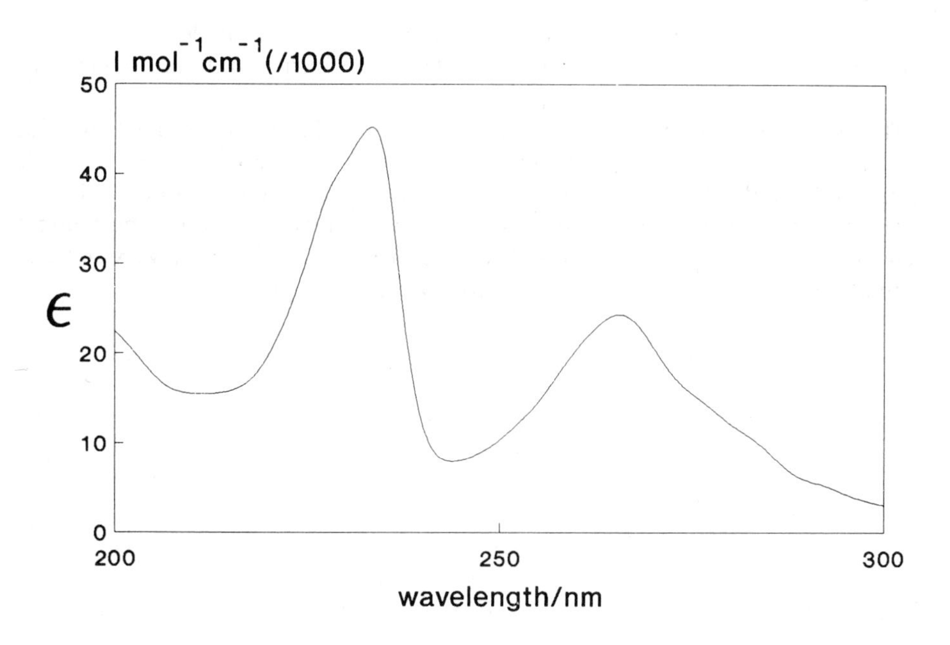

Fig. 6.10 — The electronic spectrum (ultraviolet) of 1,10-phenanthroline in hexane solution.

Within the C_{2v} point group the two nitrogen non-bonding pairs interact to give a_1 and b_2 group combination orbitals. In the free ligand molecule there are n–π^* transitions from these orbitals to anti-bonding orbitals.

There are two very intense bands, which are π–π^* type, in the spectrum of the molecule. The absorption coefficients for the bands with maximum wavelengths at 265 and 234 nm are 28 000 and 44 000 l mol^{-1}cm^{-1} respectively. Both bands have some vibrational structure. All the π–π^* transitions of the molecule are symmetry allowed.

The lowest energy transition in the free molecule (not included in the spectrum of Fig. 6.10) is from the b_2 combination of the nitrogen non-bonding pairs of electrons to the LUMO (the lowest energy π-type orbital of the delocalized system). The absorption coefficient for the transition is 560 l mol^{-1} cm^{-1}, which probably indicates that the transition is forbidden. This would be the case for a $b_2 \rightarrow b_1$ transition. There is doubt about the n–π^* transition originating in the a_1 N–N group orbital, but it is possible that it contributes to the asymmetry of the long wavelength side of the 263 nm band.

6.4 CONCLUDING SUMMARY

This chapter contains general treatments of vibrational and electronic spectroscopy.

The number of normal vibrational modes for linear and non-linear molecules are derived.

The infrared and Raman spectra of the water and carbon dioxide molecules are treated in detail. The appropriate selection rules are derived by suitable applications of group theory.

The electronic spectra of atoms are treated in terms of differences in the energies of electronic states, and the selection rules are derived. The relevance of atomic spectra to the understanding of the electronic spectra of complexes is pointed out.

The electronic spectra of molecules is dealt with in two sections, one being a discussion of six small molecules and ions, and the other being a very general treatment of polyatomic molecules using four varied examples.

7

Thermodynamics and equilibrium

Humpty Dumpty sat on a wall
Humpty Dumpty had a great fall
Not even the King's College chemists
Could put Humpty Dumpty together again

7.1 INTRODUCTION

When a chemical reaction occurs the components of the system interact so that eventually a state of **equilibrium** exists.

The passage to equilibrium by a system is characterized by two tendencies: the minimization of the energy of the system and the maximization of its entropy. This is summarized in the epigraph above. The position of equilibrium attained by a system represents a compromise between the two tendencies. The forces of attraction which exist between atoms, responsible for the production of molecules, and those between molecules which are responsible for the production of the liquid and solid states contribute such stabilization of a system recognizable as the minimization of the energy of that system. The general random nature of the gaseous state (and of the liquid state to a smaller degree) is obvious from the recognition that a gas fills any volume available to it, the molecules of the system being in constant chaotic motion.

Thermodynamics is concerned only with the initial and final states of a system, the rate and mechanism of the passage from the initial state to the final state being the province of **kinetics**.

Equilibrium is that condition of a system where there are no changes occurring in its observable properties as time proceeds. **Thermal equilibrium** is established when the system and its surroundings attain the same **temperature**. **Chemical equilibrium** is established when there is no further change of composition of the system with time. There are systems which *appear* to be in chemical equilibrium due to the reaction between the components being extremely slow.

This chapter is concerned with the understanding of the physical basis of the more important equations of thermodynamics and their relevance to chemical reactions. It is not intended to replace the more formal treatment of the subject matter, which is more properly dealt with in courses of physical chemistry.

7.2 ENERGY

There are five types of **energy** possessed by chemical systems.

(i) **Nuclear** energy resulting from the arrangements of fundamental particles in the nuclei of the atoms making up the system. Although nuclear energy is by far the largest component of the total energy of a system it has no importance with regard to chemical changes. Nuclear energy is **quantized** — there are discrete permitted energies for any given nucleus. For example, 23% of $^{238}_{92}U$ nuclei decay by α particle emission (an α particle is a $^{4}_{2}He^{2+}$ nucleus) to give a nucleus in an excited state, $^{234}_{90}Th^{*}$. The α particles have an energy of 4.147 MeV (1e V = 96.487 kJ mol^{-1}). The excited thorium nuclei decay to their lowest energy state (their **ground state**) by releasing a γ-ray photon with an energy of 0.048 MeV. The other 77% of uranium nuclei decay by emitting α particles with energies of 4.195 MeV to give the ground state of the thorium nucleus directly.
(ii) **Electronic** energy resulting from the configurations of electrons in the atoms and molecules of the system. Major changes of electronic energy occur when chemical changes take place. Electronic energy is quantized, the difference between levels being of the order of 200 kJ mol^{-1}.
(iii) **Vibrational** energy possessed by all systems except those consisting of gaseous or liquid phase **atoms**. All molecular systems possess vibrational energy even at the absolute zero of temperature (zero-point vibrational energy), the distribution of energy within the quantized levels being dependent upon the temperature of the system. Differences between vibrational levels are of the order of 25 kJ mol^{-1}.
(iv) **Rotational** energy possessed by all molecular systems and which is dependent upon the temperature of the system. It is quantized, the difference between levels being of the order of only 0.25 kJ mol^{-1}.
(v) **Translational** energy possessed by the components of gaseous and liquid phases and which is temperature dependent. Translational energy is the kinetic energy possessed by atoms and molecules as they move around in their particular system. Their energy is given by $\frac{1}{2}mv^2$ (m being their mass and v their velocity). Molecular velocities, and consequently translational energies, are not quantized, there being a continuum of possible values.

For thermodynamic purposes it is essential to consider the total energy of a system depending upon whether that system is at **constant volume** or at **constant pressure**. At constant volume the total energy of a system is the total of the five types of energy outlined above. Reactions carried out at constant volume do not carry out any work of expansion upon the surroundings, nor do the surroundings carry out any work on the system. The only quantity which may be transferred between the system and its surroundings is heat. If an **exothermic** reaction takes place then heat is released to the surroundings because they are at a lower temperature than the system. If the reaction occurring in the system is **endothermic** then heat may pass from the surroundings to the system. The temperature of the system decreases as the reaction progresses and heat is transferred from the surroundings which are at a higher temperature.

The total energy of a constant volume system is called its **internal energy** and is symbolized by U. The value of U is of little importance and would be impossible to determine. What is of importance are *changes* in the value of U in the course of any reaction. The change in the value of U in any process is symbolized by ΔU. If U_{final} and U_{initial} are the final and initial values of the internal energy of a system undergoing reaction the change in internal energy, ΔU, is given by the equation:

$$\Delta U = U_{\text{final}} - U_{\text{initial}} \tag{7.1}$$

so that ΔU would be negative for an exothermic process and positive for an endothermic one.

If an amount of energy (heat), Q_V, enters the system then:

$$\Delta U = Q_V \tag{7.2}$$

or if heat is transferred from the system to the surroundings then:

$$\Delta U = -Q_V \tag{7.3}$$

For a reaction carried out at constant pressure (which is very common, the pressure being atmospheric) the volume is variable and if an amount of heat, Q_P, enters the system there will be a change in the internal energy, ΔU, and, in addition, the volume will change by an amount, ΔV, as the system expands against the external pressure, P.

The work of expansion done by the system upon its surroundings is given by:

$$W = P\Delta V \tag{7.4}$$

so that

$$Q_P = \Delta U + P\Delta V \tag{7.5}$$

Using f and i as subscripts indicating final and initial states of the system, equation (7.5) may be written as:

$$Q_P = U_f - U_i + P(V_f - V_i) \tag{7.6}$$

which may be rearranged to give:

$$Q_P = (U_f + PV_f) - (U_i + PV_i) \tag{7.7}$$

As may be seen from the last equation, the quantity $U + PV$ is, like U and V, a **state function**, dependent upon the state of the system and not on its history. As such, $U + PV$ is given the name, **enthalpy**, and the symbol H, so that:

$$H = U + PV \tag{7.8}$$

As with the internal energy, U, it is not possible to deal with absolute values of H— only in changes in H, i.e. ΔH.

The differential form of the last equation is:

$$\Delta H = \Delta U + P\Delta V + V\Delta P \tag{7.9}$$

and since the pressure is kept constant, $\Delta P = 0$:

$$\Delta H = \Delta U + P\Delta V \tag{7.10}$$

For an amount of heat, Q_P, entering the system from the surroundings:

$$\Delta H = Q_P \tag{7.11}$$

or if Q_P is transferred from the system to its surroundings then:

$$\Delta H = -Q_P \tag{7.12}$$

7.3 ENTROPY

The **entropy** of a system is a measure of its disorder, randomness or the extent of its molecular chaos. If a spontaneous change occurs to a system the entropy of the system **and** its surroundings increases although that of the system may decrease. The best way of approaching the understanding of entropy is the concept put forward by Boltzmann and which is embodied in the Boltzmann equation:

$$S = k \ln W \tag{7.13}$$

where S = entropy, k = Boltzmann's constant and W represents the number of ways of realizing the system.

W is related to the probability of the system existing as it does, so that the more probable the system is, the higher is the related entropy. The condition of equilibrium is associated with the idea that it is achieved when the system is in its most probable state, i.e. when W is a maximum.

7.4 THE LAWS OF THERMODYNAMICS

The first and second laws of thermodynamics are essentially laws of experience and observation. There are really no formal proofs of a fundamental nature although some statistical ideas associated with the second law are worth while pursuing to a certain extent. The third law follows from the Boltzmann equation (7.13) and applies to perfectly ordered solids at a temperature of 0 K. In such a case the value of W is unity (there being only one arrangement of the constituents of a perfectly ordered solid) which results in the entropy having a zero value. The third law is that, at 0 K, the entropy of a perfectly ordered solid is zero. It forms the basis of calculated values of the entropies of substances.

The first law of thermodynamics

This is alternatively called the **law of conservation of energy** — energy may not be created nor destroyed although it may be converted from one form into another (one form of energy being mass; the two being related by the Einstein equation, $E = mc^2$, c being the velocity of electromagnetic radiation in a vacuum).

Equations (7.2), (7.5) and (7.11) are some examples of the use (and some would say statement of) of the first law. Yet another form of the law is embodied in **Hess's Law of heat summation** — if there are two chemical equations for which the respective changes in enthalpy are ΔH_1 and ΔH_2 then the combined equation will have a change in enthalpy which is given by $\Delta H_1 + \Delta H_2$. In general the first law implies that if a system in state A goes by path I to a state B, and the associated change in enthalpy is ΔH_{I}, any other path taken for the same change of state of the system, say path II, with its associated change of enthalpy, ΔH_{II}, must enforce the relationship, $\Delta H_{\text{I}} = \Delta H_{\text{II}}$.

The second law of thermodynamics

The simplest statement of the second law is that: In a spontaneous process the change in entropy is equal to or greater than zero, $\Delta S \geqslant 0$. The entropy change relates to that of the **system plus its surroundings** and can be equal to zero only if the change is carried out reversibly, otherwise (and this is relevant to reality) the entropy change **must** be positive.

When applied to the establishment of thermal equilibrium the law takes the form of saying that, in a spontaneous process, heat flows from the hotter body to the colder one until equilibrium is attained. An alternative statement of the law is that heat never flows from a colder body to a hotter one spontaneously.

A combination of the first and second laws is invaluable as a predictor of reaction direction and as an estimator of the magnitude of equilibrium constants. The use of thermodynamical principles to establish a criterion of reaction feasibility is of crucial importance in the understanding of why chemical reactions proceed in their observed directions.

A thorough understanding of the second law is essential for any of its applications to chemical and physical changes. With regard to the second law, reference has been made, and will be made in later sections, to the reversibility of a change in the system. A reversible change is one that is carried out infinitely slowly such that a system in equilibrium is altered (say, by adding an infinitesimally small amount of energy, $\mathrm{d}Q$, to the system) to another state of equilibrium. This concept of reversibility must not be confused with that of the reversibility of chemical reactions, where by adding extra product to an equilibrium mixture the reaction goes 'backwards' as written so as to restore equilibrium. Such a process would, in practice, be carried out in a non-reversible manner.

7.5 STATISTICAL APPROACH TO ENTROPY AND THE SECOND LAW — THE MAXWELL–BOLTZMANN DISTRIBUTION OF ENERGY

In this section the statistical nature of the distribution of energy in a molecular **system** is assumed. Starting with the Boltzmann equation (7.13) the criterion for predicting the direction of any change may be derived.

The derivation begins with the distribution of N molecules amongst the available energies. It is assumed that the energies are ε_1, ε_2, ε_3, ... and that the number of molecules with these particular energies are n_1, n_2, n_3, ... respectively.

Taking the Boltzmann equation as the basis for the entropy calculation it is necessary to calculate the value of W, the number of ways of achieving the postulated distribution.

The value of W is given by the formula:

$$W = \frac{N!}{n_1!n_2!n_3! \ldots} \tag{7.14}$$

The maximum probability of the system is obtained by making small changes in the n_i values (δn_i) such that:

$$\sum \delta n_i = 0 \tag{7.15}$$

and coupled with the condition that the change in W should be zero (for the maximum value of W) leads eventually to the equation:

$$n_i = \frac{N\mathrm{e}^{-\varepsilon_i/kT}}{\sum \mathrm{e}^{-\varepsilon_i/kT}} \tag{7.16}$$

The distribution of the molecules amongst the available energies as represented by equation (7.16) is known as the Maxwell–Boltzmann distribution and is used to further the understanding of entropy in this chapter and of the subject of reaction kinetics in Chapter 10.

The entropy of a system may be calculated using the Boltzmann equation and the Maxwell–Boltzmann distribution of energy. For example, if an amount of heat, $\mathrm{d}Q$, enters a system which is at equilibrium, small changes, $\mathrm{d}n_i$, occur to the occupancy numbers of the energy states, ε_i, such that:

$$\mathrm{d}U = \mathrm{d}Q = \sum \varepsilon_i \mathrm{d}n_i \tag{7.17}$$

The entropy of the system is given by:

$$S = k \ln W = k \ln \left(\frac{N!}{\prod n_i!} \right) \tag{7.18}$$

The change in entropy caused by the addition of the energy, dQ, is given by:

$$\mathrm{d}S = k \,\mathrm{d} \ln \left(\frac{N!}{\prod n_i!} \right) \tag{7.19}$$

The application of Stirling's approximation ($\ln n! = n \ln n - n$, for large values of n) allows equation (7.19) to be written as:

$$\mathrm{d}S = k \left[\mathrm{d}N \ln N - \sum n_i \mathrm{d} \ln n_i - \sum \ln n_i \mathrm{d}n_i \right] \tag{7.20}$$

The value of the first term in equation (7.20) is zero since N is a constant. The second term also has zero value since:

$$\mathrm{d} \ln n_i = \mathrm{d}n_i / n_i \tag{7.21}$$

so that the second term becomes $\Sigma \, \mathrm{d}n_i$, which is zero.

Equation (7.20) simplifies to:

$$\mathrm{d}S = -k \sum \ln n_i \mathrm{d}n_i \tag{7.22}$$

In order to simplify equation (7.22) it is necessary to use the Maxwell–Boltzmann distribution equation (7.16) where the denominator is represented by F:

$$n_i = \frac{N\mathrm{e}^{-\varepsilon_i/kT}}{F} \tag{7.23}$$

so that

$$\ln n_i = \ln N - \varepsilon_i/kT - \ln F \tag{7.24}$$

Using equation (7.24) to substitute the value for $\ln n_i$ into equation (7.22) produces:

$$\mathrm{d}S = -k \left[\ln N . \sum \mathrm{d}n_i - (1/kT) \sum \varepsilon_i \mathrm{d}n_i - \ln F . \sum \mathrm{d}n_i \right] \tag{7.25}$$

in which the first and third terms are equal to zero, so that:

$$dS = (1/T)\sum \varepsilon_i dn_i \tag{7.26}$$

and from equation (7.17) this becomes:

$$dS = dQ/T \tag{7.27}$$

Equation (7.27) applies to reversible changes and represents part of the second law of thermodynamics.

7.6 ENTROPY AND VOLUME

This section provides another (simpler) route to equation (7.27). Suppose the volume of a gaseous system *increases* from V_1 to V_2, so that the number of ways of achieving the system varies from W_1 to W_2. The probability of any one molecule occupying the initial volume is given by the ratio, V_1/V_2. The probability of all N molecules occupying the initial volume is given by $(V_1/V_2)^N$. This must represent the ratio of the number of ways of achieving the system in its two volumes:

$$W_1/W_2 = (V_1/V_2)^N \tag{7.28}$$

The entropy change, ΔS (making use of equation (7.13)), is given by:

$$\Delta S = k \ln W_1 - k \ln W_2 = kN \ln V_1 - kN \ln V_2 \tag{7.29}$$

Converting this equation to proper differential form produces:

$$dS = d(k \ln W) = kN d \ln V \tag{7.30}$$

or:

$$dS = R dV/V \tag{7.31}$$

If an expansion of the system by an amount, dV, takes place such that its pressure is held in check by an equal and opposing external pressure, P, the work done by the system upon its surroundings is given by PdV, and if the change takes place isothermally then an amount of heat, dQ, must enter the system from the outside in order that the gas should be able to expand. Since

$$dQ = PdV \tag{7.32}$$

and for one mole

$$P = RT/V \tag{7.33}$$

equation (7.32) may be written as:

$$dQ = RTdV/V \tag{7.34}$$

A comparison of equations (7.31) and (7.34) produces the conclusion that the relationship between entropy change and heat entering the system is:

$$dS = dQ/T \tag{7.35}$$

If the change in volume is carried out non-reversibly by allowing expansion into a vacuum, instead of expansion against an opposing atmosphere, the change in entropy remains as dS but dQ is **zero** since no work is done by the gas and there is no need for any heat to be absorbed from the surroundings. In such a case equation (7.35) must be modified to read:

$$dS > dQ/T \tag{7.36}$$

Equation (7.36) is the other part of the formulation of the second law — the part relating to changes carried out in a non-reversible manner. **All** changes to systems carried out under experimental conditions are inevitably of a non-reversible nature where less than the maximum amount of work is obtained from the system.

7.7 THE SECOND LAW AND EQUILIBRIUM

In section 7.4 the second law of thermodynamics was stated in the form that $\Delta S \geqslant 0$, where the change in entropy related to that of the system plus its surroundings. In sections 7.5 and 7.6 the relationship $dS \geqslant dQ/T$ has been derived, the entropy change being that of the system. In terms of differences in measurable quantities (instead of pure differentials) the relationship becomes $\Delta S \geqslant Q/T$.

Consider a spontaneous process taking place in a system which leads to a change in enthalpy, ΔH. The corresponding change in the enthalpy of the surroundings is $-\Delta H$. The change in entropy of the surroundings is given by $-\Delta H/T$. The application of the second law may be written as:

$$\Delta S_{\text{system}} + \Delta S_{\text{surroundings}} \geqslant 0 \tag{7.37}$$

and substituting the value for $\Delta S_{\text{surroundings}}$ gives:

$$\Delta S_{\text{system}} - \Delta H/T \geqslant 0 \tag{7.38}$$

Dropping the subscript and multiplying throughout by T gives:

$$T\Delta S - \Delta H \geqslant 0 \tag{7.39}$$

or

$$\Delta H - T\Delta S \leqslant 0 \tag{7.40}$$

The quantity on the left side of the inequality (7.40) is a state function and is symbolized by ΔG, such that:

$$\Delta G = \Delta H - T\Delta S \tag{7.41}$$

The G stands for the Gibbs energy (after J. W. Gibbs — one of the pioneers of thermodynamics). The change in the Gibbs energy, ΔG, is a measurable quantity for many systems and is therefore very useful. The importance of ΔG is that it allows a criterion for reaction feasibility to be enunciated. Inequality (7.40) may be written as:

$$\Delta G \leqslant 0 \tag{7.42}$$

This implies the very important conclusion that the condition for a reaction to be feasible (in the direction written) is that the value of ΔG must be negative. The magnitude of ΔG for any process may be calculated from tabulated values for the components of the reaction.

The position of equilibrium is reached when the value for ΔG becomes zero — its minimum possible value without contravening the second law.

Equation (7.31) may be integrated to give:

$$S = R \ln V + \text{constant} \tag{7.43}$$

Applied to one component of a system it is possible to introduce the concentration, c, of that component, by considering that, for one mole of the component in the volume, V, the concentration is given by $c = 1/V$ so that equation (7.43) may be written as:

$$S = \text{constant} - R \ln c \tag{7.44}$$

The integrated form of equation (7.41) is:

$$G = H - TS \tag{7.45}$$

and using equation (7.44) to substitute for the value of S gives:

$$G = H - T \times \text{constant} + RT \ln c \tag{7.46}$$

and if $G = G^{\ominus}$ when $c = 1$ then:

$$G = G^{\ominus} + RT \ln c \tag{7.47}$$

Solutes, such as real gases, behave in a non-ideal manner (except at infinite dilution) and the term c has to be replaced by a (the **activity** of the solute), where $a = \gamma c$, γ being the **activity coefficient** representing the deviation from ideality. An ideal solute has an activity coefficient of unity. For real (non-ideal) systems, equation (7.47) must be rewritten as:

$$G = G^{\ominus} + RT \ln a \tag{7.48}$$

the value of $G^{\ominus}$ being equal to G when $a = 1$ (unit activity).

The use of the symbol, $\ominus$, as a superscript indicates that the quantity (G, H, S, ΔG, ΔH or ΔS) is that relevant to a substance in its **standard state**. The standard state of a substance is defined as its physical state at 298 K and 1 atmosphere pressure.

It is conventional to rewrite equation (7.48) in the form:

$$\mu = \mu^{\ominus} + RT\ln a \tag{7.49}$$

when referring to one component of the system and to call μ the **chemical potential** of that component.

Consider the general reaction:

$$a\text{A} + b\text{B} \rightarrow c\text{C} + d\text{D} \tag{7.50}$$

where the lower case letters refer to the moles of the reactants and products. Such a reaction may, or may not, be at equilibrium. The difference in the overall chemical potential is given by:

$$\begin{aligned}\Delta\mu &= c(\mu_{\text{C}}^{\ominus} + RT\ln a_{\text{C}}) + d(\mu_{\text{D}}^{\ominus} + RT\ln a_{\text{D}}) - a(\mu_{\text{A}}^{\ominus} + RT\ln a_{\text{A}}) \\ &\quad - b(\mu_{\text{B}}^{\ominus} + RT\ln a_{\text{B}}) \\ &= c\mu_{\text{C}}^{\ominus} + d\mu_{\text{D}}^{\ominus} - a\mu_{\text{A}}^{\ominus} - b\mu_{\text{B}}^{\ominus} + RT\ln\left(\frac{a_{\text{C}}^{c}\cdot a_{\text{D}}^{d}}{a_{\text{A}}^{a}\cdot a_{\text{B}}^{b}}\right)\end{aligned} \tag{7.51}$$

The overall change in chemical potential for a reaction is then termed ΔG — the change in Gibbs energy—and the first four terms of equation (7.51) are, collectively, the change in standard Gibbs energy, $\Delta G^{\ominus}$, which results in the simplification of the equation to:

$$\Delta G = \Delta G^{\ominus} + RT\ln\left(\frac{a_{\text{C}}^{c}\cdot a_{\text{D}}^{d}}{a_{\text{A}}^{a}\cdot a_{\text{B}}^{b}}\right) \tag{7.52}$$

Equilibrium is achieved when the activities of the components of the system are such that ΔG is zero. The term in parentheses in equation (7.52) becomes an expression of the equilibrium constant, K_{eq}, for the reaction and the equation may be written in the form:

$$\Delta G^{\ominus} = -RT\ln K_{\text{eq}} \tag{7.53}$$

Equation (7.52) may be modified to read:

$$\Delta G = \Delta G^{\ominus} + RT\ln q \tag{7.54}$$

where q is the quotient in parentheses. The $\Delta G^\ominus$ term may be replaced by that expressed by equation (7.53) to give:

$$\Delta G = -RT\ln K_{eq} + RT\ln q \quad (7.55)$$
$$= RT\ln(q/K_{eq}) \quad (7.56)$$

There are two general observations which may be made from a study of equation (7.56).

(i) If $q < K_{eq}$ then ΔG is negative and the reaction should proceed in the direction as written. For q to be small the reaction mixture should have a preponderance of reactant material. This represents the general rule for the prediction of reaction feasibility. Thermodynamic feasibility of a reaction does not necessarily indicate that the reaction will proceed to equilibrium. There may be kinetic factors which prevent reaction occurring at an observable rate.

(ii) If $q > K_{eq}$ then ΔG is positive and the reaction as written will not proceed, the reverse reaction being thermodynamically feasible.

Equation (7.53) is an extremely important and useful equation, for it allows the calculation of the equilibrium constant for any reaction provided that certain data are available for the components of that reaction. The required data for very many compounds and ions are tabulated in data books, and will only be exemplified in this book. Once a basic understanding of the principles of calculation is grasped, the only book required is one full of data. The important tabulations are of values of:

(a) the standard Gibbs energy of formation, $\Delta G_f^\ominus$,
(b) the standard enthalpy of formation, $\Delta H_f^\ominus$, and
(c) the standard entropy, $S^\ominus$,

for compounds (in whatever state they happen to be at 298 K and 1 atmosphere pressure — their standard state) and for ions in aqueous solution at unit activities.

Since it is impossible to determine the absolute enthalpy and the absolute Gibbs energy of any substance (element, compound or ion) the convention is adopted that the values of $G^\ominus$ and $H^\ominus$ for **elements**, in their standard states, are **zero.**

It is possible to determine absolute values for the entropies of elements and compounds (but not ions — a subject which will be dealt with in Chapter 9) from the application of the third law of thermodynamics, and such values are tabulated in data books.

The reason for using Gibbs energies and enthalpies of formation is obvious from an example. The equation for the formation of SO_2 from its elements in their standard states is:

$$S(s) + O_2(g) \rightarrow SO_2(g) \quad (7.57)$$

the physical states corresponding to those in the standard state for each component being indicated by lower case letters in parentheses after the chemical formulae (gas, liquid or solid). The overall change in the value of the standard Gibbs energy for the reaction is, by definition, equal to the standard Gibbs energy of formation of SO_2 and is given by:

$$\Delta G_f^\ominus(SO_2, g) = G^\ominus(SO_2, g) - G^\ominus(S, s) - G^\ominus(O_2, g) \quad (7.58)$$

The application of the convention allows the last two terms on the right-hand side of equation (7.58) to have zero values so that:

$$G^\ominus(SO_2, g) = \Delta G_f^\ominus(SO_2, g) \quad (7.59)$$

and in general the standard Gibbs energy of formation of a compound is a measure of its absolute standard Gibbs energy in relation to the appropriate sum of those of its constituent elements. By a similar reasoning the absolute standard enthalpy of a substance is represented by its standard enthalpy of formation. Bearing in mind the conventions, the general equations for reactions apply:

$$\Delta G^\ominus(\text{reaction}) = \sum \Delta G_f^\ominus(\text{products}) - \sum \Delta G_f^\ominus(\text{reactants}) \quad (7.60)$$

$$\Delta H^\ominus(\text{reaction}) = \sum \Delta H_f^\ominus(\text{products}) - \sum \Delta H_f^\ominus(\text{reactants}) \quad (7.61)$$

Since the absolute standard entropies are tabulated for elements and compounds, calculations of the change in standard entropy for any reaction depend upon the equation:

$$\Delta S^\ominus(\text{reaction}) = \sum S^\ominus(\text{products}) - \sum S^\ominus(\text{reactants}) \quad (7.62)$$

in which the standard entropies of elements may not be ignored since they are non-zero quantities.

There are two methods of calculating the value of $\Delta G^\ominus$(reaction). The straightforward method is to use equation (7.60) but, if $\Delta G_f^\ominus$ data are not available, the results of equations (7.61) and (7.62) may be substituted into equation (7.60).

For the reaction (7.57) the accepted value for $\Delta G_f^\ominus(SO_2, g)$ is -300.2 kJ mol^{-1}, and since the reactants are elements in their standard states, the value for $\Delta G^\ominus$(reaction) is also -300.2 kJ mol^{-1}. The equilibrium constant for the reaction is calculated by using equation (7.53) in the form:

$$\begin{aligned} K_{eq} &= e^{-\Delta G^\ominus/RT} \\ &= \text{antiln}\,(-\Delta G^\ominus/RT) \\ &= \text{antiln}\,(300\,200/(8.31441 \times 298)) \\ &= 4.16 \times 10^{52}. \end{aligned} \quad (7.63)$$

Equation (7.63) is of general use in obtaining an idea of the magnitude of K_{eq} from the sign and magnitude of $\Delta G^{\ominus}$.

If $\Delta G^{\ominus}$ is negative then $K_{eq} \neq 1$, and the more negative the value of $\Delta G^{\ominus}$ the greater is the value of K_{eq}.

If $\Delta G^{\ominus}$ is positive then $K_{eq} < 1$, and the larger the value of $\Delta G^{\ominus}$ the smaller is the value of K_{eq}.

If the value of $\Delta G^{\ominus}$ is zero then $K_{eq} = 1$.

There are two points to note regarding the above calculation. One is that the $\Delta G^{\ominus}$ value is converted into J mol^{-1} so that when it is divided by the product of the gas constant, R (J mol^{-1} K^{-1}), and the absolute temperature, T(K), the result is a dimensionless number. The natural antilogarithm of that result gives the value of the equilibrium constant in the form of a **number**. It is important to realize this. It is not possible to take the logarithm of anything but a pure dimensionless number. The second and related point is that the equilibrium constant does not have units.

One expression of the equilibrium constant for reaction (7.57) is in terms of the partial pressures of the components of the equilibrium mixture:

$$K_{eq} = \frac{p_{SO_2}}{p_{O_2}} \tag{7.64}$$

The quotient in equation (7.64) is a dimensionless number, but there are many cases where the quotients appropriate to the reactions would have resultant units. To understand the various points involved in making sure that the quotient in an equilibrium constant expression is truly dimensionless, consider the following abstract equation:

$$\mathrm{A(s) + B(aq) + C(l) \rightarrow D(aq) + E(aq) + F(g)} \tag{7.65}$$

The equilibrium constant for this reaction would be expressed as:

$$K_{eq} = \frac{a_D \cdot a_E \cdot a_F}{a_A \cdot a_B \cdot a_C} \tag{7.66}$$

the *a*s representing (for the present) the equilibrium activities of the respective components of the equilibrium mixture. To ensure that K_{eq} is dimensionless it is necessary to express it in terms of **relative activities** rather than the **absolute activities** of equation (7.66). The relative activity of a component of the system is expressed as the ratio of the absolute activity to the activity of that component in its standard state:

$$a_{\text{relative}} = a_{\text{absolute}}/a_{\text{standard state}} \tag{7.67}$$

There are four cases to be considered when deciding upon the relative activity of a component of a system depending upon whether the component is a gas, a solid, a liquid or a solvent, or a solute.

1. Gases

The activity of a gas is usually represented by its partial pressure — defined as its actual pressure expressed as a fraction of the total pressure of the system:

$$a_{\text{relative}} = p_{\text{actual}}/P_{\text{total}} \tag{7.68}$$

This is based upon Dalton's Law of partial pressures such that the total pressure of the system is made up from the partial pressures of its components:

$$P_{\text{total}} = p_{\text{A}} + p_{\text{B}} + \ldots \tag{7.69}$$

The standard state of a gaseous component of a system is taken to be 298 K and 1 atmosphere pressure so that equation (7.69) becomes:

$$a_{\text{relative}} = p_{\text{actual}}/1 = p_{\text{actual}} \tag{7.70}$$

From this it is possible to conclude that the equilibrium constant for any gas phase reaction is dimensionless.

2. Solids

The relative activity of a solid component of a system should be taken as **unity**. Consider the equilibrium concerning the thermal decomposition of calcium carbonate:

$$CaCO_3(s) \rightarrow CaO(s) + CO_2(g) \tag{7.71}$$

The equilibrium constant is given by:

$$\begin{aligned} K_{\text{eq}} &= p_{CO_2}/p^{\ominus}_{CO_2} \\ &= p_{CO_2} \end{aligned} \tag{7.72}$$

providing that the relative activities of the solid components are taken as unity. This is merely consistent with the observation that addition of more $CaCO_3$ and/or CaO to the system does not change the equilibrium pressure of CO_2.

3. Liquids and solvents

Similar considerations apply to liquid components as to solids. Their relative activities are to be taken as unity. An exception to this rule applies when the liquid in question is a reactant or a product in the system, and is not present as the solvent (when its activity should be taken as unity). In such a case its relative activity is best represented by its **mole fraction**, x_{A}, where:

$$x_{\text{A}} = \frac{\text{no. of mols of A present}}{\text{total no. of moles in system}} \tag{7.73}$$

When the liquid concerned is a reactant or a product **and** is also the **solvent** for the system, its mole fraction is likely to be almost unity and for practical purposes should be taken as such.

4. Solutes

The concentration of a solute is expressed in terms either of its **molarity** (the number of moles contained by one litre of solution) or its **molality** (the number of moles contained by one kilogram of solvent). There are significant differences between the two expressions when:

(i) the ratio of solute/solvent is high and
(ii) the solvent has a density other than unity, as would be the case for non-aqueous solvents.

If the solvent is water, since one litre at 298 K has a mass of 1 kg, the molarity and the molality of a dilute solution of a solute, under standard conditions, are identical for all practical purposes. Only if highly concentrated solutions are used does the strict use of molality become important. In such cases it is helpful to express solubility as a mole fraction. For most normal chemical applications the relative activity of a solute is given by the ratio of its actual activity (molarity times activity coefficient, $m\gamma$) and its activity under standard conditions (unit activity at 298 K and 1 atmosphere pressure = 1 M) which is equivalent to the molarity expressed as a dimensionless number.

Returning to equation (7.66) and applying the rules concerning relative activities the equilibrium constant may now be expressed as:

$$K_{eq} = \frac{a_D \cdot a_E \cdot p_F}{a_B} \tag{7.74}$$

which is a pure number. The *a*s in the equation now represent relative activities.

7.8 APPLICATIONS OF THERMODYNAMIC PRINCIPLES TO REACTION FEASIBILITY

Two major examples of the proper application of thermodynamic principles of reaction feasibility are dealt with in this section together with some general observations.

The two areas taken to exemplify the material are:

(i) the extraction of metals from their oxides by reduction with carbon or carbon monoxide, and
(ii) the role of adenosine triphosphate (ATP) in biochemical reactions.

The general point which needs to be made is concerned with the additivity of ΔG (and $\Delta G^\ominus$) values and the absence of any connection whatsoever of such additivity with the **mechanism** of the overall process.

Consider the abstract reaction:

$$\mathrm{A + B + C \rightarrow D + E} \tag{I}$$

Although the value of $\Delta G^{\ominus}$ for this reaction may not be known it may be possible to calculate it from a knowledge of the $\Delta G^{\ominus}$ values for the reactions:

$$\mathrm{A + B \rightarrow X} \tag{II}$$
$$\mathrm{X + C \rightarrow D + E} \tag{III}$$

The first law indicates that:

$$\Delta G^{\ominus}(\mathrm{I}) = \Delta G^{\ominus}(\mathrm{II}) + \Delta G^{\ominus}(\mathrm{III}) \tag{7.75}$$

since the addition of the equations (II) and (III) produces the overall process (I), the intermediate product, X, cancelling out. Reactions (II) and (III) may very well be known to occur separately — A mixed with B gives X, and if mixed in a separate vessel, X and C give the products, D and E. Such an observation has no bearing upon the mechanism of the process.

There is the possibility that the reactions:

$$\mathrm{B + C \rightarrow Y} \tag{IV}$$
$$\mathrm{Y + A \rightarrow D + E} \tag{V}$$

could equally well be the mechanism of the reaction, or maybe the process occurs by:

$$\mathrm{A + C \rightarrow Z} \tag{VI}$$
$$\mathrm{Z + B \rightarrow D + E} \tag{VII}$$

All that thermodynamics indicates is that, for the overall process:

$$\begin{aligned} \Delta G^{\ominus}(\mathrm{I}) &= \Delta G^{\ominus}(\mathrm{IV}) + \Delta G^{\ominus}(\mathrm{V}) \\ &= \Delta G^{\ominus}(\mathrm{VI}) + \Delta G^{\ominus}(\mathrm{VII}) \end{aligned} \tag{7.76}$$

There is no possibility of discriminating between the three mechanisms from a knowledge of thermodynamics — **kinetic** studies have to be made to attempt to elucidate the mechanisms of processes. To make up an overall process by adding chemical reactions together and then conclude that something has been decided about the mechanism of the overall reaction is a serious error. The splitting up of an overall process into simpler reactions for which thermodynamic information exists is a very helpful and instructive process, but again it indicates nothing about how the overall reaction occurs.

Reaction I may have a value for $\Delta G^{\ominus}$ which is negative, indicating that it is thermodynamically feasible in the direction as written. It may also be possible to break down the overall reaction into two reactions for which the respective values of $\Delta G^\ominus$ are known:

$$A + B \rightarrow E \qquad \text{(VIII)}$$
$$C \rightarrow D \qquad \text{(IX)}$$

What must be true is that:

$$\Delta G^\ominus(\text{I}) = \Delta G^\ominus(\text{VIII}) + \Delta G^\ominus(\text{IX}) \qquad (7.77)$$

and it may be that, for instance, the value of $\Delta G^\ominus$(IX) is **positive** with $\Delta G^\ominus$(VIII) being sufficiently negative to be consistent with the value of $\Delta G^\ominus$(I) being negative. Such a possibility would imply that reaction (IX) would normally be feasible in the reverse direction and that it is being *driven* in the direction as written by reaction (VIII). That kind of interpretation is faulty and can lead to false conclusions being drawn concerning the mechanism of the overall process. As will be seen in the biochemical example later in this section it is perfectly in order for there to be a build-up of a component of a system which has a very high free energy of formation (i.e. very positive) but the overall reaction in which that component is produced **must** have a negative $\Delta G^\ominus$ value. In the abstract example dealt with above, there could be a production of the component, D, because $\Delta G^\ominus$ is negative. Just because reaction (I) may be written as the sum of the two processes, (VIII) and (IX), does not mean that those actual processes are occurring as separate chemical reactions — one being driven by the other. A proper rational interpretation would be that the Gibbs energy-rich component, D, is being produced by some mechanism (of which thermodynamics can indicate nothing) such that there is an overall reduction in the Gibbs energy of the system.

The extraction of metals from their oxides — Ellingham diagrams

One very common process for the reduction of metal oxides to yield the metal is to smelt the oxide with carbon. The carbon is usually used in the form of coke, and the processes take place at temperatures at which the oxidation of carbon occurs at a reasonable rate — up to 1000°C. Such temperatures are far removed from the thermodynamic standard state value of 298 K, and it is necessary to be able to calculate ΔG as a function of temperature to consider reaction feasibilities. Equation (7.41) is repeated here for convenience:

$$\Delta G = \Delta H - T\Delta S \qquad (7.41)$$

from which it is observed that ΔG is temperature dependent. The values of ΔH and ΔS are also slightly temperature dependent, but for general purposes such dependencies are relatively small and do not affect broad conclusions which are made by their omission. Exceptions to the previous statement arise when there are changes of state. A change from the solid state to the liquid state is accompanied by an entropy change of $\Delta S_{\text{fusion}}/T_{\text{m}}$, where T_{m} is the melting point of the component. An even larger change in entropy accompanies the change from the liquid to the gas phase: $\Delta S_{\text{evaporation}}/T_{\text{b}}$, where T_{b} is the boiling point of the component.

The oxidation of carbon can take place in two stages:

(a) $$2C(s) + O_2(g) \rightarrow 2CO(g) \quad (7.78)$$

and

(b) $$2CO(g) + O_2(g) \rightarrow 2CO_2(g) \quad (7.79)$$

with the overall process being:

(c) $$C(s) + O_2(g) \rightarrow CO_2(g) \quad (7.80)$$

In this section all reactions are written with **one** mole of dioxygen involved so that comparisons may be made of reductions of oxides of metals with different oxidation states. Similar treatment is given to the formation reactions of metal oxides so that when a metal oxide reduction is combined with any of the carbon oxidation reactions the dioxygen molecules cancel out to give the appropriate oxide-to-metal reduction equation.

For reaction (a) the data books indicate that:

$$\Delta G_{\text{f}}^{\ominus}(\text{CO,g}) = -137\ \text{kJ mol}^{-1}$$
$$\Delta H_{\text{f}}^{\ominus}(\text{CO,g}) = -111\ \text{kJ mol}^{-1}$$
$$S^{\ominus}(\text{CO,g}) = 198\ \text{J K}^{-1}\ \text{mol}^{-1}$$
$$S^{\ominus}(\text{O}_2\text{,g}) = 205\ \text{J K}^{-1}\ \text{mol}^{-1}$$
$$S^{\ominus}(\text{C,graphite}) = 5.7\ \text{J K}^{-1}\ \text{mol}^{-1}$$

and these are sufficient data to calculate the temperature dependence of $\Delta G_{\text{f}}(\text{CO,g})$ — the superscript is dropped since at temperatures other than 298 K, thermodynamic parameters are not standard. The first method is to calculate the value of $\Delta S_{\text{f}}^{\ominus}(\text{CO,g})$ as:

$$\begin{aligned}\Delta S_{\text{f}}^{\ominus}(\text{CO,g}) &= S^{\ominus}(\text{CO}) - S^{\ominus}(\text{C}) - 1/2\,S^{\ominus}(\text{O}_2) \quad (7.81)\\ &= 198 - 5.7 - 205/2\\ &= 89.8\ \text{J K}^{-1}\ \text{mol}^{-1}\end{aligned}$$

The second, more widely used, method is to make use of equation (7.41) in its standard form and write:

$$\Delta S_{\text{f}}^{\ominus} = (\Delta H_{\text{f}}^{\ominus} - \Delta G_{\text{f}}^{\ominus})/T \quad (7.82)$$

which gives

$$\begin{aligned}\Delta S_{\text{f}}^{\ominus}(\text{CO,g}) &= (-111\,000 + 137\,000)/298\\ &= 87.25\ \text{J K}^{-1}\ \text{mol}^{-1}\end{aligned}$$

The two methods produce two different answers, which does not indicate that there are faults in the theoretical equations but that there are discrepancies in published data — a very common and understandable occurrence since there are experimental errors involved with any data.

Using the results from the latter method of calculation the equation representing the variation of ΔG with temperature for equation (a) — as written — is:

$$\begin{aligned}\Delta G(\mathrm{a}) &= 2\times\Delta H_{\mathrm{f}}^{\ominus}(\mathrm{CO}) - T\times 2\times\Delta S^{\ominus}(\mathrm{CO})\\ &= -2\times 111 - T\times 2\times 87.25/1000\\ &= -222 - T\times 0.1745 \qquad (7.83)\end{aligned}$$

the divisor of 1000 being to convert the entropy units of $\mathrm{J\,K^{-1}\,mol^{-1}}$ to $\mathrm{kJ\,K^{-1}\,mol^{-1}}$ so that when multiplied by the absolute temperature, both terms in the equation have units of $\mathrm{kJ\,mol^{-1}}$. The appropriate equations for reactions (b) and (c) are:

$$\Delta G(\mathrm{b}) = -566 + T\times 0.1678 \qquad (7.84)$$
$$\Delta G(\mathrm{c}) = -394 - T\times 0.00336 \qquad (7.85)$$

Notice that the equation for ΔG(c) is the average of the other two divided by two. The three equations are plotted as $\Delta G/\mathrm{kJ\,mol^{-1}}$ against T/K in Fig. 7.1. It should be

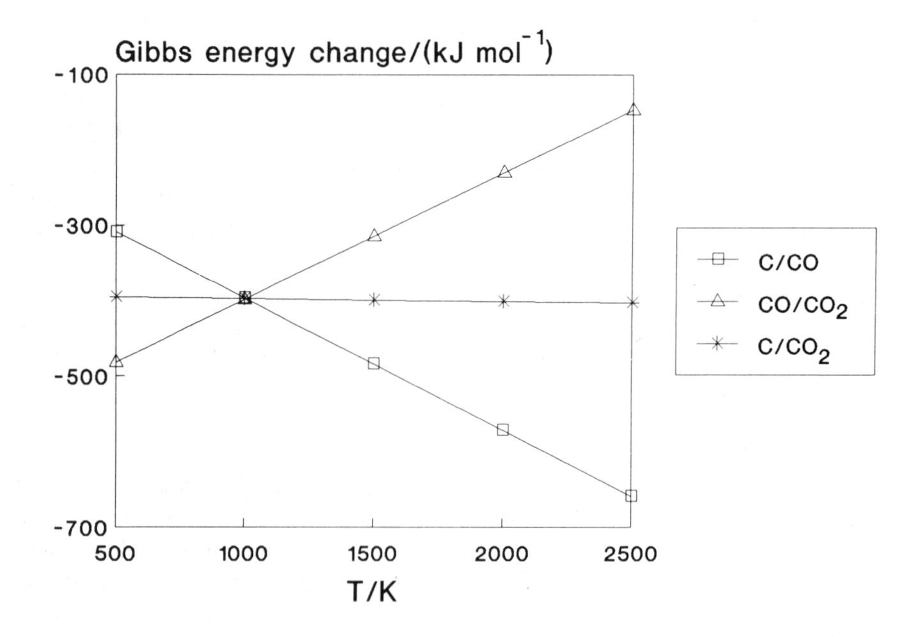

Fig. 7.1 — An Ellingham diagram for the oxidation of carbon.

noticed that all three reactions are thermodynamically feasible over the 500–2500 K temperature range.

The significance of the plots is that they form a very obvious method of comparing the lines for any of the three carbon-based reactions with those for any metal–metal oxide system in order to decide whether or not any particular metal oxide can be reduced by either carbon or carbon monoxide. If the line for the oxide to be reduced is above the carbon lines, the reaction is thermodynamically feasible and carbon or carbon monoxide may be the reducing agent. If the metal oxide line is below the carbon lines there is no possibility for its reduction to the metal by either of the carbon species. There are interesting possibilities when the metal oxide line crosses the carbon lines where the temperature and the reducing agent become important. These various possibilities are exemplified by considering the production of (i) copper, (ii) iron, (iii) zinc and (iv) aluminium. The slopes of the three lines in Fig. 7.1 are of interest. The line for the oxidation of carbon to carbon monoxide has a positive slope which derives from the **increase** in entropy in the reaction — qualitatively to be associated with there being one mole of gaseous reactant and two moles of gaseous product. The further oxidation of CO to give CO_2 (equation (7.79)) involves a change from three moles of gaseous reactants to give only two moles of gaseous products. The entropy change is much the same as in reaction (7.78) but of opposite sign. The slope of the line corresponding to reaction (7.79) in Fig. 7.1 is therefore negative. The line corresponding to reaction (7.80) in Fig. 7.1 has almost zero slope, as would be expected from a reaction which has the same number of moles of gases as reactants as there are products. The entropy change is practically zero. The lines for **all** metal-to-metal oxide reactions have, without exception, positive slopes because gaseous dioxygen is consumed in the reactions to produce solid (or sometimes liquid) oxides. Such reactions are associated with a decrease in entropy.

(i) Copper production from CuO

As may be seen from Fig. 7.2 the lines for the oxidation of copper to copper(I) oxide and for the further oxidation to copper(II) oxide are above the three carbon oxidation lines. This means that either carbon or CO may be used to reduce copper oxides to the metal. The carbon oxidation reactions have a sufficiently negative value for ΔG at almost any temperature to ensure that when added to the value for the reduction of the metal oxide (now positive since the oxidation reaction has been reversed) the overall value of ΔG is negative. It should be pointed out that this is an academic example since copper is mainly extracted from its sulphide and carbonate ores.

(ii) Iron production from Fe_2O_3

Fig. 7.3 shows the lines for the oxidation of iron to iron(II) oxide and its further oxidation to iron(III) oxide. Only the CO/CO_2 reaction (7.79) is capable of carrying out the reduction of the iron oxides at temperatures below 1000 K. The industrial production of iron is operated so that reduction of Fe_2O_3 to FeO occurs at between 500 and 700°C and the reduction of FeO to metallic iron at around 900°C. The data plotted in Fig. 7.3 indicate that the reducing agent in both areas must be carbon monoxide.

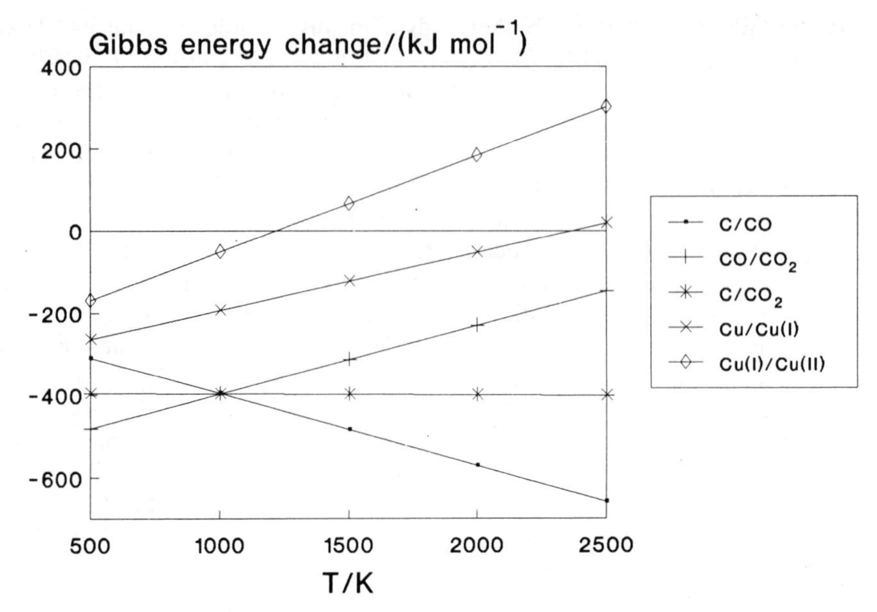

Fig. 7.2 — An Ellingham diagram for the oxidation of carbon and copper.

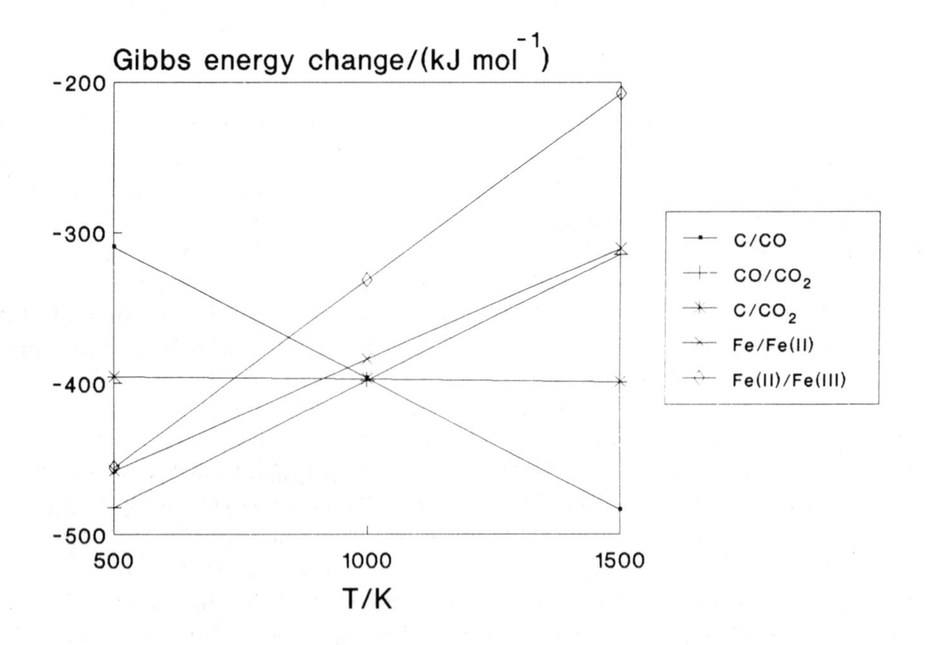

Fig. 7.3 — An Ellingham diagram for the oxidation of carbon and iron.

(iii) Aluminium production from Al_2O_3

Fig. 7.4 shows that the reduction of aluminium(III) oxide is not possible using carbon

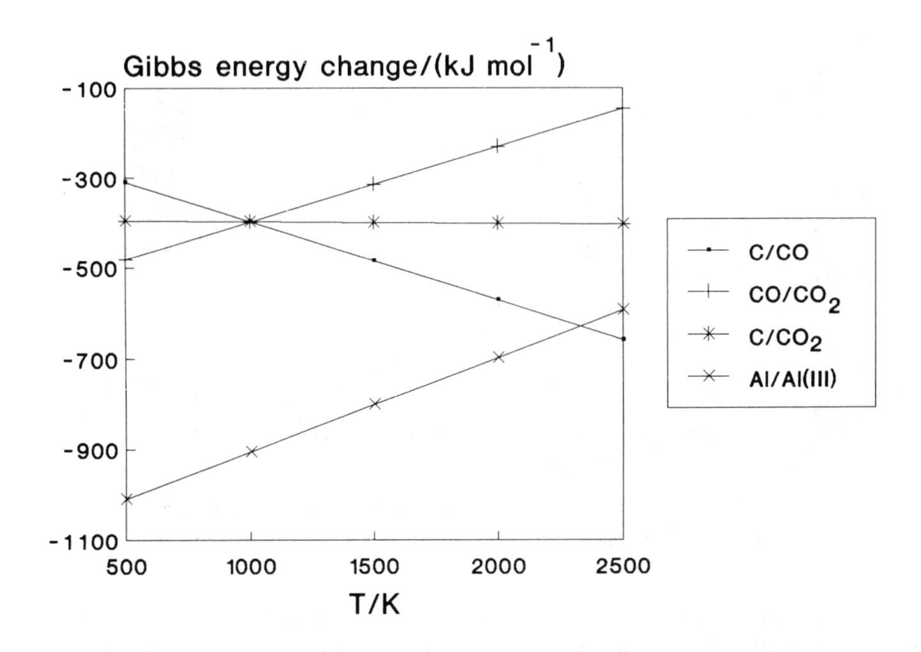

Fig. 7.4 — An Ellingham diagram for the oxidation of carbon and aluminium.

as the reducing agent unless the temperature exceeds 2200 K — a value which would make a process based upon carbon reduction uneconomic. Aluminium is produced electrolytically.

(iv) Zinc production from ZnO

Fig. 7.5 shows the data for the oxidation of zinc to ZnO. The reduction of the oxide by carbon becomes feasible at temperatures greater than 1225 K. At temperatures greater than 1400 K, all three carbon reactions may reverse the oxidation of zinc. The interesting point about zinc is that when the reactions for its production are feasible it is in the gas phase (hence the change of slope at the boiling point (1180 K)). There is an extra term to be added to the entropy change for the Zn/ZnO reaction of $\Delta H_{\text{boiling}}/T = 97.5\ \text{J K}^{-1}\ \text{mol}^{-1}$. The melting of zinc at 692 K does not produce such a large change in entropy and hardly alters the slope of the Zn/ZnO line in Fig. 7.5. That the zinc is in the gas phase when it is produced and mixed with the products of the process — CO/CO_2 — gives rise to a difficulty in separation of the zinc metal before cooling. If the products of the reaction are allowed to cool together then the reaction direction is reversed at temperatures lower than 1225 K. Industrially the separation of the zinc from the oxidation products of carbon is achieved by causing

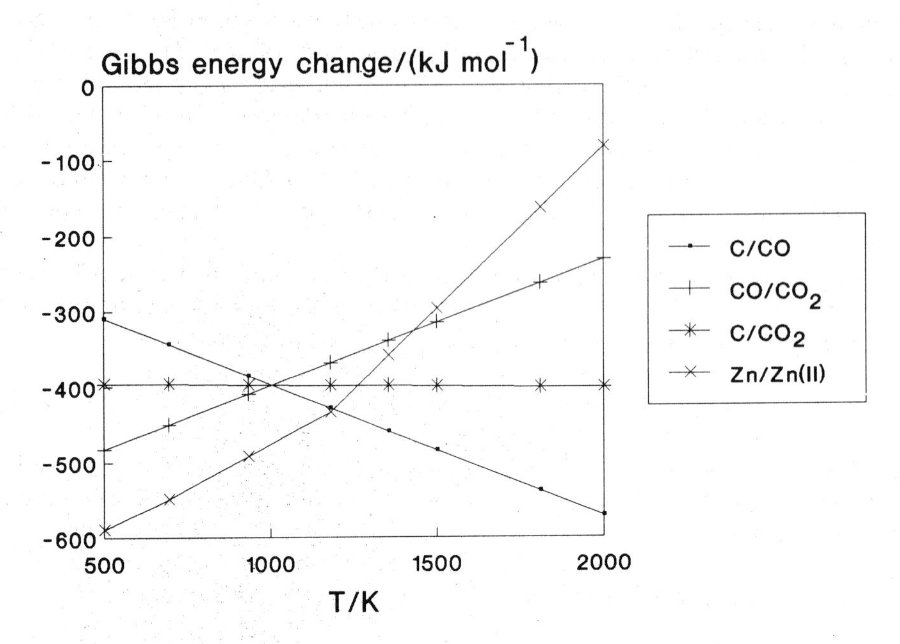

Fig. 7.5 — An Ellingham diagram for the oxidation of carbon and zinc.

the zinc to form a compound with lead in the high-temperature zone of the reactor, the lead/zinc alloy being unstable at lower temperatures when it separates into two layers.

Some biochemical processes

A great deal of misunderstanding is associated with the molecule of adenosine triphosphate (ATP) and its hydrolysis to adenosine diphosphate (ADP). The essential process is the hydrolysis of the terminal P—O—P bond of ATP to give ADP and an aqueous phosphate ion (commonly referred to as inorganic phosphate!).

At the physiological pH of 7, ATP has a charge of -4, ADP having a charge of -3 and the phosphate ion being HPO_4^{2-}, so the reaction may be simply written as:

$$ATP^{4-} + H_2O \rightarrow ADP^{3-} + HPO_4^{2-} + H^+ \qquad (7.86)$$

The essentials of the reaction are that a P—O bond has been broken, and the resulting phosphate ion has been hydrated. The hydration of the resulting components contributes to the decrease of the Gibbs energy of the system and outweighs any use of such energy in the breaking of the terminal phosphate linkage. At 37°C and pH = 7 (physiological standard state) the value of $\Delta G^{\ominus\prime}$ is -30 kJ mol^{-1}. It is very doubtful whether reaction (7.86) as formulated ever takes place in a physiological situation. There is no doubt at all that ATP and ADP feature in the **mechanisms** of many biochemical processes.

There are many reactions which involve the production of highly complicated and **ordered** molecules (e.g. peptide chains in proteins). The production of order in a system is associated with a decrease in entropy, and this tends to make the ΔG value positive. The production of one peptide linkage has been estimated to have a $\Delta G^{\ominus\prime}$ value of 17 kJ mol^{-1}. The reaction would be thermodynamically infeasible as a separate process. The involvement of the ATP-to-ADP hydrolysis with the peptide linkage reaction would produce a feasible overall process in which peptide links would be built up at the expense of converting ATP into ADP. The mechanism of such a process is extremely complicated and does not involve reaction (7.86) **as written**.

There being many building processes in biochemistry means that, if ATP is so important, that there have to be reactions in which it is actually produced so that it may take part. The reaction expressing the combustion of glucose is:

$$C_6H^{12}O_6(s) + 6O_2(g) \rightarrow 6CO_2(g) + 6H_2O(l) \tag{7.87}$$

the value of $\Delta G^{\ominus}$ being -2880 kJ mol^{-1}. Such a process takes place in living cells, but would be a waste of Gibbs energy if no use was made of it to build up essential molecules such as ATP. The production of ATP from ADP and phosphate ions requires 30 kJ mol^{-1} of Gibbs energy. It would be possible to write an equation expressing the use of the glucose oxidation to produce 38 molecules of ATP:

$$C_6H_{12}O_6 + 38ADP^{3-} + 38HPO_4^{2-} + 6O_2 \rightarrow 38ATP^{4-} + 6CO_2 + 6H_2O \tag{7.88}$$

The value of $\Delta G^{\ominus}$ for this reaction would be calculated as:

$$\begin{aligned}\Delta G^{\ominus}(7.88) &= \Delta G^{\ominus}(7.87) - 38 \times \Delta G^{\ominus}(7.86)\\ &= -2880 - 38 \times (-30)\\ &= -740 \text{ kJ mol}^{-1}\end{aligned} \tag{7.89}$$

The glucose oxidation has been harnessed to ATP production and the overall process still has a large excess of Gibbs energy for carrying out other reactions as well as contributing to the maintenance of the temperature of the system! The mechanism of reaction (7.88) is very complicated and extensive, but thermodynamics indicates nothing of this. Such matters are the province of kinetic studies.

7.9 CONCLUDING SUMMARY

This chapter contains an introduction to the following topics.

(1) The first law of thermodynamics and Hess's Law.
(2) The Boltzmann equation and the statistical approach to the second law of thermodynamics.
(3) The second law and equilibrium, Gibbs energy.
(4) The dimensionless nature of equilibrium constants.
(5) Thermodynamics and reaction feasibility criteria.
(6) Ellingham diagrams and the smelting of oxide ores.
(7) The hydrolysis of ATP and the oxidation of glucose.

8

Transition metal complexes

8.1 INTRODUCTION

Transition metal complexes represent a large area of chemistry. The material in this chapter consists of the theoretical treatment of the three main types of complex — those in which the coordination number of the metal ion (the number of ligand atoms attached to the metal) is six (octahedral, O_h) or four (square planar, D_{4h} , or tetrahedral, T_d).

In all complexes the metal-to-ligand bonding relies mainly upon pairs of electrons which are donated from the ligand to the metal. In some instances the reverse donation is important. The conventional view of the bonding describes it as **coordinate bonding** in which the electrons of the metal are of no importance. This is a great oversimplification of the apparent situation where a particular oxidation state of a metal accepts either four or six pairs of ligand electrons to allow complex formation.

The subjects which are essential to the understanding of the bonding and properties of transition metal complexes are: (i) thermodynamic stability, (ii) bonding, (iii) electronic spectra, (iv) magnetic behaviour and (v) kinetics of (a) ligand exchange and (b) oxidation-reduction reactions.

An introduction to the subject of thermodynamic stability of complexes is given in this chapter, and further treatment is given in Chapter 9. Topics (ii), (iii) and (iv) form the remainder of this chapter. Topic (v) is left until Chapter 10.

8.2 THERMODYNAMIC STABILITY OF COMPLEXES — FORMATION CONSTANTS

The thermodynamic stability of a complex is conventionally expressed as the magnitude of the equilibrium constant appropriate to the formation reaction from the metal atom or ion and the ligands from which is made. The formation reaction may be generalized as:

$$M^{n+} + xL^{l-} \rightleftharpoons ML_x^{(n-xl)+} \tag{8.1}$$

where the thermodynamic states of the components of the system have been omitted, but which are important for any particular case. The oxidation state of the metal, M, may be positive, zero or (rarely) negative. The charge on the ligand, L, is either negative or zero. Some ligands exist in acid solution as positive ions, e.g. HL^+, with the neutral ligand being protonated. The equilibrium constant appropriate to reaction (8.1) may be written as:

$$K = \frac{[ML_x^{(n-xl)+}]}{[M^{n+}][L^{l-}]^x} \quad (= \beta_x, \text{ see later}) \tag{8.2}$$

where the square brackets are an indication of the equilibrium activities of the respective species having been divided by the activity of the species in their standard states ($a^\ominus = 1$) so as to make the equilibrium constant dimensionless.

In practice so very few activity coefficients are known (or even measurable for the concentrations normally used) that actual **concentrations** are used in equations like (8.2). It should be realized that the vast number of published formation constants (the equilibrium constants for the formation) of complexes are not true thermodynamic equilibrium constants. For that reason some caution should be used in drawing too fine a conclusion about their absolute magnitudes and, in particular, in the discussion of relatively small differences between values for different complexes.

It should be noted that the term 'formation constant' is synonymous with 'stability constant'. Some treatments refer to 'instability constants', which are the reciprocals of the stability or formation constants. They refer to the reverse of the formation reactions — the dissociation reactions of complexes into their constituent metal ions and ligands.

The formation constant as described by equation (8.2) is the **overall** equilibrium constant, β_n. Overall formation constants are related to the values of step-wise formation constants, which relate to the successive additions of ligands to the central metal ion. The step-wise addition of the ammonia molecule as a ligand to copper(II) ion may be taken as an example.

The first step is written as:

$$Cu^{2+} + NH_3 \rightleftharpoons Cu(NH_3)^{2+} \tag{8.3}$$

with K_1 being given by:

$$K_1 = \frac{[Cu(NH_3)^{2+}]}{[Cu^{2+}][NH_3]} \tag{8.4}$$

The reactions concerning the formation of Cu(II)–ammine complexes occur in aqueous solution, the phase being omitted from the equations.

The second step is the addition of the second ligand to the product of the first step:

$$Cu(NH_3)^{2+} + NH_3 \rightleftharpoons Cu(NH_3)_2^{2+} \tag{8.5}$$

with the second step-wise formation constant being given by:

$$k_2 = \frac{[Cu(NH_3)_2^{2+}]}{[Cu(NH_3)^{2+}][NH_3]} \tag{8.6}$$

The complex which is normally formed when copper(II) ion, in aqueous solution, is treated with an excess of ammonia is the tetraamminecopper(II) ion, $[Cu(NH_3)_4]^{2+}$. There are four step-wise constants which determine the value of the overall formation constant. The step-wise constants (K values) may be combined to give cumulative (or overall) formation constants, denoted by β, so that, in the case of the first two steps of the formation of $[Cu(NH_3)_2]^{2+}$, $\beta_1 = K_1$, but $\beta_2 = K_1 \times K_2$. This may be seen from the operation of carrying out the multiplication of the first two step-wise constants as given by equations (8.4 and 8.6) to give:

$$\beta_2 = K_1 \times K_2 = \frac{[Cu(NH_3)^{2+}]}{[Cu^{2+}][NH_3]} \times \frac{[Cu(NH_3)_2^{2+}]}{[Cu(NH_3)^{2+}][NH_3]}$$

$$= \frac{[Cu(NH_3)_2^{2+}]}{[Cu^{2+}][NH_3]^2} \tag{8.7}$$

which is the cumulative formation constant for the first two stages of the reaction:

$$Cu^{2+} + 2NH_3 \rightleftharpoons Cu(NH_3)_2^{2+} \tag{8.8}$$

Likewise it may be written that:

$$\beta_3 = K_1 \times K_2 \times K_3 \tag{8.9}$$

and

$$\beta_4 = K_1 \times K_2 \times K_3 \times K_4 \tag{8.10}$$

such a procedure being continued as far as is necessary to describe the cumulative constants in terms of the step-wise ones.

Both types of constant are published in data books and in scientific papers, and some care has to be exercised in distinguishing between the two. One common error in the publication of the values of Ks and βs is that their values are sometimes given as the logarithm (to base 10) without actually specifying that the quoted numbers are logarithms of the actual values of the K or β in question. This arises because of the very large magnitudes of most formation constants. The K and β values concerned in the step-wise production of the tetraamminecopper(II) ion are shown in the Table 8.1.

The values in Table 8.1 are those referring to the standard conditions (298 K and

Table 8.1

i	K_i	$\log_{10} K_i$	β_i	$\log_{10} \beta_i$
1	2.0×10^4	4.3	2.0×10^4	4.3
2	4.2×10^3	3.6	8.4×10^7	7.9
3	1.0×10^3	3.0	8.4×10^{10}	10.9
4	1.7×10^2	2.2	1.4×10^{13}	13.1

1 atm. pressure). Although the β values increase as the number of ligand molecule increases, the K values decrease. The observation is a general one, and will be dealt with at length at a later stage.

In the case considered above, the subscript of the β value is equal to the number of metal ligand interactions at each stage in the formation of the complex. This is not always the case. If a **bidentate** ligand is used (such as diaminoethane, $H_2NCH_2CH_2NH_2$) which can form two linkages (via the nitrogen atoms) to a metal ion then the relationships between the K and β values are the same as in the Cu^{2+}–ammonia case but now the subscripts refer to the number of ligands bonded to the metal ion and not to the number of metal–donor atom linkages. In the extreme case of one multi-dentate ligand being used to complex with a metal ion, there could be just one value for the formation constant, K_1 being identical to β_1. That may be the case for the interaction of metal ions with edta (diaminoethane-N,N,N′,N′-tetraethanoic acid, $(HOOC)_2N(CH_2)_2N(COOH)_2$, which has two nitrogen atom and four oxygen atom donors) or with more complex ligands such as protein molecules.

A discussion of the factors which influence the magnitudes of formation constants is to be found in Chapter 10.

8.3 MOLECULAR ORBITAL TREATMENT OF THE METAL–LIGAND BOND

There are three main geometries of complexes, and these are dealt with in considerable detail in this section — (i) octahedral, O_h, (ii) square planar, D_{4h}, and (iii) tetrahedral, T_d coordination of the central metal atom or ion. Emphasis is placed upon the relative energies of the d orbitals since they lose their free-atomic five-fold degeneracy in the presence of an environment of ligands of non-spherical symmetry. This has consequences for the electronic spectra and magnetic properties of the complexes so formed.

Complexes with octahedral (O_h) symmetry

A complex such as $[FeF_6]^{3-}$ possesses true octahedral symmetry in that the ligands are monatomic. One such as $[Fe(NH_3)_6]^{3+}$ is not strictly octahedral in that the random orientations of the hydrogen atoms of the six ammonia ligands as the molecules undergo rotational and vibrational motion prevent the complex from having true octahedral symmetry. Such motion by the ligands does not seem to be important. Either the rotational and vibrational motions of the non-ligated atoms produce a statistical balancing effect or the effects upon the d orbitals are produced by the 'local' symmetry of the ligating atoms. The replacement of the six monodentate ammonia ligands of the $[Fe(NH_3)_6]^{3+}$ complex by three bidentate diaminoethane ($NH_2CH_2CH_2NH_2 = en$) ligands does not alter the 3d orbital energies significantly even though the $[Fe(en)_3]^{3+}$ complex has D_3 symmetry. It would appear that the local symmetry (O_h) of the six ligating nitrogen atoms is predominant in affecting the 3d orbitals of the iron(III) ion at the centre of the complex.

Molecular orbital treatment of ML_6 (O_h) complexes

The m.o. treatment of an ML_6 complex follows the usual pattern of classifying the orbitals of the central metal atom, and the ligand group orbitals, with respect to the O_h point group, and allowing those orbitals with the same symmetry to form bonding and anti-bonding combinations.

The orbitals of the central metal atom are, for a member of the first transition series, the 4s, 3d and 4p a.o.s. They transform, with respect to the O_h point group, as follows:

4s(M):	a_{1g}
$3d_{xy,xz,yz}$(M):	t_{2g}
$3d_{z^2,x^2-y^2}$(M):	e_g
$4p_{x,y,z}$(M):	t_{1u}

The six σ-type orbitals from the six ligands may be placed along the Cartesian axes. In Fig. 8.1 the positive ψ portions of those orbitals are shown. Their character

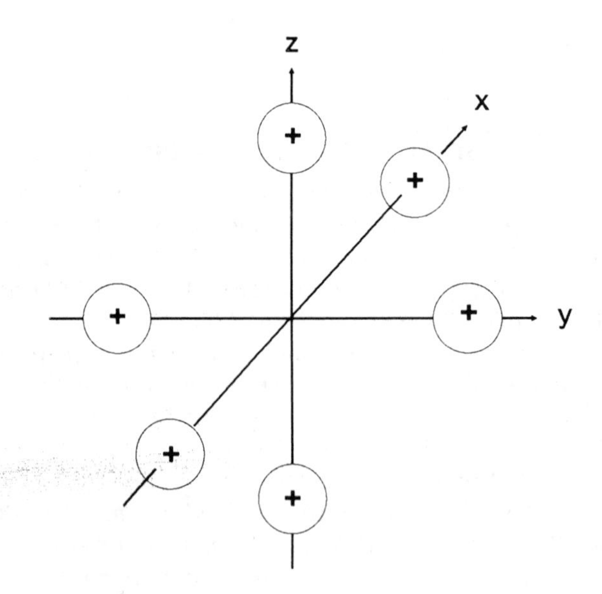

Fig. 8.1 — An octahedral arrangement of ligand 'sigma'-type orbitals in an O_h complex.

with respect to the symmetry operations of the O_h point group may be written down in the usual fashion in terms of the number of orbitals unaffected by each operation:

	E	C_3	C_2	C_4	C_2	i	S_4	S_6	σ_h	σ_d
$6 \times \sigma_L$	6	0	0	2	2	0	0	0	4	2

Some indication of the disposition of the various elements of symmetry is helpful in understanding the derivation of the above character. The C_3 axes are best visualized as passing through the centre of the triangular faces of the octahedron and the central atom of the complex. Rotation around such an axis by 120° moves all six orbitals. The C_2 axes are those which bisect any two of the Cartesian axes. Rotation around such an axis by 180° affects all six σ-type orbitals of the ligands. The C_4 axes are the three which are coincident with the Cartesian axes, and each contains the central metal atom and two diametrically opposed ligand atoms. Rotation around such an axis by 90° leaves the two ligand orbitals, centred on that axis, unchanged. The next C_2 axes are those which are coincident with the C_4 axes — rotation around them by 180° leaves two ligand orbitals unchanged.

The operation of inversion causes all six ligand orbitals to move. The S_4 axes are co-axial with the C_2 axes which bisect the pairs of Cartesian axes. The S_4 operation moves all six ligand orbitals. The S_6 axes are co-axial with the C_3 axes. The S_6 operation moves all six ligand orbitals.

Reflexion of the six ligand orbitals through any of the three horizontal planes of symmetry (represented by the *xy*, *xz* and *yz* planes) leaves four of the ligand orbitals (the ones in the particular plane) unaffected by the operation. The dihedral planes (there are six) bisect the pairs of horizontal planes (each of which contains a C_2 axis), and reflexion in any one of them causes four ligand orbitals to move, leaving two unaffected.

The character of the six σ-type ligand orbitals is reducible to the sum:

$$6 \times \sigma_L = a_{1g} + e_g + t_{1u} \tag{8.11}$$

The a_{1g} and t_{1u} combinations have the correct symmetries for interaction with the 4s and 4p orbitals, respectively, of the metal atom. Likewise the e_g orbitals of the ligands and the metal atom may combine to give bonding and anti-bonding orbitals. The remaining t_{2g} orbitals of the metal atom are non-bonding in a σ-only complex. A σ-only m.o. diagram is shown in Fig. 8.2. It is conventional in this area of m.o. theory to dispense with the Mulliken numbering of levels and to use superscript asterisks (*) to indicate anti-bonding character.

Such a bonding scheme is very helpful in explaining the spectroscopic and magnetic properties of a wide range of complexes. The main determinant of those properties is the energy difference between the anti-bonding e_g^*, and the non-bonding t_{2g}, orbitals. Such a gap in energy allows for d–d electronic transitions which happen to lie mainly in the visible region of the electromagnetic spectrum and account largely for the highly colourful chemistry observed for most transition metal complexes.

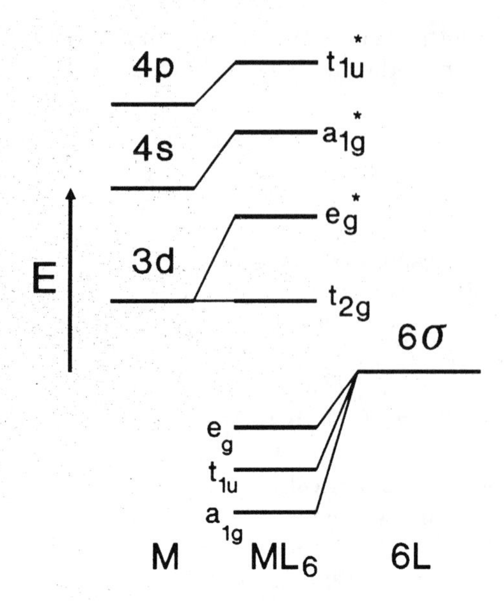

Fig. 8.2 — A 'sigma'-only molecular orbital diagram for an octahedral complex.

The energy gap referred to above is the equivalent of the $\Delta_{octahedral}$ (or sometimes 10 Dq) of crystal field theory (which treats ligands as though they are point charges).

The magnetic properties of any compound depend upon the number of unpaired electrons it possesses. Consider the hexaammineiron(III) ion, $[Fe(NH_3)_6]^{3+}$, complex in which there are five electrons supplied by the Fe(III) ion and twelve electrons by the six ammonia ligands, making a total of seventeen electrons to be placed in the m.o.s of Fig. 8.2. The electronic configuration of the complex is:

$$[Fe(NH_3)_6]^{3+}: \quad a_{1g}^2 t_{1u}^6 e_g^4 t_{2g}^5$$

providing that the five electrons of highest energy pair up in the non-bonding t_{2g} level as far as possible. Such pairing would only occur if the energy gap between the anti-bonding e_g^* level and the t_{2g} orbitals were to be sufficient to enforce pairing in the lower level. For the example in question, this does not seem to be the case because the complex has a magnetic moment which indicates that there are five unpaired electrons, rather than the single unpaired electron which would result from a t_{2g}^5 configuration. The manner in which the five electrons may be distributed so that they may occupy orbitals singly is to make use of the anti-bonding e_g^* orbitals to give the configuration $t_{2g}^3(e_g^*)^2$. In compliance with Hund's rules the five electrons have parallel spins, which is consistent with the observed magnetic moment. The energy gap between the e_g^* and t_{2g} orbitals in the complex, $[Fe(NH_3)_6]^{3+}$, must be smaller than the interelectronic repulsion energy which would be experienced in the t_{2g}^5 configuration.

The complexes of iron(III) and cobalt(III) with six ligands which are either fluoride ion, F^-, ammonia or cyanide ion, CN^-, serve to extend this discussion. A summary of the magnetic properties of the six possible ML_6 complexes is shown in Table 8.2.

Table 8.2 — Number of unpaired electrons for the metal–ligand combinations

Ligands	Fe(III), d^5	Co(III), d^6
$6F^-$	5	4
$6NH_3$	5	0
$6CN^-$	1	0

The substitution of F^- ion for NH in the Fe(III) complex has no effect upon the electron arrangement, but the substitution of CN^- causes electron pairing to be preferable to population of the e_g^* level. The pattern is somewhat different in the case of the cobalt(III) complexes, where both ammonia and cyanide complexes have a large enough e_g^*–t_{2g} energy gap to enforce electron pairing. It is clear that the energy gap depends upon the nature of the metal ion (and its oxidation state) as well as upon the nature of the ligand. Observations of their visible spectra allow estimates of the energy gap to be made, and the results are shown in Table 8.3.

Table 8.3 — e_g^*–t_{2g} energy differences for the metal–ligand combinations (/kJ mol^{-1})

Ligands	Fe(III), d^5	Co(III), d^6
$6F^-$	150	196
$6NH_3$	209	272
$6CN^-$	285	370

It is obvious from these results that electron pairing takes place (in these systems) when the e_g^*–t_{2g} gap is greater than ≈ 270 kJ mol^{-1}. Jorgensen initiated a method of combining a large number of observations which allows the e_g^*–t_{2g} energy gap to be calculated from empirically derived parameters. The gap is given by the relationship:

$$\delta(e_g^*\text{–}t_{2g}) = f \times g \qquad (8.12)$$

where f is characteristic of the ligand and g is characteristic of the metal. If the product, $f \times g$, is multiplied by 1000 the result is an estimate of the energy gap in terms of wavenumber, with units of cm^{-1}. Multiplication of $f.g$ by 11.96 gives the result in terms of kJ mol^{-1}. A selection of typical values is given in Table 8.4.

Table 8.4 — Jorgensen's f and g values for some ligands and metal ions in ML_6 complexes

Ligand	f factor	Metal ion	g factor
Br^-	0.72	Mn(II)	8.0
SCN^-	0.73	Ni(II)	8.7
Cl^-	0.78	Co(II)	9.0
F^-	0.9	V(II)	12.0
H_2O	1.0	Fe(III)	14.0
NH_3	1.25	Cr(III)	17.4
en	1.28	Co(III)	18.2
phen	1.34	Mo(III)	24.6
CN^-	1.7	Rh(III)	27.0

The chosen ligands in the order of their increasing f values are: $Br^- < SCN^-$ (ligand atom, S) $< Cl^- < F^- < H_2O < NH_3 <$ en $<$ phen $< CN^-$ (phen is the abbreviation for 1,10-phenanthroline — see Chapter 6). There are negative ligands at either end of this series and neutral ligands in between. The more basic (in terms of electron pair donation) ligands are centrally placed in the series. The series itself is called the **spectrochemical series**. This is because the values upon which it is based are derived from measurements of absorption spectra, from which the e_g^*–t_{2g} energy gaps may be determined.

The series (which can be extended to include many ligands) remains very much the same whichever metal ion is used. This means that the effect upon δ of the nature of the ligand may be discussed. Likewise it is obvious from the g values that the value of δ is dependent upon the nature of the metal ion and its oxidation state.

The variations in the e_g^*–t_{2g} energy gap may be understood in terms of the relative basicities (in terms of electron pair donation tendency) of the ligands and the type of π-type bonding which is involved. The more basic the ligand, the stronger will be the bonding and the larger the e_g^*–t_{2g} gap should be. The effects of π-type bonding explain why the cyanide ion, and the neutral CO molecule, with low basicity are at the top end of the spectrochemical series.

There are two cases of π-type bonding to be considered. These are (i) π-type donation of electrons from the ligand to the metal atom, and (ii) π-type donation of electrons from the metal atom to the ligand.

Ligand-to-metal atom π donation of electrons

The diagram in Fig. 8.3 shows the possible overlap between a metal atom d_{xy} orbital (one of the t_{2g} set) and a ligand p_x orbital. It is obvious from the signs of ψ that such an overlap would lead to m.o. formation. There is no need to resort to formal group theory to come to such a conclusion. Ligand p orbitals are usually doubly occupied so that for any π-type interaction to be energetically advantageous it is necessary for the t_{2g} orbitals of the metal ion to be either vacant or only singly occupied. Such

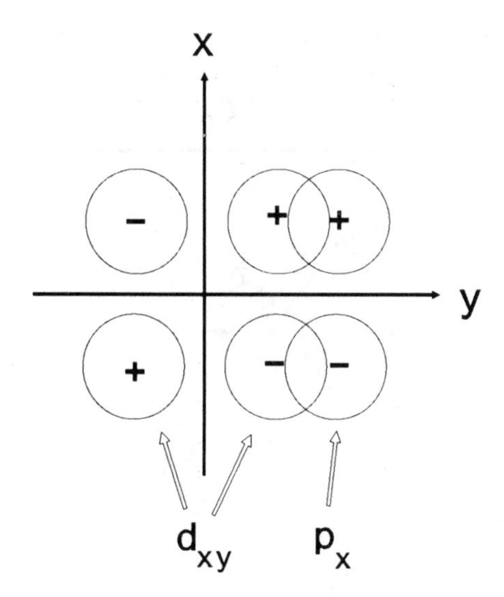

Fig. 8.3 — A diagrammatic representation of the pi-type overlap of a d_{xy} metal orbital and a p_x ligand orbital.

interaction can only be advantageous for the t_{2g}^{0-3} configurations of metals — where any electrons in excess of three must occupy the e_g^* orbitals (i.e. high-spin d^{4-5} configurations).

The energies of the ligand orbitals which engage in ligand–metal π-type bonding are normally lower than those of the t_{2g} orbitals of the metal. This means that when π-type interaction occurs the lower (bonding) orbitals have a majority contribution from the ligand orbitals. The higher (anti-bonding) combinations have energies greater than those of the original t_{2g} orbitals (which should be labelled as t_{2g}^*) and, in consequence, the e_g^*–t_{2g}^* energy gap is reduced. The effect is shown on the left-hand side of Fig. 8.4.

Metal atom–ligand π donation of electrons

The cyanide ion and the carbon monoxide molecule are isoelectronic and possess occupied π_u bonding orbitals and vacant anti-bonding π_g orbitals. The latter orbitals may interact with the metal t_{2g} orbitals to give metal-ligand bonding and anti-bonding π-type m.o.s.

The anti-bonding ligand π_g orbitals are normally of higher energies than the metal t_{2g} orbitals, so it is the latter which contribute more to the metal–ligand π-type bonding orbitals. The ligand π_g orbitals have a major contribution to the metal–ligand anti-bonding m.o.s. This has the effect of increasing the e_g^*–t_{2g} energy gap as is shown on the right-hand side of Fig. 8.4.

This ligand-to-metal donation of electrons is sometimes known as π-type **back donation**, as if to offset the normal ligand–metal donation. Another description of the same phenomenon is π-acid behaviour of the ligand, one definition of an acid being that it is a compound which has electron-pair accepting properties. The back-

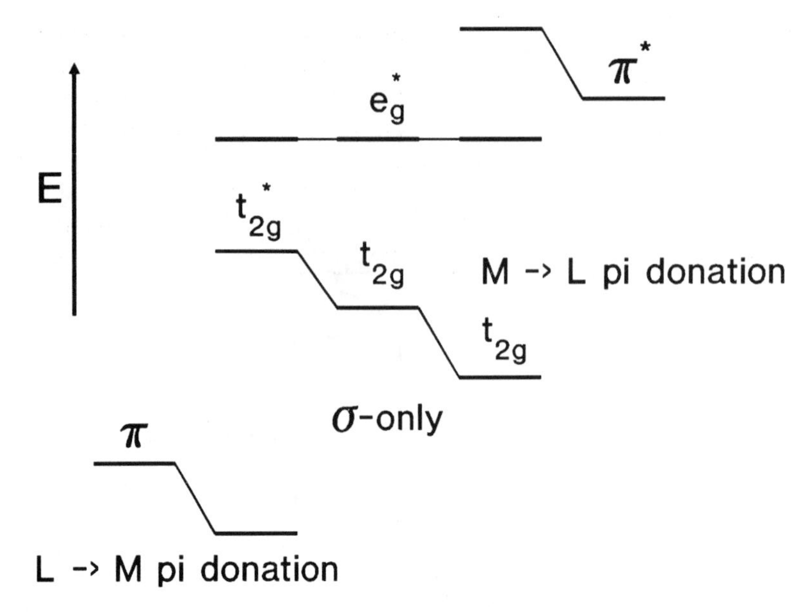

Fig. 8.4 — The effect of pi-type interaction on the energy of the t_{2g} orbitals of an octahedral complex.

donation effect explains, in electronic terms, the production of relatively strong bonds by such ligands as CN^- and CO. Their normal tendency to be weakly basic is overcome because any loss of electron pairs by ligand–metal donation is balanced, to some extent, by the metal–ligand back donation. One effect feeds the other and is known as **synergism**. The synergic effect of one type of bonding reinforcing the other contributes to very strong bonding and a very large δ value.

Complexes with back bonding are produced when the metal t_{2g} orbitals are filled so that the electrons therein are stabilized by the formation of the π-type orbitals.

The ethene molecule, C_2H_4, bonds strongly to some metal ions in a side-ways-on mode. Because of the terminal C–H bonds it does not have any σ-type orbitals which could be used in ligand–metal donation. It does, however, possess filled π_u C–C bonding orbitals, the electrons of which may be used to form a ligand–metal σ-type bond. The details of this bonding are dealt with in the section concerned with square planar complexes.

The Jahn–Teller effect

The Jahn–Teller effect applies to complexes with an odd number of electrons in the e_g^* or t_{2g} orbitals. It is more important in the e_g^* cases (because the orbitals have a greater involvement with the ligands) and, as such, it concerns the electronic configurations: high-spin d^4, low-spin d^7, and d^9, with $(e_g^*)^1$,$(e_g^*)^1$ and $(e_g^*)^3$ configurations respectively. It may be stated in the form: 'If, in a non-linear molecule, a degenerate set of orbitals is unevenly occupied, and a distortion is possible which removes their degeneracy, then such a distortion will occur and lead to a stabilization of the system'.

In the case of regularly octahedral complexes which possess the doubly degenerate e_g^* anti-bonding level the distortion which usually occurs is where two of the bonds (along the z direction) become elongated, and the four bonds in the square plane (xy) are shortened. Such a distortion causes the symmetry to change from O_h to D_{4h} and the e_g^* orbitals lose their two-fold degeneracy. The orbital with the d_{z^2} contribution becomes stabilized (the ligands are further away from the metal ion along the z axis) and is labelled a_{1g}^*. The orbital with the $d_{x^2-y^2}$ is destabilized (the ligands are nearer the metal in the xy plane) and becomes the b_{1g}^* orbital. Fig. 8.5 demonstrates these effects.

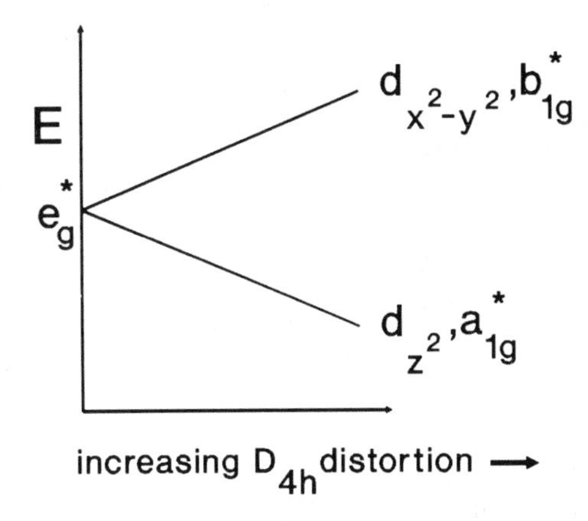

Fig. 8.5 — The Jahn-Teller effect on the energies of the e_g^* orbitals as D_{4h} distortion increases (four short bonds and two long ones).

The distortion to give the more stable a_{1g}^* orbital leads to the 'distorted' complex being more stable than the regular octahdral form when it is possible for there to be more electrons in it than occupy the destabilized b_{1g}^* orbital.

The reverse effect produced by a distortion of the two-short-bond, four-long-bond configuration of a complex does occur but only in the solid state, where crystal packing effects are an added consideration.

The case of the even filling of the e_g^* orbitals of an O_h complex may lead to stabilization by distortion to D_{4h} symmetry only if the distortion is sufficient to force electron pairing in the lower a_{1g} orbital. If the difference in energy between the b_{1g}^* and a_{1g}^* orbitals is sufficiently high then an $(e_g^*)^2$ (O_h) configuration would be more stable as (a_{1g}^*) (D_{4h}) with consequent changes in geometry and in magnetic properties. Such an effect is only observed in square planar complexes which may be thought of as distorted octahedral complexes with two infinitely long bonds along the z axis.

Molecular orbital treatment of square planar, D_{4h}, complexes

The complex is set up with its C_4 axis coincident with the z axis. Inspection of the D_{4h} character table allows the classification of the orbitals of the central metal ion as follows:

4s:	a_{1g}
$3d_{xy}$:	b_{2g}
$3d_{z^2}$:	a_{1g}
$3d_{x^2-y^2}$:	b_{1g}
$3d_{xz,yz}$:	e_g
$4p_z$:	a_{2u}
$4p_{x,y}$:	e_u

so that mixing of the 4s and $3d_{z^2}$ orbitals is possible. The reduction of the symmetry from O_h to D_{4h} causes there to be more irreducible representations to deal with, and with a consequent lowering of degeneracy, which leads to a more complicated m.o. diagram.

The four σ-type ligand orbitals lie along the metal–ligand directions (coincident with the x and y axes) and are classified by writing their character in terms of the number of orbitals unaffected by the appropriate symmetry operations of the D_{4h} point group.

	E	C_4	C_2	C_2'	C_2''	i	S_4	σ_h	σ_v	σ_d
$6 \times L_\sigma$	4	0	0	2	0	0	0	4	2	0

This character reduces to the sum:

$$4 \times L_\sigma = a_{1g} + b_{1g} + e_u \tag{8.13}$$

The four-ligand p_z (or π-type orbitals in with a z component) may be classified in a similar manner to give the representations:

$$4 \times L_\pi(z) = a_{2u} + b_{2u} + e_g \tag{8.14}$$

The other four p- or π-type ligand orbitals (those in the xy plane at right angles to the metal–ligand directions) have a character which reduces to the sum:

$$4 \times L_\pi(xy) = a_{2g} + b_{2g} + e_u \tag{8.15}$$

The m.o. diagram, for the σ-type orbitals only, is shown in Fig. 8.6. As in the case of octahedral complexes there is the possibility of π-type bonding which would alter the order of the energy levels.

The bonding of alkenes (the simplest being ethene, C_2H_4) to metals is particularly important in square planar complexes. One notable example is that of the trichloro-(η^2-(ethene)platinum(II) ion, $[(C_2H_4)PtCl_3]^-$. The ethene molecule is bonded so that its carbon–carbon axis is perpendicular to the plane containing the platinum(II) and chloride ions. The geometry is shown in Fig. 8.7. The two carbon atoms of the ethene molecule are bonded to, or are within bonding distance of, the platinum centre. It is for this reason that the nomenclature of the complex makes use

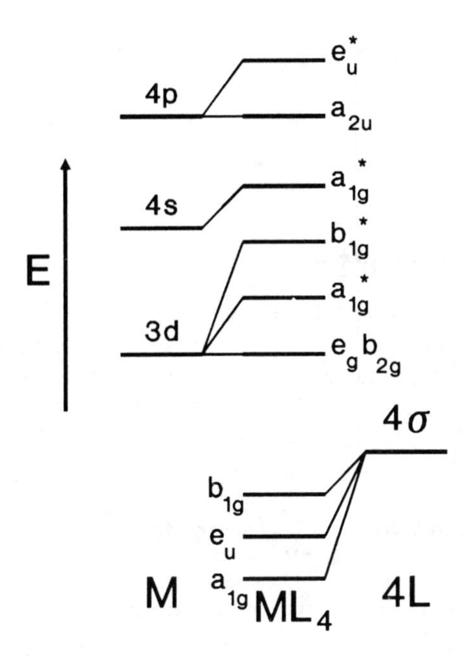

Fig. 8.6 — Molecular orbital diagram for a square planar complex (sigma orbitals only).

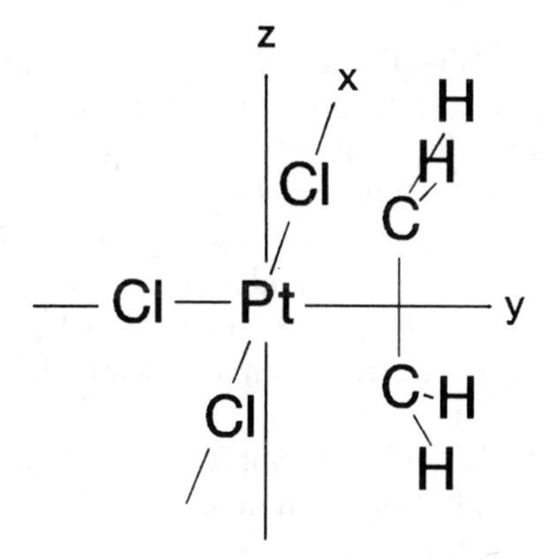

Fig. 8.7 — The structure of the trichloro-η^2-(ethene)platinum(II) ion.

of the Greek letter eta, η, (from ηαπτειν = *haptein*, to fasten), with a superscript (2 in this case) to indicate the number of bonded atoms.

The ethene molecule has a filled π bonding orbital which is in the plane which is perpendicular to the molecular plane and which contains the two carbon atoms. The vacant anti-bonding π-type orbital is in the same plane, and both are shown in Fig. 8.8. It is possible for electrons to be donated to a σ-type platinum orbital ($d_{x^2-y^2}$ is

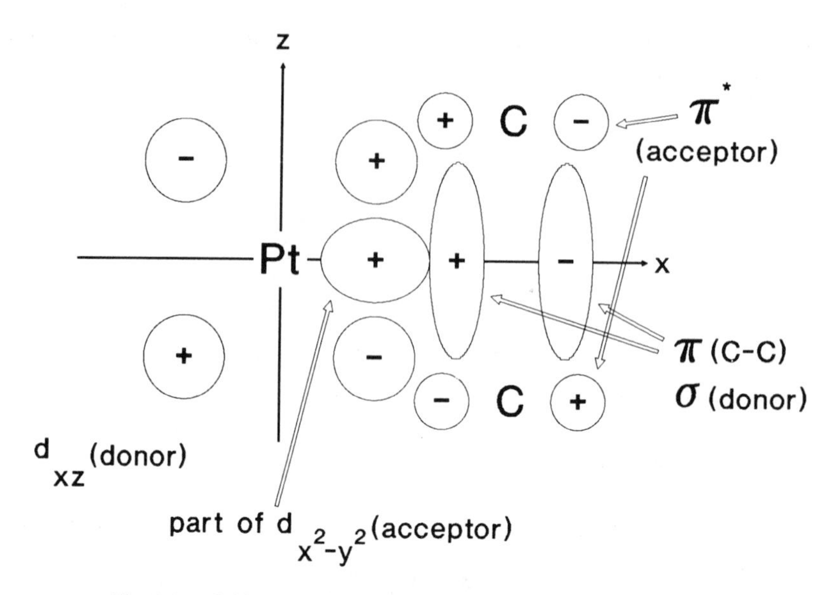

Fig. 8.8 — Orbital contributions in ethene–platinum bonding.

used in Fig. 8.8) forming an ethene-platinum bond. The ethene-to-platinum donation reduces the effectiveness of the π-bonding between the carbon atoms of the ethene ligand. The $5d_{xz}$ orbital of the platinum has the correct symmetry to form bonding and anti-bonding (Pt–ethene) combinations with the anti-bonding π-type orbital of the ethene molecule to promote π-type back bonding (donation from platinum to ethene) and produce the synergism required to ensure efficient ethene-to-platinum donation in the σ-type bonding. The promotion of an electron from $\pi \rightarrow \pi^*$ in a free ethene molecule produces a change of symmetry from D_{2h} to D_{2d} (as would be produced if one CH_2 group underwent a 90° rotation with respect to the other one). The distortion of the ethene molecule when it acts as a ligand is not consistent with such an electronic transition, the four hydrogen atoms being coplanar due to internuclear repulsion.

Alkenes (and alkynes) do not form complexes with typical acceptor metals (those with vacant π-type orbitals) but do form many complexes with metals possessing a full d^{10}. complement of electrons. In such cases the ligand-to-metal donation can only be to the s and p orbitals of the metal. Back donation of a π-type is responsible for the formation of the strong linkages which are observed.

Molecular orbital treatment of tetrahedral, T_d, complexes

In a tetrahedral environment the orbitals of a central metal atom transform as:

4s:	a_1
$4p_{x,y,z}$	t_2
$3d_{z^2, x^2-y^2}$	e
$3d_{xy, xz, yz}$	t_2

If the four σ-type ligand orbitals are arranged along the formal metal–ligand directions their character is given by:

	E	C_3	C_2	S_4	σ_d
$4 \times L_\sigma$	4	1	0	0	2

which reduces to the sum:

$$4 \times L_s = a_1 + t_2 \tag{8.16}$$

The other eight ligand orbitals (of a π-type) transform as the sum:

$$8 \times L_\pi = e + t_1 + t_2 \tag{8.17}$$

The m.o. diagram may then be constructed as in Fig. 8.9. There is a low-energy a_1

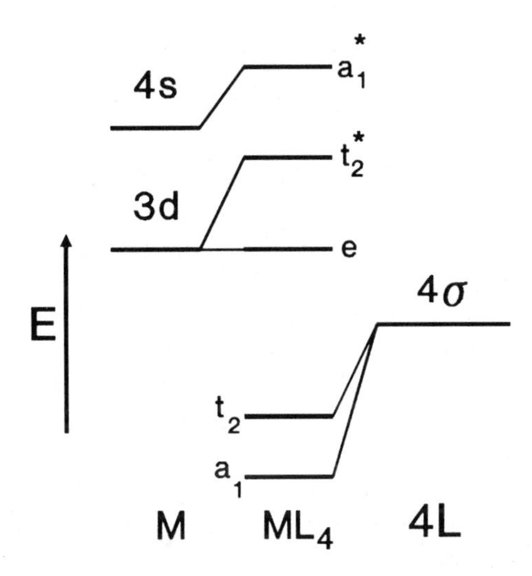

Fig. 8.9 — Molecular orbital diagram for a tetrahedral complex.

m.o. and a higher-energy t_2 set which is a mixture of σ- and π-type contributions from the ligand orbitals. The ligand t_1 set is non-bonding since there are no metal orbitals of that symmetry. Because of the general disposition of the orbitals of the metal and ligands, none of the bonding interactions is as efficient as those observed in octahedral and square planar complexes.

8.4 THE ANGULAR OVERLAP APPROXIMATION

The m.o. diagrams derived for O_h, D_{4h} and T_d symmetries of the above sections suffer from being qualitative, and even by present-day capabilities it is not possible to derive them on a fully quantitative basis. The angular overlap approximation serves to produce a semi-quantitative version of such diagrams and may be used to great effect in deciding which of several symmetries gives the lowest energy for any particular metal atom–ligand system.

The basis of the angular overlap approximation is to consider the full σ-type overlap between a metal orbital and a ligand orbital to stabilize the ligand-rich bonding orbital by one unit of energy, denoted by e_σ, and to destabilize the metal-rich anti-bonding orbital by the same amount. The situation for the overlap of a ligand orbital and the d_{z^2} orbital of a metal is shown in Fig. 8.10. As the angle

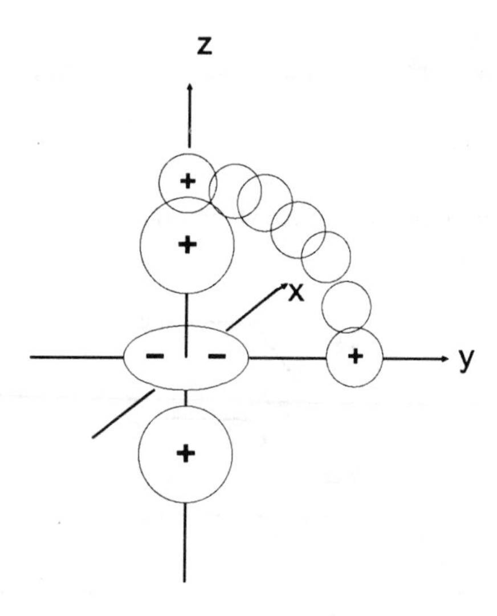

Fig 8.10 — A diagram showing the variation in the overlap of a d_{z^2} and a ligand σ-type orbital as the ligand orbital moves from the *z* direction to the *y* direction.

between the M-ligand direction and the *z* axis increases, the overlap between the two orbitals becomes less efficient such that at a 90° angle the stabilization/destabilization energy is only one quarter of that when the angle is zero.

Such position-dependent energies have been calculated for the ligand positions for the geometries considered in this chapter for the σ- and π-type overlaps of ligand orbitals and the five d orbitals of the central metal atom. The π-type overlap, being less efficient than the σ-type, causes the relevant energy term, e_π, to be smaller than e_σ. Table 8.5 is useful in the calculation of approximate m.o. diagrams for any of the symmetries O_h, D_{4h}, T_d, C_{4v} and D_{3h}. The last two are not dealt with in this section but are important in the understanding of some kinetic considerations of octahedral complexes which form part of Chapter 10.

The ligand positions referred to in Table 8.5 may be used to calculate the orbital energies for complexes with the more common symmetries. Positions 2, 3, 4 and 5 are for ligands placed along the *x* and *y* axes, and are used for square planar, D_{4h}, complexes. With the additional use of positions 1 and 2, along the *z* axis, octahedral

Table 8.5 — Angular scaling factors for e_σ and e_π

Ligand position	Metal atomic orbital				
	z^2	x^2-y^2	xz	yz	xy
1 σ	1	0	0	0	0
π	0	0	1	1	0
2 σ	1/4	3/4	0	0	0
π	0	0	1	0	1
3 σ	1/4	3/4	0	0	0
π	0	0	1	1	1
4 σ	1/4	3/4	0	0	0
π	0	0	1	0	1
5 σ	1/4	3/4	0	0	0
π	0	0	0	1	1
6 σ	1	0	0	0	0
π	0	0	1	1/0	0
7 σ	1/4	3/16	0	0	9/16
π	0	3/4	1/4	3/4	1/4
8 σ	1/4	3/16	0	0	9/16
π	0	3/4	1/4	3/4	1/4
9 σ	0	0	1/3	1/3	1/3
π	2/3	2/3	2/9	2/9	2/9
10 σ	0	0	1/3	1/3	1/3
π	2/3	2/3	2/9	2/9	2/9
11 σ	0	0	1/3	1/3	1/3
π	2/3	2/3	2/9	2/9	2/9
12 σ	0	0	1/3	1/3	1/3
π	2/3	2/3	2/9	2/9	2/9

complexes may be dealt with. Calculations involving trigonally planar arrangements of ligands would use positions 2, 7 and 8. For tetrahedral complexes, positions 9, 10, 11 and 12 are used.

Using the energy terms in Table 8.5 leads to the σ-only m.o. diagrams for O_h, D_{4h} and T_d symmetries as shown in Fig. 8.11. Such diagrams may be refined to include π-type interactions and can take into account whether the ligands have donor (leading to destabilization of the appropriate d orbitals) or acceptor (leading to stabilization of the d orbitals) properties.

The π-type interactions are of a minor nature compared to the σ-type and it is the latter which are responsible for determining the structure preferred by any metal–ligand system. The stabilization resulting from the formation of any metal–ligand interaction may be calculated for any d electron configuration. In the case of an octahedral complex it may be considered that the twelve electrons (σ-type) from the ligands are stabilized to the extent of $-12e_\sigma$ by occupying the six bonding

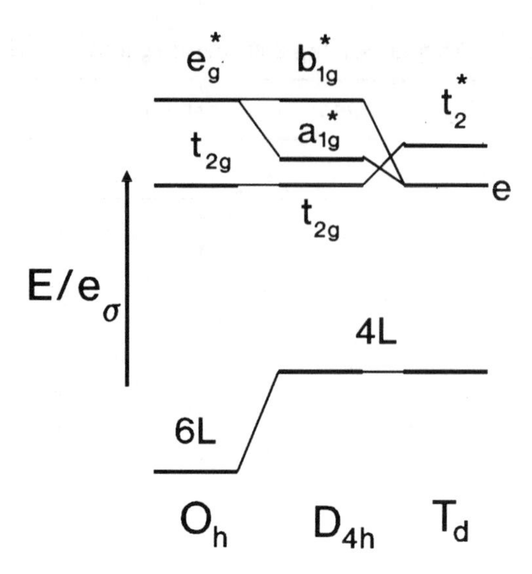

Fig. 8.11 — Sigma molecular orbital diagrams for O_h, D_{4h} and T_d complexes as derived from the angular overlap approximation.

orbitals. This stabilization is offset by the destabilizations suffered by the electrons occupying the anti-bonding d orbitals.

The resultant stability of any complex also depends upon whether the gaps between the anti-bonding orbitals are sufficient to cause electron pairing to be preferable to single occupancy. If the energy gaps are large enough to cause pairing, the resulting electronic configuration is termed **low spin**. If electron pairing is not forced upon the system, the electronic configuration is **high spin**.

It is useful to compare the stabilization energies for the low- and high-spin configurations of octahedral and square planar complexes, together with the high-spin configurations of tetrahedral complexes (there are practically no low-spin T_d complexes since the t_2^*–e energy gap is relatively low).

Calculations of the stabilization energies for the above-mentioned cases are shown in Table 8.6.

The results in Table 8.6 indicate the reason for the very great predominance of the octahedron in the chemistry of complexes. The following generalizations may be made concerning the likely complex to be formed by a metal–ligand system depending upon the number of d electrons originally possessed by the metal.

1. High-spin cases. (Small gaps between sets of d orbitals)
   (i) For d^0 to d^8 configurations the complex formed is likely to be of the ML_6, O_h, type.
   (ii) Complexes with d^9 configurations can be of either ML_6, O_h, or ML_4, D_{4h}, type. A d^9 configuration is subject to the Jahn–Teller effect which favours the production of a D_{4h}, ML_4 complex.
   (iii) Complexes with d^{10} configurations can have one of the three symmetries, O_h, D_{4h} or T_d.

Table 8.6 — Stabilization energies for O_h, D_{4h} and T_d complexes (in units e_σ)

No. of d electrons	O_h		D_{4h}		T_d
	H.S.	L.S.	H.S.	L.S.	H.S.
0	−12	−12	−8	−8	−8
1	−12	−12	−8	−8	−8
2	−12	−12	−8	−8	−8
3	−12	−12	−8	−8	$-6\frac{2}{3}$
4	−9	−12	−7	−8	$-5\frac{1}{3}$
5	−6	−12	−4	−8	−4
6	−6	−12	−4	−8	−4
7	−6	−9	−4	−7	−4
8	−6	−6	−4	−6	$-2\frac{2}{3}$
9	−3	−3	−3	−3	$-1\frac{1}{3}$
10	0	0	0	0	0

2. Low-spin cases. (Large gaps between sets of d orbitals)
   (i) Complexes with configurations d^0 to d^7 should be octahedral.
   (ii) Complexes with d^8 or d^9 configurations could be either octahedral or square planar. As in case 1(ii), the d^9 configuration is subject to Jahn–Teller distortion and is most likely to produce a D_{4h}, ML_4 complex.
   (iii) Complexes with d^{10} configurations can be either octahedral, square planar or tetrahedral.

In the tetrahedral cases it is possible for extra stability to be derived from the interaction of the 4p orbitals and the d_{xy}, d_{xz} and d_{yz} orbitals since both groups transform as t_2. Such p orbital interaction favours tetrahedral symmetry in complexes with a d^{10} configuration.

The above generalizations are subject to considerations of three other factors which may influence the preferred symmetry of a complex.

1. The Pauling electroneutrality principle.
The electroneutrality principle has as its basis the operation of Coulomb's law. The positive metal centre has an attraction for the ligand system (which is composed of negative ions or is essentially the negative end of a dipolar system) and, depending upon the strength of the attraction, will result in the stabilization of the system until the attractive forces are balanced by the repulsive forces which operate at short internuclear distances. If the metal charge is low, the position of electrostatic equilibrium will be reached by the participation of fewer negative ligands than would be the case for a highly charged metal. This ignores the other factors.

2. Lattice energy or Hydration energy
If a complex has an overall charge its stability is enhanced by the interaction with its **counter ion** (in the case of a solid) or with the solvent (in solution). Highly charged

(positive or negative) complexes are favoured by this factor (in spite of the electroneutrality principle). Metals in their zero oxidation states forming complexes with neutral ligands are exempt from this consideration.

3. Interligand repulsion and/or steric hindrance

Interligand repulsion may well influence the formula of a complex but is difficult to separate from steric hindrance (in which ligands would be closer together than indicated by their normal Van der Waals radii). There is some evidence for the ligands, Cl^- and O^{2-}, being too large in the Van der Waals sense to form ML_6 complexes. Iron(III) forms $[FeCl_4]^-$ with the larger chloride ion, but forms $[FeF_6]^{3-}$ with the smaller fluoride ion. Likewise the $[CoCl_4]^{2-}$ complex ion is tetrahedral. There are no cases with greater than four O^{2-} ligands. This may be due to steric hindrance or to the large repulsion forces in operation, or to both effects.

If we bear in mind the above calculations and factors there are only three general cases which apply to the majority of existing complexes.

1. Weak metal-ligand bonding.

If the metal–ligand interaction is weak the energy gaps between the d-type orbitals are relatively small. This allows all the anti-bonding orbitals to be occupied if necessary. It has the consequence that the complexes in this category are those with **high spin** — they have the maximum number of unpaired electrons with parallel spins. The angular overlap calculations indicate that the most likely complex to be formed by metals with d^{0-8} configurations is the octahedral ML_6 type. Square planar complexes are a possibility for d^9 configurations, while both square planar and tetrahedral complexes may be formed by metals with full d^{10} configurations.

If the complex formed is ML_6, the number of electrons in the valence shell is given by 12 (each ligand contributing two σ-type electrons) plus the number of d electrons originally possessed by the metal ion. The range is therefore from 12 (d^0) to 22 (d^{10}) electrons in the valence shell.

In the case of d^9 metals, there being no difference between the O_h and D_{4h} stabilization energies, interligand effects may favour the production of ML_4 (D_{4h}) complexes. In that case the number of valence electrons would be $9 + 8 = 17$.

In the case of a d^{10} metal, the most likely complex to be formed is a tetrahedral ML_4. Such a complex would have favourable interligand repulsion energy and p orbital participation would stabilize the T_d symmetry relative to either of the alternatives. Complexes of this type would possess 18 valence shell electrons.

Examples of weakly bonded complexes are given in Table 8.7. The formal oxidation states of the metals are indicated, together with the number of valence shell electrons (VSE).

2. Strong metal–ligand σ bonding

In a strongly bonding complex the energy differences between the various sets of anti-bonding orbitals are large enough to force electron pairing to occur and **low spin** complexes result.

The higher anti-bonding orbitals (e_g in O_h) possess too high an energy to be occupied in a stable complex. The twelve electrons from the six ligands are joined by

Table 8.7 — High-spin, weakly bonded complexes

d config.	Complex	Symmetry	VSE
0	$[TiF_6]^{2-}$	O_h	12
1	$[VCl_6]^{2-}$	O_h	13
2	$[V(C_2O_4)_3]^{3-}$	O_h	14
3	$[Cr(H_2O)_6]^{3+}$	O_h	15
5	$[Fe(C_2O_4)_3]^{3-}$	O_h	17
6	$[Fe(H_2O)_6]^{2+}$	O_h	18
7	$[Co(H_2O)_6]^{2+}$	O_h	19
8	$[Ni(H_2O)_6]^{2+}$	O_h	20
9	$[Cu(NH_3)_4]^{2+}$	D_{4h}	17
10	$[ZnCl_4]^{2-}$	T_d	18

up to six electrons from the metal so that the maximum number of electrons in the valence shell is eighteen. The angular overlap calculations indicate that metal centres with d^0–d^6 configurations should form octahedral complexes (12–18 VSE). Metals with d^7 configurations should also form octahedral complexes, but with a single e_g electron, they would be subject to the Jahn–Teller effect, and are easily oxidized to the more stable d^6 configuration. If interligand effects are important enough to restrict the coordination number to four, the angular overlap calculations indicate that for d^8 and d^9 configurations, D_{4h} complexes are the most likely alternative structures. The four ligands supply eight electrons so that the total number of valence shell electrons would be either 16 or 17 in such cases. Some examples are given in Table 8.8.

Table 8.8 — Low-spin, strongly bonded complexes

d config.	Complex	Symmetry	VSE
0	$[ZrF_6]^{2-}$	O_h	12
1	$[WCl_6]^{-}$	O_h	13
3	$[TcF_6]^{2-}$	O_h	15
4	$[OsCl_6]^{2-}$	O_h	16
5	$[PtF_6]^{-}$	O_h	17
6	$[PtF_6]^{2-}$	O_h	18
8	$[PtCl_4]^{2-}$	D_{4h}	16

Notice that the reduction of the coordination number to four in the case of $[PtCl_4]^{2-}$ is to be expected from considerations of electroneutrality **and** interligand effects, in addition to the angular overlap expectations.

3. Strong metal–ligand σ bonding with metal-to-ligand π donation

If the ligand has vacant π^* (anti-bonding) orbitals which may accept electrons from the metal centre, the ligand→metal σ-type donation is encouraged by metal→ligand π-type **back donation**. Such **synergism** enables weakly basic ligands (CN^-, CO, NO, alkenes, alkynes, arenes and numerous classes of heteroatomic aromatic molecules) to form very strongly bonded complexes, the π interaction contributing to the large energy gaps between the various sets of anti-bonding d-type orbitals. For this kind of bonding to be fully efficient the d orbitals which may engage in π-type interaction should be fully occupied. This means that metal centres should possess d^6 configurations to maximize the back-donation effect in the production of octahedral complexes with 18 valence shell electrons. If the metal contributes more than six electrons the coordination number is reduced so that the number of valence shell electrons is maintained at eighteen.

A d^8 metal centre configuration would produce a five-coordinate, ML_5, complex which would have D_{3h} symmetry. A d^{10} metal centre would produce a tetrahedral complex because in T_d symmetry **all** five d orbitals may participate in π-type bonding. Interligand effects would be consistent with T_d symmetry in that case.

This '18-electron rule' does apply to the d^7 and d^9 cases. A d^7 metal would form a ML_5 complex with 17 valence shell electrons which would be paramagnetic. Although charged paramagnetic complexes are very common, neutral ones are rare. The 17-electron ML_5 unit would gain stability by using its odd electron to form a metal–metal bond, the resulting dimer, M_2L_{10}, being diamagnetic. Each metal centre would then be sharing a valence shell electron complement of 18 electrons. A d^9 metal centre might be expected to produce a monomeric ML_4 unit with 17 valence shell electrons which would dimerize to give an M_2L_8 diamagnetic product containing a metal–metal bond.

Such dimerization does not occur with charged complexes because the electrostatic repulsion between the monomers outweighs any potential stability which would be produced by metal–metal bond formation.

There are exceptions to the 18-electron rule. One is $V(CO)_6$ with five electrons from the vanadium atom and twelve from the six ligands, making a total of 17 valence shell electrons. The compound is paramagnetic but does not form a dimer because of the steric problems which would ensue. It very easily accepts an electron to give the ion, $V(CO)_6^-$, , which does conform to the rule.

8.5 ELECTRONIC SPECTRA OF COMPLEXES

There are two types of electronic transition, characteristic of transition metal complexes, which may be observed in the visible and ultraviolet regions of the spectrum.

(1) 'd–d' transitions, which are concerned with changes in the occupation of the anti-bonding (essentially d character) molecular orbitals localized on the metal centre. These are of low intensity because of their orbital forbiddenness.
(2) Charge transfer transitions, which may be from metal to ligand (MLCT) or ligand to metal (LMCT). These are of high intensity, and are usually observed in

the ultraviolet region of the spectrum of a complex. The long-wavelength tail of such transitions can obliterate the higher-wavelength d–d low-intensity transitions.

1. d–d transitions

These transitions are essentially 'atomic' in nature, and the rules of atomic spectroscopy apply to them. Since they involve no change in the value of the *l* quantum number, the transitions are orbitally forbidden and are observed to have ε values of the order of $10\,l\,mol^{-1}\,cm^{-1}$ — a factor of about 10^3 smaller than the value for a fully allowed transition. Their small intensities are thought to be due to the influence of vibrational changes accompanying the electronic changes. This allows the incorporation of two vibrational wavefunctions in the intensity integral which could arrange for its value to be finite. Such transitions are described as **vibronic**.

The d–d spectrum of a particular complex is very much dependent upon the number of 'd' electrons in the valence shell of the metal centre. The d^1 and d^9 cases are straightforward, the other configurations being more complicated. The d^1, d^9 and d^2 cases are treated in considerable detail in this book.

The spectra of d^1 ions

In an octahedral d^1 ion, the electron in its ground state occupies the t_{2g} level. In its excited state the electron occupies the e_g^* level. In orbital terms the transition may be written as:

$$t_{2g}^1 \rightarrow (e_g^*)^1 \tag{8.18}$$

The d^1 configuration in an atom would give rise to the electronic state, ^2D, and this in an octahedral environment splits (because the orbital degeneracy breaks down) into the two states, $^2T_{2g}$ (the ground state) and ^2E_g (the excited state). In terms of changes in electronic state the transition is written as:

$$^2T_{2g} \rightarrow {}^2E_g \tag{8.19}$$

It should be remembered that such a transition is orbitally forbidden since the triple product, $g \times u \times g$, is u in character, (the dipole moment operators transforming as T_{1u}) and cannot contain the fully symmetric representation (A_{1g}).

Fig. 8.12 shows the visible absorption spectrum of the $[Ti(H_2O)]_6^{3+}$ ion, with its single absorption band with a maximum absorbance at 476 nm and a shoulder at 526 nm. The absorption band may be regarded as the sum of two overlapping bands. The two bands may be interpreted in terms of the Jahn–Teller effect. The e_g level consists of the two degenerate orbitals, d_{z_2} and $d_{x^2-y^2}$, so that the excited state, ^2E_g, in orbital terms could be either $(d_{z^2})^1$ or $(d_{x^2-y^2})^1$. The two ^2E_g states are degenerate in O_h symmetry, and so the Jahn–Teller effect applies. If a distortion of the excited state occurs such that it assumes D_{4h}. symmetry with two long axial bonds and four short bonds in the horizontal plane, the d_{z^2} orbital is stabilized and the $d_{x^2-y^2}$ orbital is destabilized with respect to their energies in O_h symmetry. The bonding along the *z* axis will be weaker so the d_{z^2}, anti-bonding orbital will have a lower energy. The

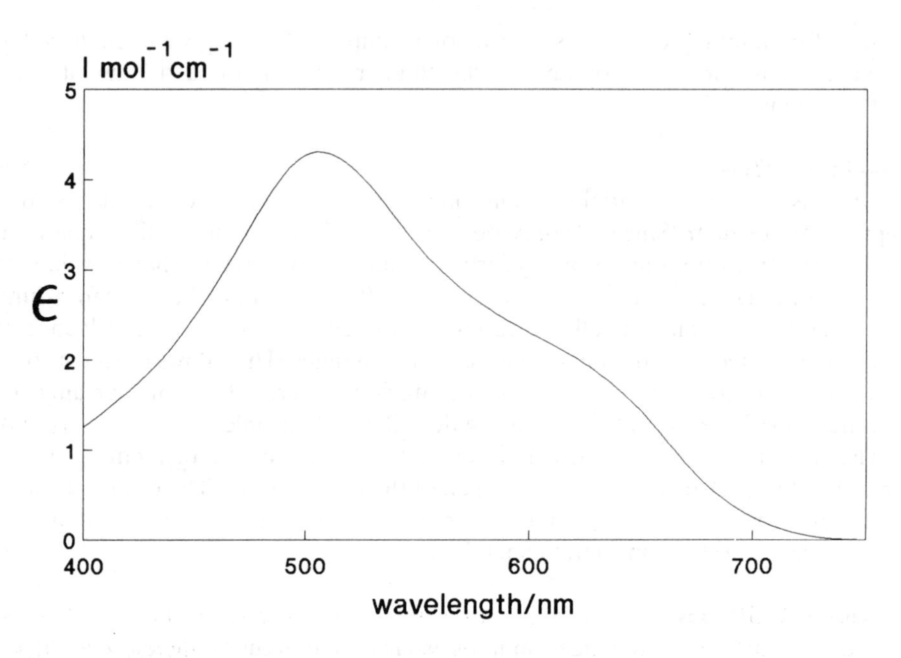

Fig. 8.12 — The electronic spectrum (visible) of the $[Ti(H_2O)_6]^{3+}$ ion.

reverse is true for bonding in the *xy* plane so that the $d_{x^2-y^2}$, anti-bonding energy is raised.

The difference in energy between the two states, ^2E_g and $^2T_{2g}$, is, in the d^1 case, equal to the difference between the energies of the e_g^* and t_{2g} orbitals (there being no interelectronic effects in the one-electron case). That difference is known as $\Delta_{octahedral}$ (shortened to Δ_o in further references) which is called the **ligand field splitting energy**. It is the splitting produced when the ligand environment interacts with the metal orbitals to produce the molecular orbitals of the complex.

If E_s represents the energy of the five-fold degenerate d orbitals in a spherical field, that of the t_{2g} orbitals will be $E_s - x$ (t_{2g} being more stable in an O_h environment) in an octahedral complex, and that of the e_g^* orbitals will be $E_s + y$ (e_g being destabilized in O_h). The difference in energy between the two orbitals in O_h is given by:

$$\Delta_o = x + y \tag{8.20}$$

For the symmetrically half-filled d^5 case the energy in a spherical field is $5E_s$ and the alteration of the ligand field to a localized O_h version does not alter the total energy so that:

$$5E_s = 3(E_s - x) + 2(E_s + y) \tag{8.21}$$

so that

$$0 = -3x + 2y \tag{8.22}$$

or

$$3x = 2y \tag{8.23}$$

Since $y = \Delta_o - x$ equation (8.23) becomes:

$$3x = 2(\Delta_o - x) \tag{8.24}$$

so that

$$x = (2/5)\Delta_o \tag{8.25}$$

and

$$y = (3/5)\Delta_o \tag{8.26}$$

Equations (8.25) and (8.26) imply that, as Δ_o increases, the energy of the t_{2g} orbitals decreases (x increases) and that of the e_g orbitals increases (y increases). These relationships are shown in Fig. 8.13, which is the Orgel diagram for the d^1 case. The

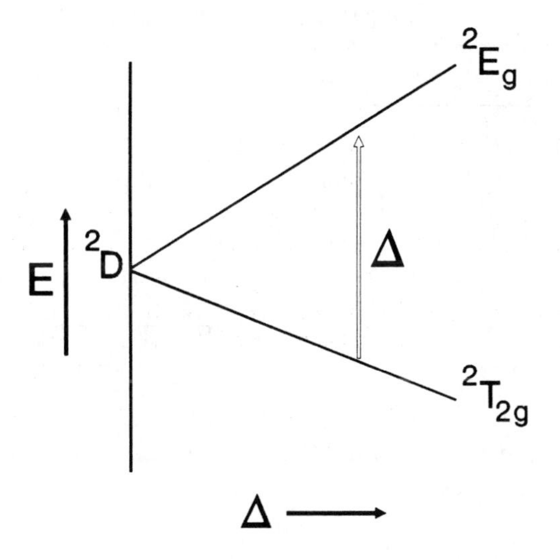

Fig. 8.13 — An Orgel diagram for d^1 complexes.

line defining the stabilization of the t_{2g} orbitals has a slope of $-(2/5)\Delta_o$ and that defining the destabilization of the e_g^* orbitals has a slope of $+(3/5)\Delta_o$, the difference between the two lines at any value of Δ_o being Δ_o. The energy of the transition observed in d^1 complexes is a direct measure of Δ_o (the actual value would be the average of the energies of the two absorptions which are caused by the operation of the Jahn–Teller effect).

The spectra of d^2 ions

Cases involving more than one electron are always much more complicated than those in which there is only one electron to consider. An isolated atom (or ion) with a d^2 configuration gives rise to 45 microstates, which may be sorted into the following terms: ^3F, ^1D, ^3P, ^1G and ^1S.

In an O_h environment, atomic states **split** as a result of the differences in energy between the orbitals. The previous section deals with the splitting of a D state into T_{2g} and E_g states. Just as d orbitals transform as t_{2g} and e_g in O_h symmetry, the splitting of any electronic state may be read off from the appropriate character table (whatever the symmetry of the environment of the central atom). All states arising from d orbitals (which have g symmetry) are g type. In the cases of those orbitals which are u type the 'u' must be changed to 'g'.

For any states with L values larger than 3 (F states) the above procedure is not possible, and the basic wavefunctions have then to be classified. The results for some useful states in O_h symmetry are given in Table 8.9.

Table 8.9 — Splitting of atomic states in O_h symmetry

Atomic state	States in an O_h environment
S	A_{1g}
P	T_{1g}
D	$E_g + T_{2g}$
F	$A_{2g} + T_{1g} + T_{2g}$
G	$A_{1g} + E_g + T_{1g} + T_{2g}$
H	$E_g + 2T_{1g} + T_{2g}$
I	$A_{1g} + A_{2g} + E_g + T_{1g} + 2T_{2g}$

Application of the results of Table 8.9 to the atomic states of the d^2 configuration produces the following states:

$$^3F \rightarrow {}^3A_{2g} + {}^3T_{1g} + {}^3T_{2g}$$
$$^1D \rightarrow {}^1E_g + {}^1T_{2g}$$
$$^3P \rightarrow {}^3T_{1g}$$
$$^1G \rightarrow {}^1A_{1g} + {}^1E_g + {}^1T_{1g} + {}^1T_{2g}$$
$$^1S \rightarrow {}^1A_{1g}$$

When the metal atom is placed in an O_h environment by bonding with six ligands, the anti-bonding d orbitals have different energies, the t_{2g} set being lower than the e_g

set. This allows for there being three configurations of different energies: t_{2g}^1, $t_{2g}^1(e_g^*)^1$ and $(e_g^*)^2$. The lowest-energy t_{2g}^2 configuration gives rise to the states A_{1g}, E_g, T_{1g} and T_{2g} (leaving out their individual multiplicities). Their multiplicities are decided at a later stage. What is certain at this stage is that there are fifteen microstates which represent the t_{2g}^2 configuration. These arise from the six ways of placing one electron into three orbitals (taking the spin of the electron into account), and the five ways of placing the second electron. Such a consideration would result in thirty ways of placing the two electrons but, because of the indistinguishability of electrons, this is reduced to fifteen. If the multiplicities of the states arising from the t_{2g}^2 configuration are represented by w, x, y and z respectively, the total degeneracy (number of microstates) is given by:

$$15 = w + 2x + 3y + 3z \tag{8.27}$$

x being multiplied by 2 because an E_g state is doubly degenerate, and y and z are multiplied by three because the T_{1g} and T_{2g} states are triply degenerate. Since there are only two electrons to be considered, the multiplicities of the states (w, x, y and z) must be either 1 or 3. The possible solutions are given in the following table:

	w	x	y	z
(a)	1	1	1	3
(b)	1	1	3	1
(c)	3	3	1	1

Reference to these solutions is made later.

The doubly excited $(e_g^*)^2$ configuration gives rise to the states:

$$e_g \times e_g = A_{1g} + A_{2g} + E_g \tag{8.28}$$

and is represented by $(4 \times 3)/2 = 6$ microstates. If p, q and r are their respective multiplicities the following equation is produced:

$$6 = p + q + 2r \tag{8.29}$$

and the only solutions are:

	p	q	r
(d)	1	3	1
(e)	3	1	1

implying that the latter state is ^1E_g, the decision about the values of p and q being left until later.

The singly excited $t_{2g}^1(e_g^*)^1$ configuration gives rise to the following states:

$$t_{2g} \times e_g = 2T_{1g} + 2T_{2g} \tag{8.30}$$

consistent with there being $6 \times 4 = 24$ microstates. The two electrons may not occupy the same orbital so that both singlet and triplet, T_{1g} and T_{2g}, states are produced, as shown in Table 8.10.

Table 8.10

State	No. of microstates		
$^1T_{1g}$	1×3	=	3
$^3T_{1g}$	3×3	=	9
$^1T_{2g}$	1×3	=	3
$^3T_{2g}$	3×3	=	9
	Total	=	24

The correlation between the atomic states (as split by their O_h environment) and the states produced from the t_{2g}^2, $t_{2g}^1(e_g^*)^1$ and $(e_g^*)^2$ configurations is shown in Fig. 8.14. The correlation diagram is constructed by the proper application of two rules. One is the rather trivial one that the correlation lines must represent only one state, the other being the **non-crossing rule**. The latter implies that correlation lines of identical states must not cross. The reason for this is that if two states of identical symmetry approach each other (in energy) then they will interact so that they form two new states which are further apart. The same principle is used in the construction of m.o. diagrams.

Starting the consideration of Fig. 8.14 at the lowest energy correlation it is seen that the lowest energy T_{1g} state must be a triplet state, $^3T_{1g}$. That enforces the conclusion that option (a) above contains the correct assignments of multiplicities for the states arising from the t_{2g}^2 configuration.

The $^3T_{2g}$ and $^3A_{2g}$ correlations are unique and are not in doubt. The latter enforces the conclusion that option (d) gives the correct multiplicities for the states arising from the $(e_g^*)^2$ configuration. The $^1T_{2g}$ state arising from the atomic ^1D state correlates with the lowest $^1T_{2g}$ state arising from the t_{2g}^2 configuration (otherwise some forbidden crossing would occur). The two lower ^1E_g states are correlated by the same reasoning.

The second $^3T_{1g}$ states are correlated, there being no alternatives.

The lower $^1A_{1g}$ state, arising from the ^1G atomic state, correlates with that coming from the t_{2g}^2 configuration.

The remaining correlations ($^1T_{2g}$, $^1T_{1g}$, ^1E_g and $^1A_{1g}$) are in no doubt.

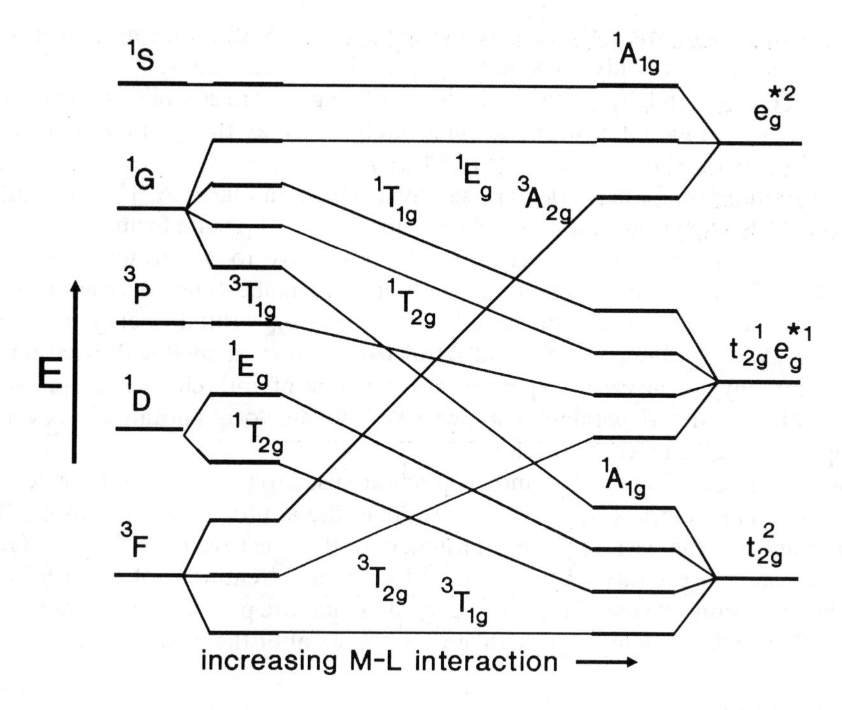

Fig. 8.14 — A diagram showing the splitting of the atomic states of a d^2 configuration in an octahedral environment and the correlation of their energies with the states arising from the t_{2g}^2, $t_{2g}^1(e_g^*)^1$ and $(e_g^*)^2$ configurations appropriate to a strong metal–ligand interaction.

Fig. 8.14 may be used to interpret the spectrum of any d^2 complex. Whatever the strength of the interaction between the metal centre and the six ligands in the O_h complex, the ground state must be $^3T_{1g}$. Any orbitally allowed transitions can only be to higher-energy triplet states. Transitions involving a change in multiplicity are spin forbidden and would not be expected to be observed. That restricts the allowed transitions to three:

(a) $^3T_{1g} \rightarrow {}^3T_{2g}$
(b) $^3T_{1g} \rightarrow {}^3T_{1g}$
(c) $^3T_{1g} \rightarrow {}^3A_{2g}$

The hexaaquavanadium(III) ion is an example of a d^2 complex. In aqueous solution, transition (a) is observed at a wavelength of 588 nm and transition (b) at 400 nm, but transition (c) is overlain by an intense (allowed) charge transfer transition. Transition (c) may be observed in crystalline V_2O_3 (where no charge transfer is possible) at 263 nm.

Diagrams such as the one in Fig. 8.14 are available for all d configurations, but they are not normally used for the interpretation of spectra. There are two modified forms of the diagrams which are to be found in the literature. These are (i) Orgel diagrams and (ii) Tanabe–Sugano diagrams.

An Orgel diagram for the d^2 case is shown in Fig. 8.15. The singlet states are omitted since they are only relevant to spin-forbidden transitions. The relationship between Figs 8.14 and 8.15 is clear, the former being a more complicated version of the latter, only the states with the same multiplicity as the ground state being included.

The appropriate Tanabe–Sugano diagram is shown in Fig. 8.16. The basis of the diagram is that the ground state correlation line for the $^3T_{1g}$ state forms the x axis of the diagram, with the other correlations being relative to the ground state. The quantities which are plotted against each other in a Tanabe–Sugano diagram — the energy of the state and the magnitude of the $e_g^* - t_{2g}$ orbital energy difference (known as $\Delta_{octahedral}$) — are each divided by what is known as the Racah B parameter which, for any given case, expresses the amount of interelectronic repulsion. Specification of the B parameter allows one diagram to quantitatively cover all examples for a given d configuration.

Both Orgel and Tanabe–Sugano diagrams are used to (i) interpret the spectra of transition metal complexes, and (ii) to estimate the appropriate value of Δ_o. The latter quantity is the basis of the establishment of the spectrochemical series. Orgel diagrams are more easily understood, and a general treatment of them follows. Simplified versions of some Tanabe–Sugano diagrams are presented in Appendix II. The further understanding of such diagrams follows after the next section.

Orgel diagrams

All Orgel diagrams depend for their understanding on the principles outlined above and, in particular, on the details of the d^1 case. The ground state of the d^2 configuration, ^3F, splits into the three states, $^3T_{1g}$, $^3T_{2g}$ and $^3A_{2g}$, in an octahedral environment, which correlate with the configurations, t_{2g}^2, $t_{2g}^1\ (e_g^*)^1$ and $(e_g^*)^2$, respectively. Such information allows the **slopes** of the correlation lines to be inferred from those for one t_{2g} electron and one e_g^* electron. The line for the ground term, $^3T_{1g}(t_{2g}^2)$, will have a slope which is $-2 \times (2/5)\Delta_o = -(4/5)\Delta_o$, because of there being **two** t_{2g} electrons. The line for the first excited state, $^3T_{2g}(t_{2g}^1(e_g^*)^1)$ (one electron in each orbital), has a slope given by:

$$\text{Slope}(^3T_{2g}) = -(2/5)\Delta_o + (3/5)\Delta_o = +(1/5)\Delta_o \tag{8.31}$$

The line for the second excited state, $^3A_{2g}(e_g^*)^2$, has a slope given by:

$$\text{Slope}(^3A_{2g}) = +2 \times (3/5)\Delta_o = +(6/5)\Delta_o \tag{8.32}$$

These lines are shown in Fig. 8.15 together with the line representing the variation with Δ_o of the $^3T_{1g}$ state which is derived from the atomic ^3P state. The ^3P state is unaffected by the value of Δ_o, and is shown with a positive slope which represents the increasing difference between the atomic ^3P and ^3F states as the metal–ligand interaction increases.

Since there are two $^3T_{1g}$ states there is the possibility that they may interact with the stabilization of the lower state and the destabilization of the upper one. The extent of the interaction depends upon the difference in energy between the two

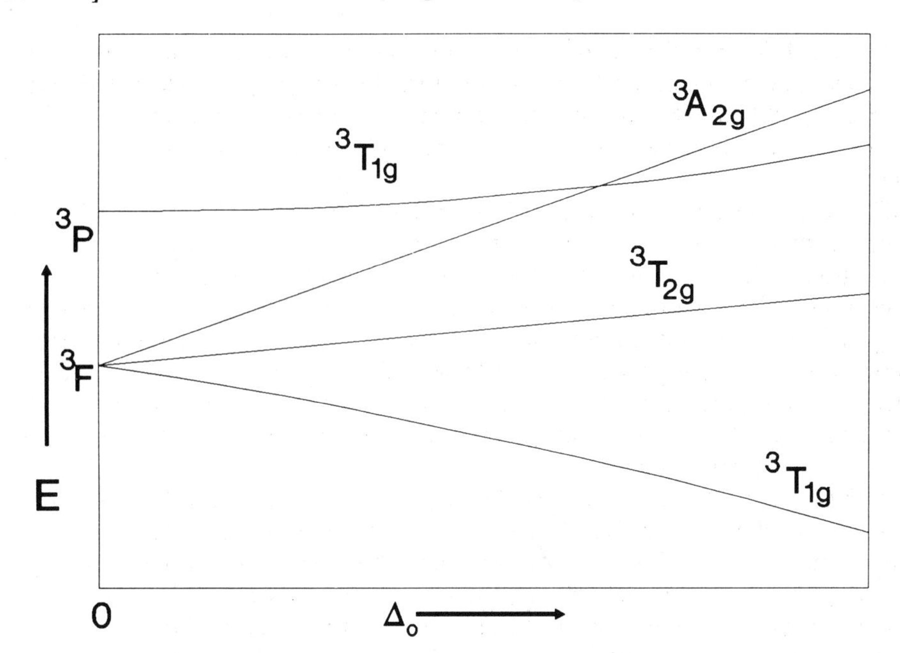

Fig. 8.15 — An Orgel diagram for d^2 octahedral complexes.

states. The effects of the interaction cause both $^3T_{1g}$ states to be curved, as shown in Fig.8.15. There are no such complications with the $^3T_{2g}$ and $^3A_{2g}$ states, and this is important since the difference between them amounts to $(6/5)\Delta_o - (1/5)\Delta_o = \Delta_o$ at any value of Δ_o. This means that Δ_o can be estimated as the difference in energy between the two transitions $^3T_{1g} \rightarrow {^3T_{2g}}$ and $^3T_{1g} \rightarrow {^3A_{2g}}$, which may be the first and second transitions or the first and third transitions (in terms of increasing energy, decreasing wavelength), depending upon whether Δ_o is large enough to allow the $^3A_{2g}$ state to lie above the higher of the two $^3T_{1g}$ states.

The two diagrams in Fig 8.13 and 8.15 may be extended in a relatively simple way to cover almost all the cases of d electron configurations in either octahedral or tetrahedral complexes. The d^1 case in a tetrahedral coordination is the opposite of that in octahedral coordination in that the e* orbitals have lower energy than do the t_2. The value of $\Delta_{tetrahedral}$ (Δ_t in the following text) has the value $(4/9)\Delta_o$ for a given ligand, as may be confirmed by carrying out an angular overlap calculation for the two symmetries. The ground state for the T_d environment is ^2E, the excited state being $^2T_2$. As Δ_t increases, the ^2E state is stabilized, with the line having a slope of $-(3/5)\Delta_t$. The excited state is similarly destabilized, with a slope of $+(2/5)\Delta_t$.

The d^9 (O_h) case is very similar to that of the $d^1(T_d)$ configuration. The ground state, $^2E_g(t_{2g}^6(e_g^*)^3)$, is associated with a slope of stabilization given by:

$$\text{Slope}(^2E_g) = -6 \times (2/5)\Delta_o + 3 \times (3/5)\Delta_o$$

$$= -(3/5)\Delta_o \tag{8.33}$$

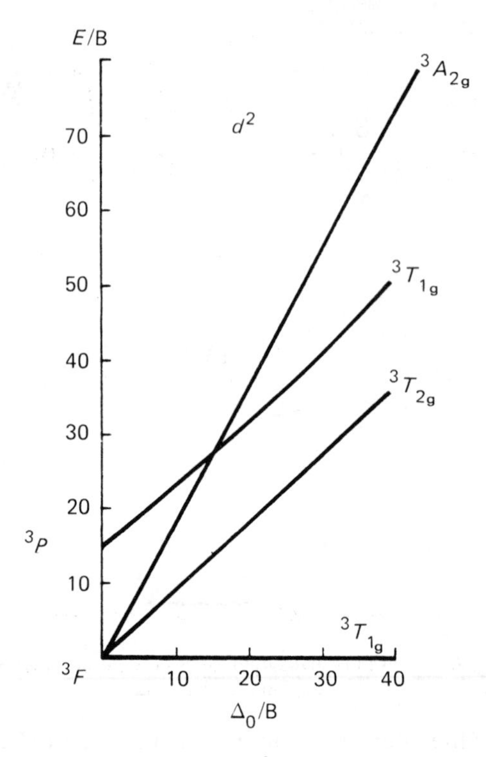

Fig. 8.16 — A Tanabe–Sugano diagram for d^2 octahedral complexes. (After Tanabe and Sugano, *J. Phys. Soc. Japan*, **9**, 753, (1954).)

The excited state, $^2T_{2g}(t_{2g}^5(e_g^*)^4)$, has a slope of destabilization given by:

$$\text{Slope}(^2T_{2g}) = -5 \times (2/5)\Delta_o + 4 \times (3/5)\Delta_o$$

$$= +(2/5)\Delta_o \tag{8.34}$$

It should be noticed that the Orgel diagrams for the $d^9(O_h)$ and d^1 (T_d) cases are (apart from the scale and the absence of the g subscripts in the T_d case) the inverse of that for the $d^1(O_h)$ configuration. The splittings in O_h and T_d environments of atomic states (the ingredients for the appropriate Orgel diagrams) and whether the d configurations are d^n or d^{10-n} have the following general relationships:

$$d^n(O_h) = d^{10-n}(T_d) = \text{inverse of } d^{10-n}(O_h) \tag{8.35}$$

The majority of spectra may be interpreted via the diagrams in Figs 8.17 and 8.18, which are based upon the above generalizations. The cases involving spin pairing induced by high values of Δ_o are considered in the next section.

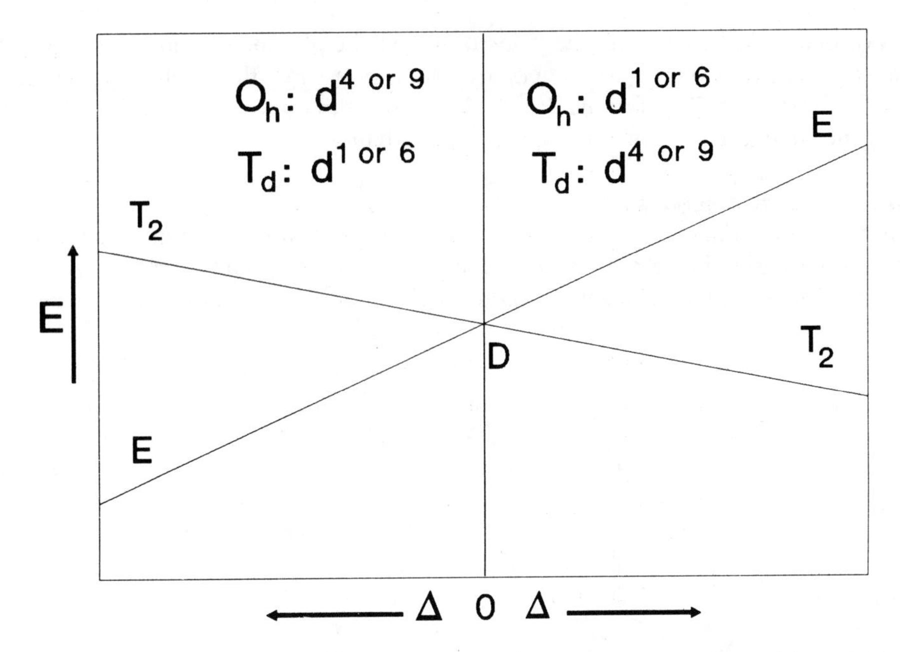

Fig. 8.17 — An Orgel diagram for O_h and T_d complexes with one, four (high spin), six (high spin) or nine d electrons.

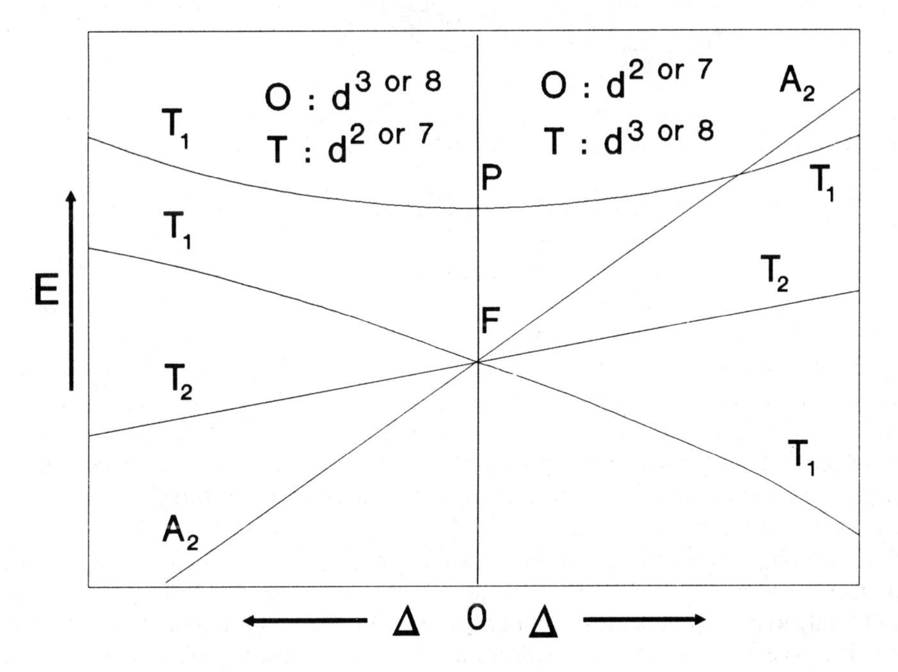

Fig. 8.18 — An Orgel diagram for O_h and T_d complexes with two, three, seven (high spin) or eight d electrons.

For octahedral complexes the value of Δ_o may be obtained as the energy of the lowest energy (longest wavelength) transition in the following cases: $d^{1,3,4,6,7\,(\text{high-spin}),8\text{ and }9}$. For the $d^{2,7\,(\text{low-spin})}$ cases, the value of Δ_o is obtained as the difference in energy between the first and third bands.

Tanabe–Sugano diagrams

Tanabe–Sugano diagrams for 3, 5, 6, 7 and 8 d electrons are shown in Appendix II. Those for 1 and 9 d electrons are too similar to the appropriate Orgel diagrams to include. The d^4 Tanabe—Sugano diagram is shown in Fig. 8.19 and is selected as it is

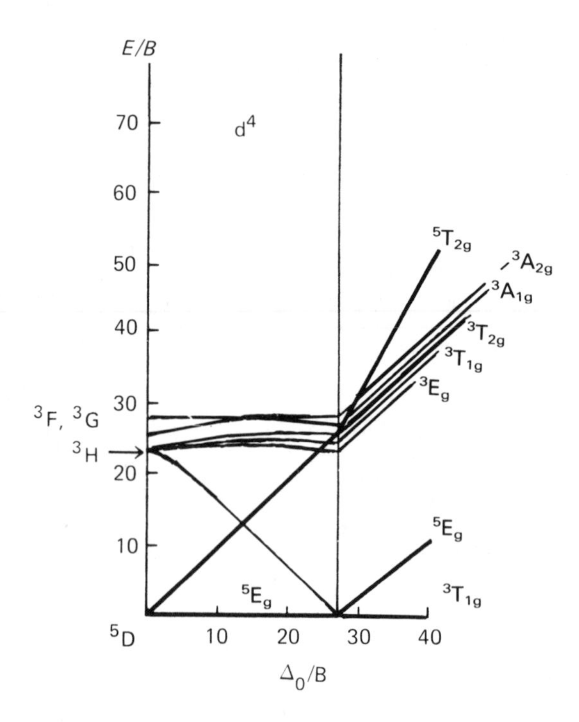

Fig. 8.19 — A Tanabe–Sugano diagram for d^4 octahedral complexes. (After Tanabe and Sugano, *J. Phys. Soc. Japan*, **9**, 754, (1954)).

the simplest example which includes a change of ground state multiplicity as Δ_o increases. At large values of Δ_o spin pairing is induced, causing the ground state of the complex to be based upon the t_{2g}^4 configuration instead of the $t_{2g}^3(e_g^*)^1$ configuration, which is more stable at lower values of Δ_o. The $t_{2g}^3(e_g^*)^1$ configuration produces an atomic ground state, ^5D, which splits into ^5E_g and $^5T_{2g}$ states in an octahedral environment, the former being the lower in energy. Higher-energy states arise from triplet and singlet atomic states, and retain those multiplicities in O_h symmetry. At sufficiently high values of Δ_o (indicated on the diagram by a vertical line) to cause some electron pairing, the $^3T_{1g}$ state (derived from the atomic ^3H state) becomes the ground state. At these high values of Δ_o the spectrum of a d^4 complex consists of triplet–triplet transitions. The singlet states and higher triplet states are omitted from the simplified diagram. They have no relevance to the observed spectra of complexes.

In the midst of all the colourful chemistry of the transition elements there is the virtually colourless chemistry of the commonest (II) oxidation state of manganese. Manganese(II) ions have a d^5 configuration and are normally of high-spin $t_{2g}^3(e_g^*)^2$ so that the ground states are ^6S. Any electronic transition **must** alter the multiplicity of the state and is therefore spin forbidden.

2. Charge transfer spectra

The term 'charge transfer' implies the movement of an electron from one part of a complex to another. Charge transfer transitions occur as the result of the absorption of a photon of suitable energy to cause such a movement of an electron. The movement of charge may be either (i) from an orbital largely localized on the ligands to one largely localized on the metal (LMCT) or (ii) from the reverse process (MLCT).

In a complex there are bonding σ-type orbitals largely composed of ligand group orbitals with only a minor contribution from any metal orbitals. In addition there is the possibility of there being π-type ligand orbitals which may interact with the t_{2g} orbitals of the metal to give bonding and anti-bonding combinations. In such a case the t_{2g} orbitals become anti-bonding (with respect to the metal-ligand interaction). If the ligands possess anti–bonding π-type orbitals, interaction with the t_{2g} metal orbitals causes the stabilization of the latter.

Ligand–metal charge transfer transitions

The diagram of Fig. 8.20 shows the four types of possible LMCT transition, which may be classified as follows.

(i) $L(\sigma \rightarrow M(t_{2g})$
(ii) $L(\sigma) \rightarrow M(e_g^*)$
(iii) $L(\pi) \rightarrow M(t_{2g})$
(iv) $L(\pi) \rightarrow M(e_g^*)$

Classes (i) and (iii) are impossible if there is a t_{2g}^6 configuration and are limited to those complexes with fewer than six t_{2g}electrons. None of the transitions is possible for d^{10} complexes. The transitions are usually of high intensity since the ligand group orbital combinations contain u-type representations.

Metal–ligand charge transfer transitions

The diagram of Fig. 8.20 shows the two classes of possible metal–ligand charge transfer transition:

(i) $M(e_g^*) \rightarrow L(\pi^*)$
(ii) $M(t_{2g}) \rightarrow L(\pi^*)$

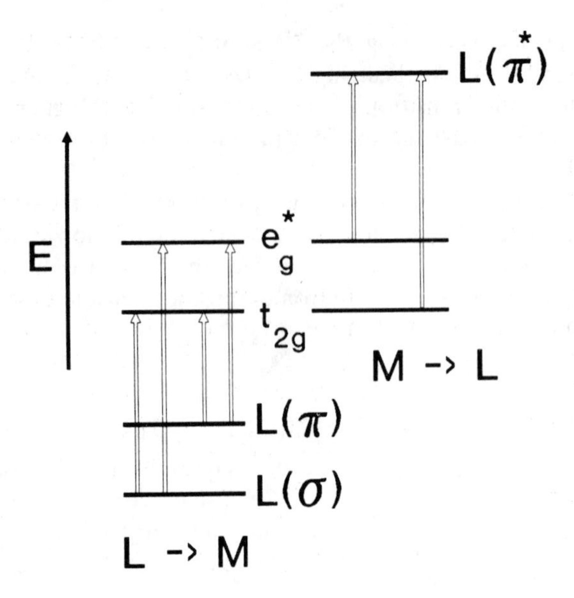

Fig. 8.20 — Possible charge transfer transitions.

which may occur. These transitions are available in those complexes in which there is π-type back bonding (unsaturated ligands). The existence of class (i) depends upon the anti-bonding e_g^* orbitals being occupied in the ground state. As the π^*-type group orbitals contain u-type representations, the MLCT transitions are allowed and are observed to have high intensities.

The detailed interpretations of charge transfer spectra are complicated and are not pursued any further in this book. The complications arise from two sources. One is that the ligand orbitals (π andor π^*) may form group orbitals so that several states are produced. The other is that some of the charge transfer bands may be obscured by π–π^* transitions between various ligand orbitals. Two examples of CT spectra follow.

The hexaiodoosmium(IV) ion, OsI_6^{2-}, exhibits. intense absorptions at 862–538 nm, 373–281 nm and 224 nm, which are thought to be π–t_{2g}, π–e_g^* and σ–e_g^* LMCT transitions respectively. The tris-1,10-phenanthrolineiron(II) ion, $[Fe(phen)_3]^{2+}$, exhibits intense absorptions at 520 nm, 320 nm and 290 nm. They are MLCT transitions. Any higher-energy MLCT transitions are obscured by the two intra-ligand π–π^* transitions of the ligand system which are observed at 229 and 263 nm (compare these absorptions with those given for the free ligand in Fig. 6.10).

8.6 MAGNETIC PROPERTIES OF COMPLEXES

The operation of molecular orbital theory allows the prediction of the number of unpaired electrons to be made for any complex. The actual number for any d configuration depends upon the value of Δ_o (for octahedral complexes) relative to the **pairing energy**, P, which represents the interelectronic repulsion energy which has to be overcome for electron pairing to occur. If $P > \Delta_o$ then pairing will not occur and a **high spin** complex results. A **low spin** complex is formed if $\Delta_o > P$. For octahedral complexes there are significant differences observable for the d^{4-7} configurations. High-spin and low-spin complexes have different d–d absorption spectra, as has been dealt with above. The other area of experimentation in which differences are observable is the determination of magnetic moments of such complexes.

The magnetic moment of a compound depends upon the number of unpaired electrons (with parallel spins) and whether their orbital contributions are quenched or not. The orbital contribution may be quenched by an octahedral environment. The orbital contribution arises from the availability of another orbital (to the one containing the electron) which the electron may occupy (without violating the Pauli exclusion principle) by a rotatory motion. A single electron in one of the t_{2g} orbitals can undertake such motion, and a t_{2g}^1 configuration would be expected to give an orbital contribution to the magnetic moment of the complex. A t_{2g}^2 configuration would also give an orbital contribution. A t_{2g}^3 configuration would not be able to contribute orbitally to the magnetic moment since rotation from one orbital to another by one electron would violate the Pauli principle. By similar arguments it may be decided that t_{2g}^4 and t_{2g}^5 configurations contribute, but the t_{2g}^6 one does not. High-spin configurations, $t_{2g}^4(e_g^*)^2$ and $t_{2g}^5(e_g^*)^2$, may also contribute to the orbital fraction of the magnetic moment. The magnetic moment, μ, of a compound is given by the equation:

$$\mu = g.\mu_B.(S(S+1) + L(L+1))^{1/2} \tag{8.36}$$

where g is the **magnetogyric** or **gyromagnetic ratio**, μ_B is the Bohr magneton ($eh/4\pi m$), and S and L are the total spin and angular momentum quantum numbers of the state (m is the rest mass of the electron). g has the value:

$$g = 1 + \left(\frac{J(J+1) - L(L+1) + S(S+1)}{2J(J+1)}\right) \tag{8.37}$$

where J is the appropriate value for the interaction of L and S for the state.

For a fully orbitally quenched situation the value of g reduces to 2 (strictly 2.00023 because of a relativistic effect) and this, together with the absence of the L term from equation (8.36), produces the spin-only formula:

$$\mu = 2\mu_B(S(S+1))^{1/2} \tag{8.38}$$

and since S is given by $n/2$, n being the number of unpaired electrons, the equation may be modified to:

$$\mu = \mu_B(n(n+2))^{1/2} \tag{8.39}$$

The expected spin-only values for the magnetic moments of complexes with one to five unpaired electrons are shown in Table 8.11 below (in terms of Bohr magnetons).

Table 8.11 — Spin-only magnetic moments

No. of unpaired electrons	μ/μ_B
1	1.73
2	2.83
3	3.87
4	4.90
5	5.92

If there are orbital contributions to the magnetic moment, the equation becomes:

$$\mu = (1 - \lambda/\Delta_o)\mu_B(n(n+1))^{1/2} \qquad (8.40)$$

where λ is the **spin-orbit coupling parameter**. The value of λ for a given complex may be either positive (if the electron shell is less than half full) or negative (if the electron shell is more than half full). The orbital contribution is distinguishable from the spin-only contribution by its temperature dependence, the spin-only moment being independent of temperature.

8.7 CONCLUDING SUMMARY

This chapter contains extensive treatments of the following topics.

(1) The thermodynamic stability of complexes.
(2) The application of molecular orbital theory to the bonding in octahedral, square planar and tetrahedral complexes.
(3) The extension of molecular orbital theory in the form of the angular overlap approximation to produce predictions of possible symmetries and stoichiometries for any given d orbital configuration of the metal atom.
(4) The d–d (Orgel and Tanabe–Sugano diagrams) and charge transfer electronic spectroscopy of complexes.
(5) The magnetic properties of complexes.

9

Thermodynamics of ions in solution

9.1 INTRODUCTION

This chapter covers the factors which influence the chemical form and stability of ions (including complex ions) in solution with respect to their formation and with respect to their reactivities with either water or other ions. Stability with respect to formation is discussed in terms of $\Delta G^\ominus$ and $E^\ominus$, the appropriate **standard reduction potential**. The reactivity of ions is discussed in terms of $E^\ominus$ only.

9.2 THERMODYNAMICS AND ELECTRODE POTENTIALS

This section deals with the important definitions, equations and conventions, which are necessary for the understanding of the stabilities of ions in solution.

The basis of the equation:

$$\Delta G^\ominus = -RT \ln K \tag{9.1}$$

is dealt with fully in Chapter 7, and is of primary importance.

It is convenient to express the change in standard Gibbs energy, $\Delta G^\ominus$, for any reaction, as the difference between the two standard reduction potentials, for the two 'half-reactions' which, when added together, produce the overall equation. For example, the overall equation:

$$Cu^{2+}(aq) + Zn(s) \longrightarrow Cu(s) + Zn^{2+}(aq) \tag{9.2}$$

may be considered to be the sum of the two half-reactions:

$$Cu^{2+}(aq) + 2e^- \longrightarrow Cu(s) \tag{9.3}$$

and

$$Zn(s) \rightarrow Zn^{2+}(aq) + 2e^- \tag{9.4}$$

where the 'state' of the electrons is not defined — they cancel out when the two half-reactions are summed.

Reaction (9.2) forms the basis of the classical Daniell cell (Daniell was the first professor of chemistry at King's College London) with its copper anode in contact with aqueous copper(II) ions separated by a porous membrane from the zinc cathode in contact with aqueous zinc(II) ions. A diagram of the cell is shown in Fig. 9.1. The

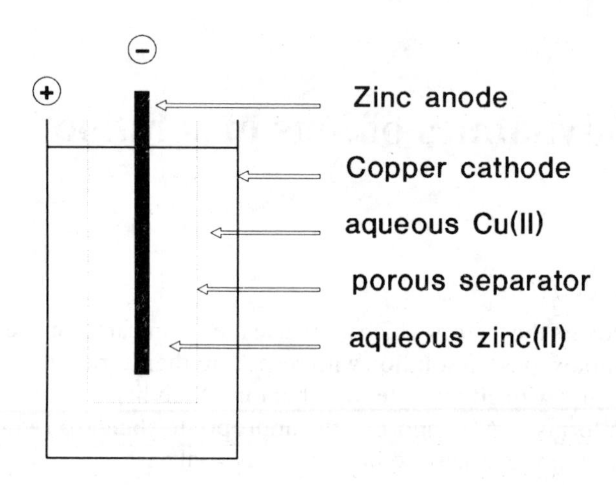

Fig. 9.1 — A diagrammatic representation of a Daniell cell.

maximum voltage of a Daniell cell is 1.1 V. Although there is the possibility of what is known as a **junction potential** it is conventional to divide the cell potential into contributions from the two electrode systems.

The vital equation to remember as the basis of understanding the subject of electrode potentials is that which expresses the relationship between the standard electrode potential, $E^{\ominus}$, and the change in standard Gibbs energy for any reaction:

$$\Delta G^{\ominus} = -nFE^{\ominus} \tag{9.5}$$

where n is the number of electrons involved in the process, and F is the Faraday — the conversion factor with the value: 96487 C mol^{-1}, (one volt-coulomb = one joule).

The value of $\Delta G^{\ominus}$ for reaction (9.2) is -212.3 kJ mol^{-1} corresponding to the observed maximum voltage for the Daniell cell of 1.1 V. The standard electrode potential for the reaction may be considered to be made up from the **difference** between the values of the standard **reduction** potentials of the two half-reactions, (9.3) and (9.4):

$$E^{\ominus} = E^{\ominus}(Cu^{2+}/Cu) - E^{\ominus}(Zn^{2+}/Zn) \tag{9.6}$$

the appropriate reductions being shown for the two half-reaction potentials.

Consider the meaning of equation (9.6) in terms of $\Delta G^\ominus$ values. The value of $\Delta G^\ominus$ for the reduction of Cu^{2+} to Cu ((9.3)) is:

$$\Delta G^\ominus(Cu^{2+}/Cu) = -2 \times F \times E^\ominus(Cu^{2+}/Cu) \quad (9.7)$$

and the value for the reduction of Zn^{2+} to Zn (the reverse of equation (9.4)) is:

$$\Delta G^\ominus(Zn^{2+}/Zn) = -2 \times F \times E^\ominus(Zn^{2+}/Zn) \quad (9.8)$$

the values of $n=2$ being included because the half-reactions are two-electron reductions.

Equation (9.2) may be constructed from its constituent half-reactions (9.3) and (9.4), the appropriate value for its $\Delta G^\ominus$ being the sum of the $\Delta G^\ominus$ values for the half-reactions:

$$\begin{aligned} \Delta G^\ominus\ (9.2) &= \Delta G^\ominus\ (9.3) + \Delta G^\ominus\ (9.4) \\ &= -2F \times E^\ominus(Cu^{2+}/Cu) + 2F \times E^\ominus(Zn^{2+}/Zn) \\ &= -2F(E^\ominus(Cu^{2+}/Cu) - E^\ominus(Zn^{2+}/Zn)) \end{aligned} \quad (9.9)$$

It is most important to notice that the **sign** of $E^\ominus(Zn^{2+}/Zn)$ in equation (9.9) is the reverse of that in equation (9.8) since equation (9.9) is relevant to the oxidation of Zn to Zn^{2+}.

The standard potential for the overall reaction (9.2) is calculated from equation (9.9) (applying equation (9.5)) as:

$$-\Delta G^\ominus\ (9.2)/2F = E^\ominus(Cu^{2+}/Cu) - E^\ominus(Zn^{2+}/Zn) \quad (9.10)$$

— the difference between the two standard reduction potentials.

Since neither of the two half-reaction potentials is known it is necessary to propose an arbitrary zero relative to which all half-reaction potentials may be quoted. The half-reaction chosen to represent the arbitrary zero is:

$$H^+(aq) + e^- \longrightarrow \tfrac{1}{2}\,H_2(g) \quad (9.11)$$

where the state of the electron is not defined. The standard reduction potential for the reaction is taken to be zero:

$$E^\ominus(H^+/H_2) = 0 \quad (9.12)$$

Equation (9.12) applies strictly to the standard conditions where the activity of the hydrogen ion is unity and the pressure of the hydrogen gas is 1 atmosphere at 298 K. It is conventional to quote **reduction** potentials for the general half-reaction:

$$\text{Oxidized form} + ne^- \rightarrow \text{Reduced form} \quad (9.13)$$

The $E^{\ominus}$ value for the reaction:

$$H^{+}(aq) + e^{-}(aq) \longrightarrow \tfrac{1}{2} H_2(g) \tag{9.14}$$

with the state of the electron being defined as aqueous may be estimated to be 2.6 V (see Chapter 10 for the calculation). If reaction (9.14) is reversed and added to reaction (9.11) the result is:

$$e^{-} \longrightarrow e^{-}(aq) \tag{9.15}$$

the appropriate value of $E^{\ominus}$ being -2.6 V, so allowing the hydrated electron to be compared to any other reducing agent on the arbitrary scale.

The values of $E^{\ominus}$ for half-reactions on the hydrogen scale are determined either electrochemically or by thermochemical methods, and are tabulated in data books. This allows the storage of a great amount of thermodynamic data, for n half-reactions may be arranged in pairs to give $n(n-1)/2$ different reactions — fifty values could describe at least $(50 \times 49)/2 = 1225$ reactions, for example.

The accepted values of the standard reduction potentials for the Cu^{2+}/Cu and Zn^{2+}/Zn **couples** are 0.34 V and -0.76 V respectively. The standard potential for the Daniell cell is thus:

$$E^{\ominus}\text{ (Daniell cell)} = 0.34 - (-0.76) = 1.1 \text{ V} \tag{9.16}$$

The overall standard potential for the reaction:

$$2Cr^{2+} + Cu^{2+} \longrightarrow Cu + 2Cr^{3+} \tag{9.17}$$

is calculated from the standard reduction potentials for the constituent half-reactions, equation (9.3), and:

$$Cr^{3+} + e^{-} \longrightarrow Cr^{2+}, \ E^{\ominus} = -0.41 \text{ V} \tag{9.18}$$

as:

$$E^{\ominus}\text{ (9.17)} = 0.34 - (-0.41) = 0.85 \text{ V} \tag{9.19}$$

the chromium half-reaction potential having its sign reversed because the half-reaction itself is reversed to make up the overall equation (9.17). Notice that the 2:1 stoichiometry of equation (9.17) has no effect upon the calculation of the overall $E^{\ominus}$ value. This is because $E^{\ominus}$ values are values of the changes in standard Gibbs energy

per electron, and since the number of electrons cancels out in an overall reaction it has no influence upon the calculation. This is not so for the calculation of the overall change in the standard Gibbs energy which, for reaction (9.17), is given by:

$$\Delta G^{\ominus}\ (9.17) = -nFE^{\ominus}\ (9.17) = -2F \times 0.85\ \text{V} = -164\ \text{kJ mol}^{-1} \quad (9.20)$$

the negative value indicating that the. reaction is **thermodynamically feasible**. Thermodynamic feasibility for any process is indicated by either a **negative** $\Delta G^{\ominus}$ value or a **positive** $E^{\ominus}$ value. A combination of equations (9.1) and (9.5) allows the equilibrium constant of a reaction to be expressed as:

$$K_{eq} = e^{nFE^{\ominus}/RT} \quad (9.21)$$

which makes the point that a positive value of $E^{\ominus}$ gives rise to a value of K_{eq} which is greater than unity.

It is often necessary to calculate the $E^{\ominus}$ value of a half-reaction from two or more values for other half-reactions. In such cases the above considerations do not apply, and care must be taken to include the number of electrons associated with each half-reaction. For example, for the two half-reactions:

$$Cr(VI) + 3e^- \longrightarrow Cr^{3+} \quad (9.22)$$

equation (9.18) has $E^{\ominus}$ values of 1.38 V and -0.41 V respectively. When added together they give the half-reaction:

$$Cr(VI) + 4e^- \longrightarrow Cr^{2+} \quad (9.23)$$

but its $E^{\ominus}$ value is not the simple sum of the two $E^{\ominus}$ values for equations (9.18) and (9.22). The calculation of $E^{\ominus}$ (9.23) via the value of $\Delta G^{\ominus}$ illustrates why this is so.

$$\begin{aligned}\Delta G^{\ominus}\ (9.23) &= \Delta G^{\ominus}\ (9.18) + \Delta G^{\ominus}\ (9.22)\\ &= -F \times E^{\ominus}\ (9.18) - 3F \times E^{\ominus}\ (9.22)\\ &= -F(E^{\ominus}\ (9.18) + 3 \times E^{\ominus}\ (9.22)) \quad (9.24)\end{aligned}$$

The $E^{\ominus}$ (9.23) value is then given by:

$$\begin{aligned}E^{\ominus}\ (9.23) &= -\Delta G^{\ominus}\ (9.23)/4\text{F}\\ &= (E^{\ominus}\ (9.18) + 3 \times E^{\ominus}\ (9.22))/4\\ &= (-0.41 + 3 \times 1.38)/4\\ &= 0.93\ \text{V}\end{aligned}$$

The final $E^{\ominus}$ value is the sum of the individual $E^{\ominus}$ values suitably weighted by their respective numbers of electrons. This has to be so because the electrons do not cancel out in such cases.

The above facility allows for even more equations than $n(n-1)/2$ for any value of n to be represented by a table of n $E^{\ominus}$ values!

On the hydrogen scale the range of values for standard reduction potentials varies from about 3.0 to -3.0 V for solutions in which the activity of the hydrogen ion is 1 M. It is also conventional to quote values for standard reduction potentials in solutions with unit activity of hydroxide ion $-E_B^\ominus$. Typical values range from about 2.0 to -3.0 V.

Examples of half-reactions at each end of the $E^\ominus$ and $E_B^\ominus$ ranges are:

$$\tfrac{1}{2}F_2(g) + e^- \longrightarrow F^-(aq), \quad E^\ominus = 2.65 \text{ V} \tag{9.25}$$

$$Li^+(aq) + e^- \longrightarrow Li(s), \quad E^\ominus = -3.04 \text{ V} \tag{9.26}$$

$$O_3(g) + H_2O(l) + 2e^- \longrightarrow O_2(g) + 2OH^-(aq), \quad E_B^\ominus = 1.24 \text{ V} \tag{9.27}$$

$$Ca(OH)_2(s) + 2e^- \longrightarrow Ca(s) + 2OH^-(aq), \quad E_B^\ominus = -3.03 \text{ V} \tag{9.28}$$

Bearing in mind the general equation (9.13) for a half-reaction and considering the thermodynamic implications of $E^\ominus$ values it is possible to state the following generalizations.

(1) A highly negative value of $E^\ominus$ implies that the reduced form of the couple is a good reducing agent.
(2) A large positive value of $E^\ominus$ implies that the oxidized form of the couple is a good oxidizing agent.

9.3 FORMS OF IONS IN AQUEOUS SOLUTION

The form of any ion in aqueous solution depends largely upon the size and charge (oxidation state) of the central atom. These two properties influence the extent of any interaction with the solvent molecules. A large singly charged ion is likely to be simply hydrated—surrounded by a small number (4 to 10) of water molecules some of which may be formally bonded to it (as in the case of a transition metal ion). The interaction of gaseous ions with water to give their hydrated versions causes the liberation of **enthalpy of hydration**, ΔH_{hyd}, the negative value, $-\Delta H_{hyd}$, being quoted as the **hydration energy** (a positive quantity).

Forms of ions other than the simply hydrated ones may be understood in terms of the increasing interaction between the central ion and its hydration sphere as its oxidation state increases and its size decreases.

The first stage of **hydrolysis** may be written as:

$$M(H_2O)^{n+} \longrightarrow MOH^{(n-1)+} + H^+ \tag{9.29}$$

implying that a metal–oxygen bond is formed with the release of a proton into the solution. As the value of n (the oxidation state of M) increases, the next stage of hydrolysis would be expected to occur:

$$MOH(H_2O)^{(n-1)+} \longrightarrow M(OH)_2^{(n-2)+} + H^+ \tag{9.30}$$

There is then the possibility that water could be eliminated from the two OH^- groups:

$$M(OH)_2^{(n-2)+} \longrightarrow MO^{(n-2)+} + H_2O \tag{9.31}$$

Further hydrolyses involving more pairs of water molecules would yield the oxo-ions, $MO_2^{(n-4)+}$, $MO_3^{(n-6)+}$ and $MO_4^{(n-8)+}$.

It may be envisaged that water molecules could be eliminated **between** two hydrolysed ions to give a dimeric product:

$$2MOH^{(n-1)+} \longrightarrow MOM^{2(n-1)+} + H_2O \tag{9.32}$$

Further hydrolysis of the dimeric product could yield ions such as $O_3MOMO_3^{2(n-7)+}$. In some cases it is possible that further linkage takes place to give polymeric ions.

Examples of these forms of ions are to be found in the transition elements and in those main group elements which can exist in higher oxidation states. As may be inferred from equations (9.29) and (9.30), alkaline conditions encourage hydrolysis, so that the form an ion takes is dependent upon the acidity of the solution. Highly acid conditions tend to depress the tendency of an ion to undergo hydrolysis.

Table 9.1 contains some examples of ions of different form, the form depending upon the oxidation state of the central element.

Table 9.1 — Some ionic forms of vanadium

Oxidation state	Acid solution	Basic solution
(II)	V^{2+}	VO
(III)	V^{3+}	V_2O_3
(IV)	VO^{2+}	$V_2O_5^{2-}$
(V)	VO_2^+	VO_4^{3-}

In acidic solution the (II) and (III) oxidation states are simple hydrated ions, the (IV) and (V) states being oxocations. In basic solution the (II) and (III) states form neutral insoluble oxides (the lattice energies of the oxides giving even more stability than any soluble form in these cases) and the (IV) and (V) states exist as oxyanions (dimeric in the (IV) case).

Simple cations such as $Fe(H_2O)^{3+}$ undergo a certain amount of primary hydrolysis depending upon the pH of the solution. The ion is in the hexaaqua-form, only at pH values lower than 2.0. Above that value the hydroxypentaaquairon(III) ion is prevalent. Further increase in the pH of the solution produces more hydrolysis until $Fe(OH)_3$ is precipitated.

With some ions the extent of polymer formation is high, even in acidic solution, e.g. vanadium(V) can exist as $V_{10}O_{28}^{6-}$.

The above ideas are not limited to species with central metal ions. They apply to the higher oxidation states of non-metallic elements. Many simple anions do exist with primary solvation spheres in which the positive ends of dipoles are attracted to the central negative charge. Examples of ions which may be thought about in terms of the hydrolysis of the parent hydrated ions are SO_3^{2-}, SO_4^{2-}, $S_2O_7^{2-}$, ClO^-, ClO_2^-, ClO_3^-, ClO_4^-, and the vast number of different polymeric silicate ions.

9.4 FACTORS AFFECTING THE MAGNITUDE OF $E^{\ominus}$

This section contains considerations of the trends in $E^{\ominus}$ values for:

(i) the reduction of Group 1 unipositive cations to the metal, and a comparison with the reduction of silver(I) ion to silver(O).
(ii) the reduction of Group 17 elements to their uninegative anions,
(iii) the reduction of Na^+, Mg^{2+} and Al^{3+} cations to their metals, and
(iv) the reduction of M^{3+} to M^{2+} for the elements V to Co.

(i) Group 1 — trend in $E^{\ominus}$ down a typical group

The accepted values for the standard reduction potentials for the Group 1 unipositive cations being reduced to the solid metal are given in Table 9.2, together with some

Table 9.2 — Data for Group 1 elements and M^+

Element	Li	Na	K	Rb	Cs
$E^{\ominus}(M^+/M)$	−3.04	−2.71	−2.92	−2.92	−2.92
$\Delta H_a(M_s/M_g)$	161	109	90	86	79
$I_1(M \rightarrow M^+)$	519	494	418	402	376
$\Delta H_{hyd}(M^+)$	−523	−419	−331	−314	−285
$r_{ionic}(M^+)$	68	98	133	149	165

relevant thermochemical data and the appropriate ionic radii. The units of all the energies are kJ mol^{-1}, and those of the radii are pm.

The $E^{\ominus}$ values imply that all the elements are very powerful reducing agents (M^+ being difficult to reduce to M) and that lithium is the most powerful, with sodium being the least powerful. The reducing powers of the elements K, Rb and Cs are very similar and intermediate between those of Li and Na.

The calculations which follow are carried out in terms of enthalpy changes, with entropy changes being ignored. This is because the entropy changes are (a) difficult to estimate, (b) expected to be fairly similar for the reactions of the elements of the same periodic group, and (c) relatively small compared to the enthalpy changes.

The standard enthalpy change for the reduction of $M^+(aq)$ to M(s) may be estimated from the data in Table 9.2 by using Hess's Law (first law of thermodynamics) on the reactions:

$$M^+(aq) \longrightarrow M^+(g) \tag{9.33}$$

$$M^+(g) + e^- \longrightarrow M(g) \tag{9.34}$$

$$M(g) \longrightarrow M(s) \tag{9.35}$$

their sum being:

$$M^+(aq) + e^- \longrightarrow M(s) \tag{9.36}$$

The general expression for the standard enthalpy change for reaction (9.36) is:

$$\Delta H^\ominus\ (9.36) = -\Delta H^\ominus_{hyd} - I_M - \Delta H^\ominus_a \tag{9.37}$$

The reduction enthalpy as calculated from equation (9.37) may be compared with that calculated for the reference reduction of the hydrated proton to molecular dihydrogen. The latter enthalpy change is calculated from the $\Delta H^\ominus$ values for the reactions:

$$H^+(aq) \longrightarrow H^+(g) \tag{9.38}$$

$$H^+(g) + e^- \longrightarrow H(g) \tag{9.39}$$

$$H(g) \longrightarrow \tfrac{1}{2}\, H_2(g) \tag{9.40}$$

their sum being:

$$H^+(aq) + e^- \longrightarrow \tfrac{1}{2}\, H_2(g) \tag{9.41}$$

The standard enthalpy change for reaction (9.41) is given by:

$$\begin{aligned} \Delta H^\ominus\ (9.41) &= -\Delta H^\ominus_{hyd}(H^+) - I_H - \Delta H^\ominus_a \\ &= 1101 - 1312 - 218 \\ &= -429\ \text{kJ mol}^{-1} \end{aligned} \tag{9.42}$$

The reduction enthalpies for the Group 1 cations are obtained by subtracting -429 kJ mol^{-1} from the results of substituting the appropriate data into equation (9.42), and are given in Table 9.3, together with their conversion into volts:

The calculated values are similar to those observed, the small differences being attributable to the ignored entropy terms. The calculation of reduction enthalpy shows that the value is the resultant, relatively small quantity, of the interaction between two large quantities; the ionization energy of the gaseous metal atom and the hydration energy of the cation, the heat of atomization being a relatively small contribution. The trends in the values of the three contributing quantities are understandable in terms of the changes in size of the ions (hydration enthalpy decreases with increasing ionic radius), the electronic configuration of the atoms

Table 9.3 — Reduction enthalpies annd calculated $E^{\ominus}$ values for Group 1 cations

Element	Reduction enthalpy	Calculated $E^{\ominus}$
Li	272	−2.82
Na	245	−2.54
K	252	−2.61
Rb	255	−2.64
Cs	259	−2.68

(ionization energy decreases as Z increases) and the strength of the metallic bonding (bonding becomes weaker as atomic size increases).

The aqueous silver(I) ion, with its $3d^{10}$ outer electronic configuration, is more easily reduced to its metallic state than the unipositive ions of the elements of Group 1, the Ag^+/Ag standard reduction potential being 0.8 V. The calculated value for this quantity, using the above approach, is 1.33 V, the discrepancy between theory and calculation being greater than those of the Group 1 cases. The calculated value arises from the greater standard enthalpy of atomization of silver (289 kJ mol^{-1}) due to the stronger metallic bonding (silver metal bonding is enhanced by overlap of the 3d and 4s bands), and a higher first ionization energy (732 kJ mol^{-1}). The hydration enthalpy of the silver(I) ion (−464 kJ mol^{-1}) is not a major factor in determining the differences between the above potentials.

(ii) Group 17

The Group 17 elements are the most electronegative of their respective periods, and the reduction of the elements to their uninegative ions are thermodynamically very feasible, as can be seen from the values for $E^{\ominus}$ given in Table 9.4. The table contains

Table 9.4 — Data and calculations of $E^{\ominus}$ for Group 17 elements

Element	F	Cl	Br	I
$E^{\ominus}$ (observed)	2.87	1.36	1.07	0.54
$\Delta H_a^{\ominus}$	158	244	224	212
Electron affinity	348	364	342	314
$\Delta H^{\ominus}(X^-)$	−490	−356	−310	−255
$r_i(X^-)$	133	181	196	220
Reduction enthalpy	−330	−169	−119	−34
$E^{\ominus}$ (calculated)	3.42	1.75	1.23	0.35

the data for the calculation of the appropriate reduction enthalpies, the results of the calculations and the calculated $E^\ominus$ values.

The appropriate equation for the calculation of the enthalpy change for the reaction:

$$1/2\ X_2\ (g, l\ or\ s) + e^- \longrightarrow X^-(aq) \qquad (9.43)$$

is:

$$\text{Reduction enthalpy} = \Delta H_a^\ominus(X_2) - E_X + \Delta H_{hyd}^\ominus(X^-) + 429 \qquad (9.44)$$

where E_X is the electron affinity (the energy released upon ion formation) of the gaseous X atom, the 429 term being the 'oxidation enthalpy' for the reference reaction.

The discrepancies between the observed and calculated values for $E^\ominus$ are attributable to the missing entropy term, but are not large.

Fig. 9.2 shows the trends in the hydration energies, electron affinities, enthalpies of atomization and in the ionic radii.

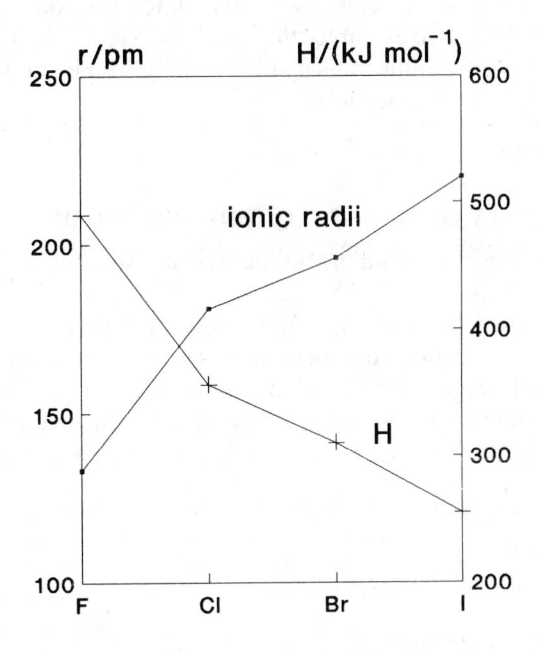

Fig. 9.2 — The variation of ionic radii and hydration energies of Group 17 uninegative ions.

(iii) $E^\ominus$ values of sodium, magnesium and aluminium

As the second short period is traversed, the reducing powers of the elements lessen. This is shown for the elements Na, Mg and Al, appropriate data being presented in

Table 9.5. The reduction enthalpies are calculated as in the above cases, with the modification for the varying number of electrons involved.

Again the calculations are in keeping with the observed trends, and it should be noticed that the major factor responsible for the lessening reducing power of the elements is the increasing total ionization energy which is not offset by the smaller increase in hydration enthalpy.

(iv) $E^{\ominus}$ values for some first row transition elements (M^{3+}/M^{2+})

Not all the first row transition elements have the same pairs of stable oxidation states' and the elements from V to Co are chosen for this example since they all have (II) and (III) states for which data are available. The data and the calculated $E^{\ominus}$ values are shown in Table 9.6.

In these calculations the omission of the entropy terms has caused some large differences between the observed and calculated values for the M^{3+}/M^{2+} potentials. Two hydrated ions are involved in each couple and the associated entropy changes (dependent upon differences in ion radii) would be expected to be larger than in the previously considered cases. In spite of the omissions, the trend in the calculated values is the same as that which is observed. There is a tendency for the M(III) states of V and Cr to be more stable than their M(II) states, whereas the reverse is the case for the elements Mn, Fe and Co, whose M(III) states are oxidizing agents.

It is clear from the above examples that quite simple calculations can identify the major factors which govern the values of $E^{\ominus}$ for any couple, and that major trends may be accounted for and understood.

9.5 THE STABILITY OF IONS IN AQUEOUS SOLUTION

This section is not concerned with the thermodynamic stability of ions with respect to their formation. It is concerned with whether or not a given ion is capable of existing in aqueous solution without reacting with the solvent. Hydrolysis reactions are dealt with above; the only reactions discussed in this section are those in which water is either oxidized to dioxygen or reduced to dihydrogen.

The limits of stability for water at different pH values may be defined by the Nernst equations for the hydrogen reference half-reaction and that for the reduction of dioxygen to water.

The Nernst equation

Equation (7.54) is repeated here:

$$\Delta G = \Delta G^{\ominus} + RT \ln q \qquad (9.45)$$

and represents the change in Gibbs energy for non-standard conditions of a system. The quotient, q, is the usual product of the activities of the products divided by the product of the reactant activities, taking into account the stoichiometry of the overall reaction. It may be converted into the **Nernst equation** by using equation (9.5), so that:

Table 9.5 — $E^\ominus$ data and calculations for Na, Mg and Al

Element	Na	Mg	Al
$E^\ominus$ (observed)	-2.71	-2.37	-1.66
$\Delta H_a^\ominus(M)$	109	150	314
I_1	494	740	580
I_2	—	1500	1800
I_3	—	—	2700
$\Delta H_{hyd}^\ominus$	-419	-1942	-4697
$E^\ominus$ (calculated)	-2.54	-2.12	-2.04

Table 9.6 — $E^\ominus$ data and calculations for the elements V to Co

Element	V	Cr	Mn	Fe	Co
$E^\ominus$ (observed)/V	-0.26	-0.42	1.5	0.77	1.92
$-\Delta H_{hyd}^\ominus(M^{3+})$/(kJ mol^{-1})	4408	4626	4596	4487	4713
I_3/(kJ mol^{-1})	2828	2987	3248	2957	3232
$-\Delta H_{hyd}^\ominus(M^{2+})$/(kJ mol^{-1})	1896	1926	1863	1959	2080
$E^\ominus$ (calc.)/V	-1.17	-1.47	0.89	0.0	1.76

$$E = E^\ominus - \frac{RT}{nF} \ln q \tag{9.46}$$

The Nernst equation may be applied to the hydrogen standard reference half-reaction and gives:

$$E(H^+/H_2) = -(RT/F) \ln (1/a_{H^+}) \tag{9.47}$$

considering the activity of dihydrogen to be unity ($E^\ominus$ $(H^+/H_2) = 0$). In order to obtain an equation which relates E to pH it is necessary to convert equation (9.42) to one in which decadic logarithms are used:

$$E(H^+/H_2) = -2.303 \times (RT/F)\log_{10}(1/a_{H^+}) \tag{9.48}$$

which, at 298 K, becomes:

$$E(H^+/H_2) = -0.059 \log (1/a_{H^+}) \tag{9.49}$$

The theoretical definition of pH is given by the equation:

$$\mathrm{pH} = -\log a_{H^+} \tag{9.50}$$

and substitution into equation (9.44) produces:

$$E(H^+/H_2) = -0.059 \times pH \tag{9.51}$$

The equation for the half-reaction expressing the reduction of dioxygen to water in acid solution is:

$$O_2(g) + 4H^+(aq) + 4e^- \rightarrow 2H_2O(l) \tag{9.52}$$

the $E^\ominus$ value being 1.23 V when the activity of the hydrogen ion is unity. The Nernst equation for the dioxygen/water couple is:

$$\begin{aligned} E(O_2/H_2O) &= 1.23 - (RT/4F) \ln (1/(a_{H^+})^4) \\ &= 1.23 - (RT/F) \ln (1/a_{H^+}) \\ &= 1.23 - 0.059 \times pH \end{aligned} \tag{9.53}$$

It has been assumed that the activities of dioxygen and water are unity.

Equations (9.51) and (9.53) are plotted in Fig. 9.3 with pH values varying from 0

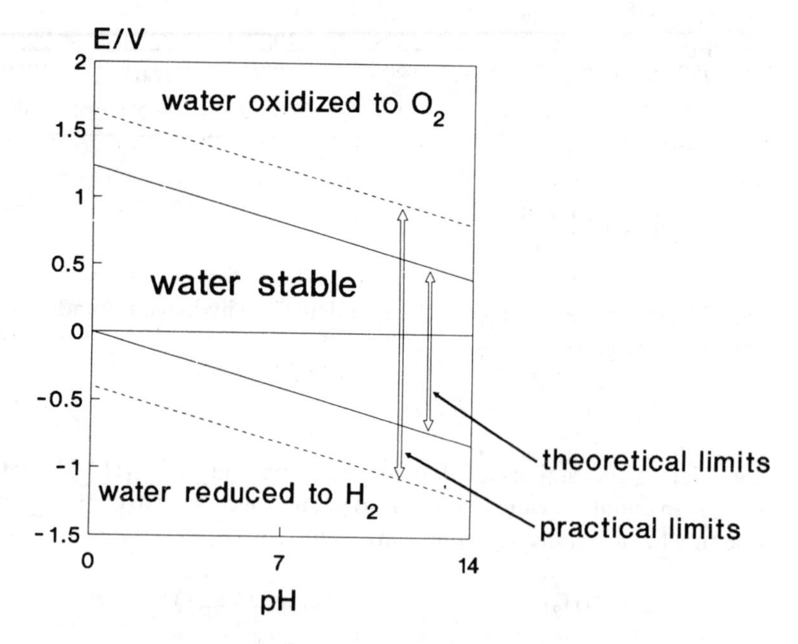

Fig. 9.3 — A diagram showing the Nernst equations for (i) the reduction of dioxygen to water (first full line) and (ii) the reduction of water to dihydrogen (second full line). The dotted lines represent the practical limits for the two reductions.

to 14. Theoretically the oxidized form of any couple whose E value lies above the oxygen line should be unstable with respect of its potential for oxidizing water to dioxygen. Likewise the reduced form of any couple whose E value lies below the hydrogen line should be unstable and should be capable of reducing water to dihydrogen.

In practice it is found that wherever gases are evolved (in particular the necessity for atoms to come together to form diatomic molecules) there is a barrier to the process (sometimes known as overpotential) so that the E value needs to be around 0.4 V lower than the hydrogen line or above the oxygen line for water to be reduced or oxidized. These practical limits are shown in Fig. 9.3. The oxidized and reduced forms of couples with E values anywhere between the two practical limits are stable in aqueous systems with respect to either the reduction or the oxidation of the solvent.

Couples with E values outside of the practical limits of stability do not necessarily cause the destruction of the solvent. Some reactions, although they may be thermodynamically feasible, are kinetically very slow. The Co^{3+}/Co^{2+} couple has an $E^{\ominus}$ value of 1.92 V and Co^{3+} should not exist in aqueous solution but its oxidation of water to oxygen is very slow and solutions containing $Co(H_2O)_6^{3+}$ give off dioxygen slowly. The $E^{\ominus}$ value for the Al^{3+}/Al couple is -1.66 V and indicates that aluminium should dissolve in acidic aqueous solutions, but the reaction is extremely slow.

9.6 POURBAIX DIAGRAMS

Pourbaix diagrams are plots of E versus pH for the various couples in the oxidation of an element. They are useful in defining 'areas' of stability for any particular oxidation state of that element, and may include the Nernst equations for the reduction and oxidation of water to dihydrogen and dioxygen respectively. The diagram for iron is derived as follows.

The chemistry of iron in acid solution ($a_{H^+} = 1$) may be summarized by what is called a Latimer diagram (in its simplest form this is a list of the various oxidation states of an element arranged in descending order (left to right), with the appropriate reduction potentials placed between each pair of states):

$$Fe^{3+} \xrightarrow{0.77\ V} Fe(OH)_2 \xrightarrow{-0.44\ V} Fe$$

The Latimer diagram for the oxidation states in a solution with a unit activity of OH^- ions

$$Fe^{3+} \xrightarrow{-0.56\ V} Fe(OH)_2 \xrightarrow{-0.887\ V} Fe$$

Latimer diagrams are available for all the elements which exhibit more than one oxidation state and form an excellent and concise summary of the aqueous chemistry of such elements. They are usually written down for the extreme conditions for either $a_{H^+} = 1$ (theoretical pH = 0) or $a_{H^-} = 1$ (theoretical pH = 14).

Pourbaix diagrams correlate the Latimer diagrams of the two extremes of the pH scale and take into account the **speciation** of the oxidation states of the element. The stages in the construction of the Pourbaix diagram for iron now follow.

1. The half-reaction:

$$Fe^{3+} + e^- \longrightarrow Fe^{2+} \tag{9.54}$$

does not feature the proton and so has a Nernst equation which is simply:

$$E(Fe^{3+}/Fe^{2+}) = E^{\ominus}(Fe^{3+}/Fe^{2+}) = 0.77 \tag{9.55}$$

and is a horizontal line on the Pourbaix diagram. However, it is only of relevance at pH values up to 1.8, above which the hydroxide becomes insoluble.

2. At a pH value above 1.8 the appropriate half-reaction for the Fe(III)/Fe(II) couple is:

$$Fe(OH)_3 + 3H^+ + e^- \longrightarrow Fe^{2+} + 3H_2O \tag{9.56}$$

The Nernst equation for reaction (9.56) is:

$$E(Fe(OH)_3/Fe^{2+}) = E^{\ominus}(Fe(OH)_3/Fe^{2+}) - 3 \times 0.059 \times pH \tag{9.57}$$

the factor of 3 in the slope being due to the three protons in equation (9.56). The value of $E^{\ominus}(Fe(OH)_3/Fe^{2+})$ is estimated from the reactions and their $\Delta G^{\ominus}$ values given in Table 9.7.

Table 9.7

Reaction	$E^{\ominus}$/V	$\Delta G^{\ominus}$/(kJ mol^{-1})
$Fe^{3+} + e^- \longrightarrow Fe^{2+}$	0.77	−74.3
$3OH^- + 3H^+ \longrightarrow 3H_2O$	—	−239.7
$Fe(OH)_3 \longrightarrow Fe^{3+} + 3OH^-$	—	207.6
$Fe(OH)_3 + 3H^+ + e^- \longrightarrow Fe^{2+} + 3H_2O$	1.1	−106.4

The $E^{\ominus}$ value for the $Fe^{3+}Fe^{2+}$ couple is converted into its $\Delta G^{\ominus}$ value. The $\Delta G^{\ominus}$ value for the formation of water from its ions is known. The $\Delta G^{\ominus}$ value for the dissociation of $Fe(OH)_3$ is calculated from the solubility product of the compound by the equation:

$$\Delta G^{\ominus} = -RT \ln SP_{Fe(OH)_3} \tag{9.58}$$

The solubility product is essentially the equilibrium constant for the dissociation of $Fe(OH)_3$.

A Hess's law calculation then allows the $\Delta G^\ominus$ for equation (9.56) to be estimated, the appropriate $E^\ominus$ being derived by making use of equation (9.5). Equation (9.57) may now be written as:

$$E(\text{Fe(OH)}_3/\text{Fe}^{2+}) = 1.1 - 3 \times 0.059 \times \text{pH} \qquad (9.59)$$

The lines represented by equations (9.55) and (9.59) cross each other when the pH is:

$$\text{pH} = (1.1 - 0.77)/(3 \times 0.059) = 1.86 \qquad (9.60)$$

3. A similar calculation must be carried out to take into account that at a pH of 6.6 the stable form of Fe(II) is the insoluble hydroxide, $Fe(OH)_2$. The reactions and data are given in Table 9.8.

Table 9.8

Reaction	$E^\ominus$/V	$\Delta G^\ominus$/(kJ mol^{-1})
$Fe(OH)_3 + 3H^+ + e^- \longrightarrow Fe^{2+} + 3H_2O$	1.1	− 106.4
$3H_2O \longrightarrow 3H^+ + 3OH^-$		239.7
$Fe^{2+} + 2OH^- \longrightarrow Fe(OH)_2$		− 84.4
$Fe(OH)_3 + e^- \longrightarrow Fe(OH)_2 + OH^-$	− 0.51	− 48.9

The value of $\Delta G^\ominus$ for the third reaction in Table 9.8 is calculated from the solubility product of $Fe(OH)_2$:

$$\Delta G^\ominus = -RT \ln (1/\text{SP}_{\text{Fe(OH)}_2}) \qquad (9.61)$$

The Nernst equation for the half-reaction:

$$\text{Fe(OH)}_3 + \text{e}^- \longrightarrow \text{Fe(OH)}_2 + \text{OH}^- \qquad (9.62)$$

is:

$$\begin{aligned} E(9.62) &= -0.51 - (RT/F) \ln a_{\text{OH}^-} \\ &= -0.51 + 0.059 \times \text{pH} \\ &= -0.51 + 0.059(14 - \text{pH}) \\ &= -0.51 + 0.826 - 0.059 \times \text{pH} \\ &= 0.316 - 0.059 \times \text{pH} \end{aligned} \qquad (9.63)$$

Equations (9.59) and (9.63) are univalued when:

$$1.1 - 3 \times 0.059 \times pH = 0.316 - 0.059 \times pH$$

which is when the pH is given by:

$$pH = (1.1 - 0.316)/(2 \times 0.059) = 6.6 \quad (9.64)$$

Equation (9.59) is only relevant to the Pourbaix diagram between the pH values 1.8 and 6.6, and at pH values greater than 6.6 equation (9.63) is appropriate.
4. The Nernst equation for the half-reaction:

$$Fe^{2+}(aq) + 2e^- \longrightarrow Fe(s) \quad (9.65)$$

is the horizontal line:

$$E(Fe^{2+}/Fe) = -0.44 \quad (9.66)$$

5. At pH values greater then 6.6 the stable form of iron(II) is the insoluble hydroxide, $Fe(OH)_2$, and by considerations similar to the ones of the previous cases the Nernst equation for the half-reaction:

$$Fe(OH)_2 + 2e^- \longrightarrow Fe(s) + 2OH^- \quad (9.67)$$

is derived:

$$E(Fe(OH)_2/Fe) = -0.06 - 0.059 \times pH \quad (9.68)$$

6. The appropriate sections of the equations (9.55), (9.59), (9.63), (9.65) and (9.68) form the graphs on the Pourbaix diagram for the Fe/Fe(II)/Fe(III) system shown in Fig. 9.4. Vertical lines are added to indicate the phase transitions. The graphs define five areas of the diagram which are the areas of stability for the species, $Fe^{3+}(aq)$, $Fe(OH)_3$, $Fe^{2+}(aq)$, $Fe(OH)_2$ and Fe(s).

Reaction feasibilities can be predicted by overlaying Pourbaix diagrams for any two systems. The Pourbaix diagram for the As(III)As(V) (dotted line) and iron (full lines) systems is shown in Fig. 9.5. The region above the dotted line is the field of stability for As(V), the region below it being the field of stability for As(III). Fe(III) has the potential to oxidize As(III) to As(V) in the pH regions 0–5.5 and 9–14, but that in the region 5.5–9 the reverse reaction is feasible. Such predictions must be qualified with the possibility that kinetic barriers may cause the reactions to be very slow (which is the case in this particular example).

9.7 THE EFFECTS OF COMPLEXATION UPON $E^\ominus$ VALUES

The above discussions concern aquated ions, but the value of $E^\ominus$ is very dependent upon the ligands surrounding any metal ion centre. Some typical values are given for Fe(II)Fe(III) systems in Table 9.9.

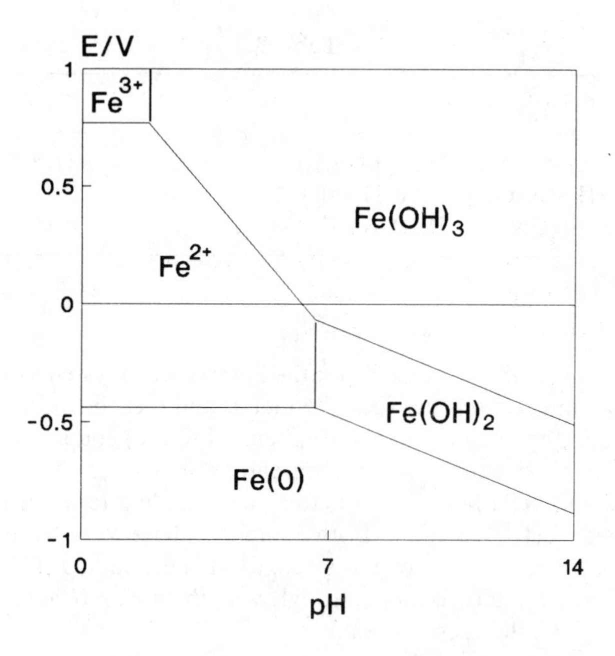

Fig. 9.4 — A Pourbaix diagram for the zero, II and III oxidation states of iron.

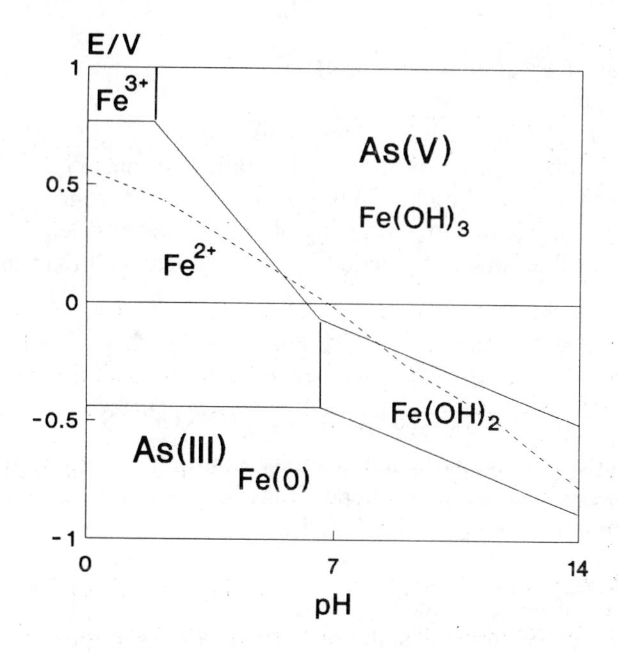

Fig. 9.5 — The Pourbaix diagram for iron (full lines) overlain with that for the III and V states of arsenic (dotted line).

Table 9.9

Couple	$E^{\ominus}$/V
$[Fe(phen)_3]^{3+}/[Fe(phen)_3]^{2+}$	1.14
$(Fe(H_2O)_6]^{3+}/[Fe(H_2O)_6]^{2+}$	0.77
$[Fe(CN)_6]^{3-}/[Fe(CN)_6]^{4-}$	0.36

The variations in the value of $E^{\ominus}$ for the Fe(III)Fe(II) systems can be understood in terms of the interactions between the metals and their ligands. Compared to the aquated system, the phen (phen = 1,10-phenanthroline) couple has a higher potential which is mainly due to the relatively greater stabilization of the Fe(II) state. The lower charge on Fe(II) facilitates a better back bonding to occur compared to the higher-charged Fe(III) complex. Both complexes have very large values for their formation constants because of the strength of the bonding. Cyanide ion, being negatively charged, interacts more strongly with the higher oxidation state, Fe(III), both ions being stabilized considerably.

Another striking example is the effect of replacing water by ammonia ligands on the $E^{\ominus}$ values for the Co(III)/Co(II) couple. The values of $E^{\ominus}$ for the two half-reactions:

$$Co(H_2O)_6^{3+} + e^- \longrightarrow Co(H_2O)_6^{2+} \tag{9.69}$$

$$Co(NH_3)_6^{3+} + e^- \longrightarrow Co(NH_3)_6^{2+} \tag{9.70}$$

are 1.84 and 0.11 V respectively. The complexation by ammonia ligands causes a great loss of oxidizing power of the Co(III) state. This may be explained in general terms by considering the bonding in the (III) state to be stronger with the ammonia ligands, which contributes to a lowering of the effective nuclear charge so that the electron responsible for the reduction is attracted less well than in the case of the hydrated ion.

9.8 FACTORS AFFECTING STABILITY CONSTANTS

The magnitude of any particular stability constant (or formation constant) is governed by several factors, some depending upon the nature of the central metal ion and others upon the nature of the ligands.

1. The nature of the metal ion

Variations in the magnitude of stability constants for a series of complexes of metals ions with a common set of ligands may be discussed in terms of the relative sizes and charges of the metal ions.

(i) Effect of ionic size

Table 9.10 contains the values of the ionic radii (crystal radii), the standard

Table 9.10 — Thermodynamic data for complexes, M(II)

Element	r_i/pm	$-\Delta H^{\ominus}_{hyd}/(\text{kJ mol}^{-1})$	$\log_{10} K_1$ (en)
Mn	97	1845	2.7
Fe	92	1920	4.3
Co	88	2054	5.9
Ni	83	2106	7.6
Cu	87	2100	10.7
Zn	88	2044	5.9

enthalpies of hydration and the decadic logarithms of the stability constants for the formation of $M(en)^{2+}$ complexes for the (II) oxidation states of the elements from Mn to Zn. The radius and formation constant data are shown graphically in Fig. 9.6.

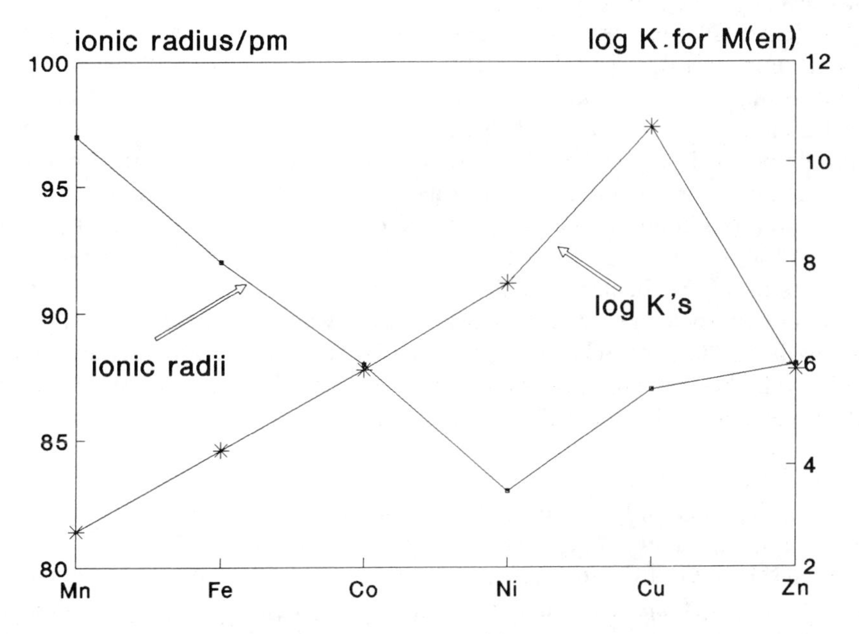

Fig. 9.6 — The variations of ionic radii and log K_1 values of M(en) complexes for the II oxidation states of the elements Mn to Zn.

The crystal radius used for copper(II) is the accepted one for octahedral coordination. The Jahn-Teller effect is important in the d^9 case, and is probably responsible for the extraordinary stability of the $Cu(en)^{2+}$ complex. The trends in the values of

$\Delta H^\ominus_{hyd}$ and $\log_{10}K_1$(en) are those which would be expected from the variations in the radii of the central metal ions. The smaller ions produce the more stable complexes.

The general shape of the graph of the $\log_{10}K_1$ values for the diaminoethane(en) 1:1 complexes is an example of the general trend for the stability constants for any particular ligand. The general trend in stability for the complexes of the M(II) cases:

$$\mathrm{Mn < Fe < Co < Ni < Cu > Zn}$$

is known as the Irving–Williams series.

(ii) Effect of the oxidation state of the central metal ion

As the charge on the central metal ion increases, the stability constant for a given metal with the same ligand environment would be expected to show an increase. One example suffices to make the general point; the β_6 values for the complexes, $[Fe(CN)_6]^{4-}$ and $[Fe(CN)_6]^{3-}$, are 10^{34} and 10^{44} respectively, implying a very large increase in stability as the formal charge on the metal ion increases from +2 to +3.

2. The nature of the ligand

There are three main effects of the nature of the ligand upon the stability constants of complexes.

(i) The donor atom

The effect of the donor atom upon complex stability is mainly summarized by the spectrochemical series, which is the order of ligands in terms of the differences in energy between the e_g and t_{2g} orbitals produced for a given central metal ion. Bearing in mind that the series takes into account not only σ-type bonding but π-type bonding **and** anti-bonding effects, it is to be expected that a large $e_g - t_{2g}$ energy difference would be associated with a large value for the stability constant of the complexes of any particular metal ion and a variety of ligands. There are minor deviations from such a generalization which apply when metal–ligand π-type bonding is possible. That kind of bonding produces an increase in stability (compared to a purely σ-type bonded complex) but reduces the magnitude of the $e_g - t_{2g}$ energy difference.

(ii) Denticity

In general the magnitude of the stability constant increases with the denticity (the number of donor atoms) of the ligand. For example, K_1 for the $[Cu(en)]^{2+}$ complex is observed to be 5×10^{10} whereas the β_2 value for $[Cu(NH_3)_2]^{2+}$ is only 5×10^7 – factor of 10^3 less stable than the complex with the bidentate ligand with the same donor atoms. This general observation is known as the chelate effect — a chelate involving the formation of a ring structure.

The chelate effect is due to differences in the enthalpy and entropy changes of formation of the complexes. The data for the formation reactions of the diammine and diaminoethane complexes:

$$[Cu(H_2O)_4]^{2+} + 2NH_3 \longrightarrow [Cu(NH_3)_2]^{2+} + 2H_2O \qquad (9.71)$$

$$[Cu(H_2O)^{2+}] + en \longrightarrow [Cu(en)]^{2+} + 4H_2O \tag{9.72}$$

are given in Table 9.11. The respective $\Delta G^{\ominus}$ values indicate the greater stability of

Table 9.11 — Data for reactions (9.71) and (9.72)

Reaction	(9.71)	(9.72)
$\Delta G^{\ominus}/(kJ\ mol^{-1})$	−43.9	−61.1
$\Delta H^{\ominus}/(kJ\ mol^{-1})$	−46.4	−54.4
$\Delta S^{\ominus}/(J\ K^{-1}\ mol^{-1})$	−8.4	22.6

the Cu(II)–en combination. The contribution to the greater stability of 17.2 kJ mol^{-1} is made up from a greater exothermicity of 8.0 kJ mol^{-1} and an advantage of 9.2 kJ mol^{-1} from the $T\Delta S^{\ominus}$ terms. There is a small decrease in entropy in the formation of the diammine complex compared to a comparatively large increase in entropy of formation of the diaminoethane complex. This difference may be understood in terms of there being no change in the number of species taking part in reaction (9.71) whereas there is an increase in the number of species in reaction (9.72). The translational freedom afforded to one of the water molecules in reaction (9.72) is mainly responsible for the entropy increase observed.

The greater exothermicity of the chelate formation reaction is due mainly to a change in the relative solvation energies of one diaminoethane molecule compared to the two water molecules it replaces in reaction (9.72). The en molecule, with its two CH_2 groups, is not particularly hydrophilic and does not interact strongly with the water solvent. The two water molecules it displaces are very hydrophilic (!) and their entry into the bulk solution is a major factor in determining the enthalpy advantage of the chelation reaction.

(iii) Ring size

In general a five-membered ring (metal plus appropriate ligand atoms) confers a greater stability to a complex than a six- (or more) membered ring. Table 9.12 below

Table 9.12 — $\log_{10}K_1$ values for Mn(II) complexes

Ligand	$\log_{10}K_1$
Ethandioate (oxalate)	2.93
Malonate	2.30
Succinate	1.26

contains data for the Mn(II) complexes with ethandioate (oxalate), ($C_2O_4^{2-}$), malonate ($CO_2C_2CO_2^{2-}$) and succinate ($CO_2(CH_2)_2CO_2^{2-}$) ions. Here and in many other examples the five-membered ring is preferred to any other number.

9.9 CONCLUDING SUMMARY

This chapter contains a discussion of the thermodynamics of ions in solution, and includes the following topics.

(1) Thermodynamics and electrode potentials.
(2) Hydrolysis and the forms of ions in solution — speciation.
(3) The factors influencing the values of standard reduction potentials.
(4) The stability of ions in aqueous solution, the application of the Nernst equation in the construction of Pourbaix diagrams.
(5) The factors affecting the values of stability constants.

10

Kinetics and mechanics of reactions

10.1 INTRODUCTION

The sign of $\Delta G^{\ominus}$ for a reaction is an indication of its feasibility. A negative value of $\Delta G^{\ominus}$ indicates that the position of equilibrium favours the products. Although that criterion may be fulfilled there are **kinetic barriers** which may cause a reaction to be extremely slow. For instance, the reaction:

$$H_2(g) + Cl_2(g) \longrightarrow 2HCl(g) \qquad (10.1)$$

has a value of $\Delta G^{\ominus}$ of $-184.7\,kJ\,mol^{-1}$ but, at 298 K and in the absence of light, it is immeasurably slow. Fig. 10.1 shows an idealized reaction energy profile for an exothermic reaction (negative ΔH) and with a kinetic barrier of magnitude, E, above the initial state. It shows the portions of the reaction which are governed by thermodynamics (ΔH) and which are governed by kinetics (E), and that there is no connexion between the two.

The subject of **reaction kinetics** covers two main areas of interest. One is the attempt to understand the factors which affect the rates of chemical processes. The other is the attempt to understand such processes at the molecular level in terms of the **mechanism** of the reaction.

All kinetic studies are dependent upon some form of measurement of the **rate** of a reaction. It is important to define exactly what is meant by the rate of a reaction.

The reaction:

$$2Fe^{2+}(aq) + H_2O_2(aq) \longrightarrow 2Fe^{3+}(aq) + 2OH^-(aq) \qquad (10.2)$$

could be followed by measuring the concentration of the hydrogen peroxide at various times. One mathematical expression for the rate is:

$$\text{Rate} = -\,d[H_2O_2]/dt \qquad (10.3)$$

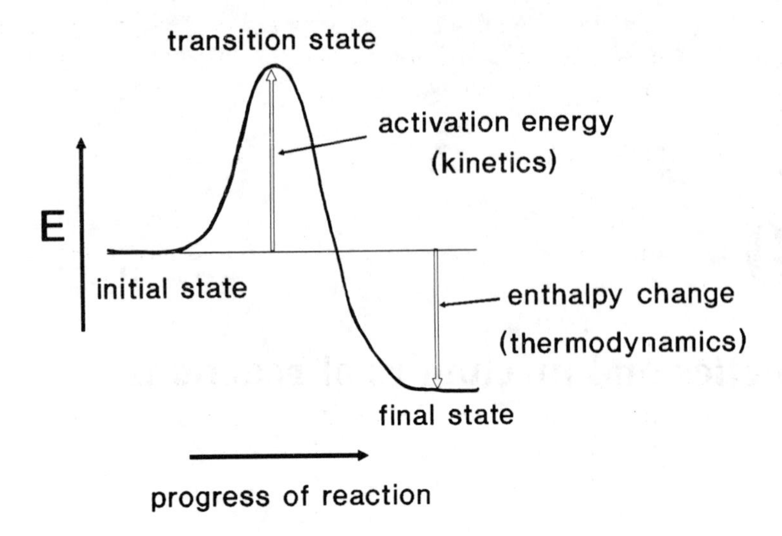

Fig. 10.1 — A diagram showing the reaction profile which emphasizes the spheres of influence of kinetics and thermodynamics.

the negative sign indicating that the concentration decreases with time.

If the reaction were to be followed by measuring the variation of the concentration of iron(II) ion with time the rate would be expressed as:

$$\text{Rate} = -\,\mathrm{d}[\mathrm{Fe}^{2+}]/\mathrm{d}t \qquad (10.4)$$

The two rates given by equations (10.3) and (10.4) are not equal since for every hydrogen peroxide molecule destroyed there are two iron(II) ions oxidized. The stoichiometry of the reaction must be taken into consideration in the strict definition of its rate.

The relationship between the two rates is given by:

$$-\,\mathrm{d}[\mathrm{H_2O_2}]/\mathrm{d}t = -\,\mathrm{d}[\mathrm{Fe}^{2+}]/\mathrm{d}t \times \tfrac{1}{2} \qquad (10.5)$$

The primary quest in an investigation of the kinetics of a reaction is to establish the form of the right-hand side of the equation:

$$-\,\mathrm{d}[C]/\mathrm{d}t = ? \qquad (10.6)$$

where C is some reactant. Equation (10.6), when complete, is known as the rate law.

This may be carried out by systematically varying the reaction conditions such as the concentrations (or partial pressures in the case of a gas phase reaction) of the components of the system (reactants, products and catalysts), the sizes of particles (in a heterogeneous reaction), and the radiation intensity (in photochemical or radiolysis reactions).

The rate of reaction is usually obtained from a plot of a function (usually the logarithm) of the concentration of either a reactant or a product against time.

One form of a rate law is given by:

$$-\mathrm{d}[C]/\mathrm{d}t = k[C] \tag{10.7}$$

the rate being proportional to the concentration of C at any time, the proportionality factor, k, being known as the **rate constant**. Equation (10.7) may be transformed into a more manageable form by rearrangement to:

$$-\mathrm{d}[C]/[C] = k\,\mathrm{d}t \tag{10.8}$$

and by integration to give:

$$-\ln[\mathrm{C}] = kt + \text{constant} \tag{10.9}$$

If the concentration of C at time zero is expressed as $[C_0]$ the equation (for $t = 0$) becomes:

$$-\ln[C_0] = \text{constant} \tag{10.10}$$

which identifies the integration constant. A combination of equations (10.9) and (10.10) gives:

$$\ln[C_0] - \ln[C] = kt \tag{10.11}$$

or

$$\ln([C_0]/[C]) = kt \tag{10.12}$$

Equation (10.8) is an example of **first order** kinetics in that the rate is proportional to the first power of the concentration term, $[C]$. Applied to a single concentration term the order of the reaction is defined by the equation:

$$-\mathrm{d}[C]/\mathrm{d}t = k[C]^{\text{order}} \tag{10.13}$$

Although most reactions have a more complicated rate law than that expressed by equation (10.8) it is usual to attempt to arrange the reaction conditions so that the equation does apply. In a reaction which is first-order in two components (second order overall) the rate law may be forced to follow a simple first-order law by having a large excess (e.g. ten-fold) of one of the reactants. If in the reaction:

$$\mathrm{A} + \mathrm{B} \longrightarrow \text{Products} \tag{10.14}$$

the rate law was observed to be:

$$-\mathrm{d}[\mathrm{A}]/\mathrm{d}t = -\mathrm{d}[\mathrm{B}]/\mathrm{d}t = k[\mathrm{A}][\mathrm{B}] \tag{10.15}$$

— first order in A and B — by having a large excess of the component B the rate law could be arranged to be:

$$-\,d[A]/dt = k'[A] \tag{10.16}$$

where the observed rate constant, k', is equal to k[B], since [B] is virtually constant.

In the reaction of iron(II) ion with hydrogen peroxide the technique may be used to force the reaction to exhibit first-order kinetics with respect to Fe^{2+} by having an excess of hydrogen peroxide in the reaction mixtures. Similarly, with an excess of Fe^{2+} in the mixtures, the kinetic law indicates a first-order dependence upon the H_2O_2 concentration. The reaction is second order overall — a conclusion which is not necessarily obvious from the stoichiometry of the reaction! The only method of obtaining the rate law for a reaction is by experiment — its prediction from the overall stoichiometry should not be attempted.

Some examples which illustrate the above ideas follow.

(a) The reaction of diiodine with propanone is given by the equation:

$$I_2 + CH_3COCH_3 \longrightarrow CH_3COCH_2I + H^+ + I^- \tag{10.17}$$

The reaction may be carried out such that the propanone and acid concentrations are relatively much (say ten times) higher than that of the iodine. The plot of $[I_2]$ versus time is observed to be a straight line, indicating that the reaction is zero order with respect to the diiodine concentration.

While keeping an excess of propanone and acid it is possible to vary their concentrations and to decide how such variations affect the slope (which is the zero-order rate constant) of the $[I_2]$ versus time plots. This allows the order of the reaction with respect to propanone and H^+ to be observed to be unity in both cases. The rate law is observed to be:

$$-\,d[I_2]/dt = k[\text{propanone}][H^+] \tag{10.18}$$

The reaction is an example of homogeneous catalysis by the H^+ ion.

(b) The reaction:

$$BrO_3^- + 5Br^- + 6H^- \longrightarrow 3Br_2 + 3H_2O \tag{10.19}$$

has a rate law which is:

$$d[Br_2]/dt = k[Br^-][BrO_3^-][H^+]^2 \tag{10.20}$$

i.e. first order in Br^- and BrO_3^-, and second order in H^+. Such a conclusion would be impossible to predict from equation (10.19).

(c) The ostensibly simple reaction between dihydrogen and dibromine:

$$H_2 + Br_2 \longrightarrow 2HBr \tag{10.21}$$

has the very complicated rate law:

$$d[HBr]/dt = \frac{k[H_2][Br_2]^{1/2}}{1 + k'[HBr]/[Br_2]} \tag{10.22}$$

Such rate laws may be interpreted in terms of the smallest number of **bimolecular** stages to give the accepted mechanisms of the reactions. That is a statement of the **principle of maximum simplicity**—which must be the guiding principle of any kinetic investigation. It is an alternative statement of **Occam's Razor**: 'don't multiply things unnecessarily'.

The mechanism of a reaction is elucidated by a consideration of the results of experimental investigation of its kinetics, together with any relevant thermodynamic information. Before anything can be concluded about a mechanism it is essential to know the rate law for the reaction. The rate law expresses the effect upon the rate of the reaction of concentrations of reactants, products and catalysts. It contains rate constants which have a temperature dependence. It may (in the case of photochemical reactions) contain a radiation intensity term.

It is most important to set up alternative mechanisms, for any process, which are consistent with the experimental data. After this is done, tests may be suggested which could discriminate between the various alternatives. Such a procedure is an excellent example of Popper's approach to the development of scientific theory by attempting to falsify current ideas to enhance their credibility.

The subject of kinetics may be divided into two main sections:

(1) the study of single stages of a reaction or one stage processes and
(2) the study of multi-stage processes.

The first section is concerned with the intimate mechanism whereby a single-stage reaction occurs and the factors which determine its rate, and forms a major part of physical chemistry. Further treatment of this topic is left to textbooks of physical chemistry.

The second section is concerned with the elucidation of the mechanism of a multi-stage process — in the identification of the various stages which can explain the overall rate law of the reaction. Most of the examples used are inorganic reactions, but the principles apply to any reaction.

This chapter contains an account of the conclusions from the basic theory of single-stage reactions — the theoretical treatment of the rate constant. This is followed by examples of multi-stage processes:

(1) ligand substitution reactions
(2) redox reactions
(3) heterogeneous reactions
(4) enzyme reactions
(5) bacterial growth

(6) trioxygen (ozone) photochemistry
(7) water radiolysis.

10.2 RATE CONSTANTS AND ACTIVATION ENERGY

The rate constant of any process is an experimentally determined value. Rate constants are temperature dependent and for most simple processes follow the Arrhenius law:

$$K = A\mathrm{e}^{-E/RT} \tag{10.23}$$

Equation (10.23) was derived empirically from observations, and A represents a constant (the Arrhenius constant or the pre-exponential factor), E being the **activation energy** for the process, R is the gas constant and T is the temperature (in kelvin). It is customary to obtain values of k at four or five different temperatures and to determine the value of E from a plot of the natural logarithm of k versus $1T$. The logarithmic version of equation (10.23) is:

$$\ln k = \ln A - E/RT \tag{10.24}$$

indicating that the above plot should yield a straight line with a slope of $-E/R$ and with an intercept of ln A. The normal experimental errors experienced in kinetic measurements have the consequence that A values are subject to large errors.

10.3 THEORY OF THE RATE CONSTANT AND ACTIVATION ENERGY

There are two theories of the rate constant which should be mentioned — collision theory and transition state theory — both of which have use in understanding the progress of chemical reactions. Brief treatments are given in this section.

Collision theory

The rate constant for reactant molecules colliding with each other is known as the **collision number**, Z, and can be calculated from kinetic theory. In molar terms its value is around $10^{10}\,\mathrm{mol}^{-1}\,\mathrm{sec}^{-1}$. Although some rate constants are observed to be of such a magnitude, the majority have values which are far lower than that figure. The reason for this is that only those molecules which collide with an energy equal to, or greater than, that deemed to be the activation energy will actually react to give products. The rate of any individual reaction is very rapid and equivalent to that of the breakage of a chemical bond. That takes place within the time of a typical stretching vibration — in the region of 10^{-15} s. The bulk observed rate is much lower in that it represents only the very small fraction of energetically suitable collisions occurring in the system.

A collision with energy below that of the activation energy will not lead to reaction. The colliding molecules will simply recoil from the collision, with the possibility of some energy being transferred from one to the other. To produce a realistic equation for the rate constant, the collision number must be multiplied by an expression representing the fraction of energetically favourable collisions. The required expression is known as a Boltzmann factor.

The Maxwell–Boltzmann equation (Chapter 7) implies that the fraction of molecules in a system at a given temperature which possess an energy ε_i is given by:

$$\frac{n_i}{N} = \frac{e^{-\varepsilon_i/kt}}{\sum e^{-\varepsilon_i/kT}} \tag{10.25}$$

the summation in the denominator being over all possible values of i. In the next equations, that summation will be denoted by Σ. If energy level i is that which corresponds to the minimum activation energy needed for reaction it is possible to calculate the fraction of molecules in the system which have energies equal to, or greater than, ε_i, as:

$$n_i/N = e^{-\varepsilon_i/kT}/\Sigma \tag{10.26}$$

$$n_{i+1}/N = e^{-(\varepsilon_i + x)/kT}/\Sigma \tag{10.27}$$

$$n_{i+2}/N = e^{-(\varepsilon_i + 2x)/kT}/\Sigma \tag{10.28}$$

The term x in the above equations represents the difference in energy between any two adjacent levels. Equations (10.26)–(10.28) can be extended infinitely, and the sum of the equations is the required fraction of those molecules with energies equal to, or greater than, ε_i. Their right-hand sides contain a common factor, $e^{-\varepsilon_i/kT}$, as well as Σ.

The sum of the equations reduces to:

$$n_{\geq\varepsilon_i}/N = e^{-\varepsilon_i/kT}(1 + e^{-x/kt} + e^{-2x/kT} + \ldots)/\Sigma \tag{10.29}$$

which, since the infinite series in parentheses is equal to Σ, simplifies to:

$$\frac{n_{\geq\varepsilon_i}}{N} = e^{-\varepsilon_i/kT} \tag{10.30}$$

This important result, combined with the rate constant for reaction at every collision, Z, gives a general equation for the rate constant:

$$k = Ze^{-E/RT} \tag{10.31}$$

The activation energy, E, is expressed in molar terms, with the gas constant, R, replacing the Boltzmann constant.

Equation (10.31) is only moderately successful in explaining observed rate constants, in that the experimentally determined values of the pre-exponential factor are usually considerably lower than the appropriately calculated Z value. The discrepancies are usually excused as 'steric' factors, and there is some justification for this. If two atoms collide, their spherical symmetries would preclude the necessity for a steric factor. If two molecules collide then their relative orientations (irrespective of the energy of the collision) would have to be suitable for a particular reaction to occur efficiently. In such cases there is a logic behind the inclusion of a steric factor or probability factor into equation (10.31) to give:

$$k = PZe^{-E/RT} \tag{10.32}$$

The value of P represents the probability of an energetically satisfactory collision leading to the production of products.

Although the steric factor can be estimated from experimental data there is no way in which it may be predicted for a particular reaction. It is of no help in advancing the theory of kinetic processes. There are cases where the value of P is found to be greater than unity, which is another difficulty for the theory.

Transition state theory

Transition state theory is more satisfactory than collision theory. It is based upon the idea that the initial state of the system has to pass through a transition state before reaching the final state. The transition state is that intermediate state of minimum energy which allows the reorganization of the system (bond breakage and formation, reassignment of electrons to new molecular orbitals) so that it may proceed to the final state. Although the mathematical form of the expression for the rate constant may be derived from statistical thermodynamics, the treatment in this section is not so rigorous but yields the same expression.

The underlying idea of transition state theory is that there is a pseudo-equilibrium between the reactants and the transition state, the latter either yielding products or reverting to the initial state. This may be written as:

$$\mathrm{A + B \rightleftharpoons TS \xrightarrow{k_2} Products} \tag{10.33}$$

The equilibrium constant for the production of the transition state may be written as $K^{\dagger} = [\mathrm{A}][\mathrm{B}]/[\mathrm{TS}]$, and the rate of production of the products is given by:

$$\mathrm{d}[P]/\mathrm{d}t = k_2[\mathrm{TS}] = k_2K^{\dagger}[\mathrm{A}][\mathrm{B}] \tag{10.34}$$

so that the observed rate constant, k, has the value:

$$k = k_2K^{\dagger} \tag{10.35}$$

$K^{\dagger}$ may be treated like a true equilibrium constant and is related to the change in Gibbs energy of activation, $\Delta G^{\dagger}$, by the equation:

$$\ln K^{\dagger} = -\Delta G^{\dagger}/RT \tag{10.36}$$

so that:

$$K^{\dagger} = \mathrm{e}^{-\Delta G^{\dagger}/RT} \tag{10.37}$$

which substitutes into equation (10.035) to produce:

$$k = k_2 \mathrm{e}^{-\Delta G^{\dagger}/RT} \tag{10.38}$$

Since $\Delta G^{\dagger} = \Delta H^{\dagger} - T\Delta S^{\dagger}$, equation (10.38) becomes:

$$k = k_2 \mathrm{e}^{\Delta S^{\dagger}/R} \mathrm{e}^{-\Delta H^{\dagger}/RT} \tag{10.39}$$

Equation (10.39) has the same form as equation (10.32) so that if $\Delta H^{\dagger}$ is equated to the activation energy, E, the pre-exponential factor PZ is given by:

$$PZ = k_2 \mathrm{e}^{\Delta S^{\dagger}/R} \tag{10.40}$$

The term k_2, is the equivalent of Z, the collision number, and although it is represented as a rate constant itself it must be realized that it is the rate constant for a reaction with zero activation energy. That leaves the probability factor as given by:

$$Z = \mathrm{e}^{\Delta S^{\dagger}/R} \tag{10.41}$$

This is a more satisfactory expression for Z, as it relates the value of Z to the entropy of activation, $\Delta S^{\dagger}$. If the production of the transition state is accompanied by a decrease in entropy then Z has a value which is less than unity. Two molecules coming together to produce a transition state would be associated with a loss of translational freedom, with $\Delta S^{\dagger}$ being negative. If, in addition, there were particular steric restrictions necessary for the production of the transition state, the value of $\Delta S^{\dagger}$ would be considerably more negative, causing the Z factor to be very small. On the other hand, if the production of the transition state was accompanied by an increase of entropy, Z values greater than unity could be accounted for.

This is the case for reactions between oppositely charged ions in solution. Although there is a loss of entropy as two ions form the transition state, the entropy change due to the increased translational freedom of a considerable number of solvent molecules (the solvation spheres of both ions) can be very positive. The transition state either has either a reduced charge or is neutral and does not restrict the translational freedom of as many solvent molecules as do the two reactant ions. In such cases the Z values may be greater than unity and allow an explanation for those values of A which are greater than expected from purely collisional grounds.

The factor $\mathrm{e}^{-\Delta H^{\dagger}/RT}$ is identical in form with the Boltzmann factor of previous equations, $\Delta H^{\dagger}$ being the activation energy for the reaction. The identity is strictly true for reactions carried out at constant pressure and has to be modified for those carried out at constant volume.

10.4 LIGAND SUBSTITUTION REACTIONS

There are three classes of ligand substitution reactions:

(i) Associative (A) reactions in which the **rate-determining step** is that which involves the association of the two reactants — the complex and the incoming ligand. The first stage of the reaction is the production of an **intermediate** (not a transition state) where the metal centre has a coordination number one more than it has in the initial state.

(ii) Interchange (I) reactions, which are single-stage processes not involving an intermediate of any kind. In this kind of process there is a synchronous interchange so that as the incoming ligand approaches the complex, the outgoing ligand leaves. It has been necessary to define two subsections of interchange reactions. One is the I_a process, in which the rate depends upon the **nature**, of the incoming group as well as its concentration. In such a case, bond formation in the transition state is important in determining the magnitude of the activation energy. The other subsection consists of I_d processes in which the rate is independent of the nature of the incoming group but may depend upon its concentration. The activation energy is largely determined by bond breaking in the transition state. In such a process the incoming ligand enters the outer sphere of the complex as the leaving group dissociates (hence I_d) from the central atom.

(iii) Dissociative (D) reactions in which there is a detectable intermediate with a coordination number which is lower than that of the reactant complex.

The above classification derives from the ideas of Langford and Gray, and is in general use in this branch of kinetic investigation.

Only two general examples of ligand substitution reactions are dealt with in this book: (i) square planar platinum(II) complexes and (ii) octahedral complexes.

(i) Ligand substitution reactions of platinum(II) complexes

Platinum(II) complexes are in general square planar and conform to the associative (A) mechanism. The incoming ligand is able to approach the complex without experiencing much steric hindrance. There are complications with the involvement of the solvent. The observed rate laws for such reactions as:

$$PtL_3X + Y \longrightarrow PtL_3Y + X \qquad (10.42)$$

in which X represents the outgoing ligand have the form:

$$\text{Rate} = (k_S + k_Y[Y])[PtL_3X] \qquad (10.43)$$

in which k_S and k_Y are rate constants. This is a case where the reactions are carried out under pseudo-first-order conditions, with an excess of Y (compared to the concentration of the platinum complex). The observed pseudo-first-order rate constant, k_{obs}, is found to have the form:

$$k_{obs} = k_S + k_Y[Y] \qquad (10.44)$$

Whenever the observed rate constant has a form such as this, the interpretation is that there must be two competing pathways for the observed reaction. Equation (10.43) may be multiplied out to give:

$$\text{Rate} = k_S[PtL_3X] + k_Y[PtL_3\,X][Y] \qquad (10.45)$$

the two terms representing the rates of the competing pathways.

The second term is that which would be expected for a simple bimolecular association between the two reactants. The first term is independent of the concentration of the incoming ligand and may be thought to be typical of a dissociative process. When experiments are carried out with different solvents it becomes clear that the first term represents a pathway in which k_S is determined by the nature of the solvent. The rate-determining step does not involve the incoming ligand, and is a slow process in which X is possibly replaced by a solvent molecule:

$$PtL_3X + S \longrightarrow PtL_3S + X \qquad (10.46)$$

the solvent molecule being itself replaced by a faster (and therefore not rate limiting) step:

$$PtL_3S + Y \longrightarrow PtL_3Y + S \qquad (10.47)$$

(ii) Ligand substitution reactions of octahedral complexes

The discussion of these reactions is limited to the replacement of coordinated water molecules by water molecules (exchange reactions) or by other ligands. The mechanism of reactions of six-coordinate complexes is dominated by the dissociative (D) or dissociative interchange (I_d) processes. It is very improbable that an associative process with the production of a seven-coordinate intermediate (or transition state) would be able to compete with the dissociative alternatives.

An important possibility in these reactions is the formation of an outer-sphere complex between the six-coordinate complex and the incoming ligand:

$$ML_5X + Y \longrightarrow ML_5X,Y \qquad (10.48)$$

in which the incoming ligand, Y, replaces one of the water molecules in the outer sphere of the complex, ML_5X. The second stage of the process is an I_d type of ligand substitution:

$$ML_5X,Y \longrightarrow ML_5Y,X \qquad (10.49)$$

The outgoing ligand, X, then equilibrates with the bulk solution phase thus completing the substitution reaction.

$$ML_5Y,X + H_2O \longrightarrow ML_5Y + X \quad (10.50)$$

If an equilibrium is established between the complex, C, and the incoming ligand, Y, to give an outer-sphere complex, OS:

$$C + Y \longrightarrow OS \quad (10.51)$$

the equilibrium constant, K_{os}, is given by:

$$K_{os} = [OS]/[C][Y] \quad (10.52)$$

If the rate-determining step is the interchange reaction:

$$\text{Rate} = k_i[OS] \quad (10.53)$$

then

$$\text{Rate} = k_i K_{os}[C][Y] \quad (10.54)$$

implying a second-order process.

The total concentration of the complex undergoing substitution is given by:

$$T = [C] + [OS] \quad (10.55)$$

$$= [C] + K_{os}[C][Y] \quad (10.56)$$

and solving for [C] gives:

$$[C] = T/(1 + K_{os}[Y]) \quad (10.57)$$

Substitution in equation (10.54) gives:

$$\text{Rate} = k_i K_{os} T[Y]/(1 + K_{os}[Y]) \quad (10.58)$$

Equation (10.58) reduces to one implying second-order kinetics if K_{os} is such that $K_{os} \ll 1[Y]$ so that:

$$\text{Rate} = k_i K_{os} T[Y] \quad (10.59)$$

(under these conditions $T \approx [C]$).

For reactions carried out under pseudo-first-order conditions where $[Y] \gg T$ the observed rate constants are given by:

$$k_{obs} = k_i K_{os}[Y]/(1 + K_{os}[Y]) \quad (10.60)$$

which may be inverted to give:

$$1/k_{obs} = 1/k_i K_{os}[Y] + 1/k_i \quad (10.61)$$

in which form a plot of $1/k_{obs}$ against $1/[Y]$ gives a straight line with a slope of $1/k_i K_{os}$ and an intercept of $1/k_i$. Determination of the slope and intercept from experimental data gives values for both k_i and K_{os}.

Rates of water molecule replacement

The above formulation for the kinetics of ligand replacement reactions at an octahedral centre are modified in the case of water replacement reactions taking place in aqueous solution because of the very high concentration of the incoming ligand, which is 55.5 M. This has the consequence that $K_{os}[H_2O] \gg 1$ so that equation (10.60) becomes:

$$k_{obs} \approx k_i \quad (10.62)$$

The observed rate constants for water exchange in a series of transition metal hexaaqua-complexes are given in Table 10.1. All the aqua-complexes, are of the

Table 10.1 — Observed rate constants for water exchange

Central metal	d electrons	$\log (k_i/s^{-1})$
Ca(II)	0	8.5
V(II)	3	1.8
Cr(II)	4	9.0
Mn(II)	5	7.5
Fe(II)	6	6.5
Co(II)	7	6.4
Ni(II)	8	4.4
Cu(II)	9	9.0
Zn(II)	10	7.5

high-spin variety (water does not bond particularly strongly with +2 transition elements). There is an enormous variation in the water exchange rates of these complexes, ranging from the relatively very slow V(II) to the very fast Cr(II) and Cu(II) cases.

The reason for this variation cannot be based upon any consideration of ionic size. There is a variation of ionic radius from the largest (Ca^{2+}, $r_i = 100$ pm) to the smallest (Cu^{2+}, $r_i = 57$ pm) which is irrelevant compared to the variation in rate constant from the highest (Cr(II) = Cu(II), $k_i = 10^9 s^{-1}$) to the lowest (V(II), $k_i = 63 s^{-1}$) — a factor of 1.58×10^7. In the Cu(II) case there is no doubt an involvement of the Jahn–Teller effect which tends to ensure that the axial ligands are very weakly bound. In all the cases except for Ca(II) and V(II) there are e_g anti-bonding electrons present which weaken the metal–ligand binding, and in the V(II) case the symmetrical filling of the t_{2g} orbitals (non-bonding) is consistent with the low rate of exchange.

The above data, together with an enormous quantity to be found in data books, give rise to the classification of complexes as being either **inert** or **labile** with respect to ligand substitution. These terms belong to descriptions of **kinetic** behaviour and must not be confused with **thermodynamic** terms such as **stable** or **unstable**, which are used when discussing the stability or otherwise of complexes with respect to their constituent metal and ligands.

10.5 KINETICS AND MECHANISMS OF REDOX PROCESSES

This section is restricted to a discussion of (a) the bromide ion–bromate ion reaction, (b) the reaction between iron(II) ion and hydrogen peroxide (both referred to in the introductory section), and (c) redox processes between transition metal complexes.

(a) The reaction between Br^- and BrO_3^-

The rate law of reaction is expressed by equation (10.20). The rate-determining step involves four ions — bromide ion, bromate ion and two protons — and as such represents a fourth-order reaction. A four-body collision process would be extremely improbable. The influence of the two protons is crucial in that they can (by associating with the bromide and bromate ions) produce two **neutral** reactants, HBr and $HBrO_3$, which may react together in a simple bimolecular manner:

$$H^+ + Br^- \longrightarrow HBr \qquad (10.63)$$

$$H^+ + BrO_3^- \longrightarrow HBrO_3 \qquad (10.64)$$

$$HBr + HBrO_3 \longrightarrow HBrO + HBrO_2 \qquad (10.65)$$

The products of equation (10.65) are somewhat speculative but they are known compounds and the reaction is a simple oxygen atom transfer. Such reactions are very typical of the redox chemistry of main group elements where oxidation states usually change by two units. The mechanistic ideas are completed by the postulated fast reactions (since they do not influence the rate):

$$HBr + HBrO_2 \longrightarrow 2HBrO \qquad (10.66)$$

and

$$3HBr + 3HBrO \longrightarrow 3Br_2 + 3H_2O \qquad (10.67)$$

Although the above postulated mechanism is consistent with the observed rate law it must not be considered to be proved. Further evidence must be sought such as the detection of the intermediates (not to be confused with transition states) HBrO and $HBrO_2$.

(b) The oxidation of iron(II) ion by hydrogen peroxide

The observed rate law for the reaction of iron(II) ion with hydrogen peroxide in acidic solution is:

$$\text{Rate} = k[Fe^{2+}][H_2O_2] \tag{10.68}$$

and the rate-determining step is the bimolecular reaction between the two reactants:

$$Fe^{2+} + H_2O_2 \longrightarrow Fe^{3+} + OH^- + OH \tag{10.69}$$

with the production of the hydroxyl free radical (the hydroxide ion reacts with a proton to give a water molecule in a very rapid step). Equation (10.69) represents a one-electron transfer, which is the most probable (one-photon–two-electron transfers in spectroscopy are highly improbable). The electron is transferred from the iron(II) ion to the electronegative hydrogen peroxide molecule (it would momentarily occupy an anti-bonding orbital of $H_2O_2^-$), causing its destruction. The very reactive intermediate hydroxyl radical reacts rapidly with another iron(II) ion to complete the stoichiometry of the reaction:

$$Fe^{2+} + OH \longrightarrow Fe^{3+} + OH^- \tag{10.70}$$

— another simple bimolecular one-electron transfer process.

The above mechanism is consistent with the rate law, and further confirmation comes from the observation that, when carried out in the presence of acrylonitrile, the polyacrylonitrile product contains terminal OH groups (as shown by infrared absorption spectroscopy).

(c) Redox reactions between transition metal complexes

There are three classes of electron transfer reaction to be discussed:

(i) one-electron exchange processes between metal centres involving different oxidation states of the same metal (exchange reactions),

(ii) one-electron redox processes between complexes of two different metal centres, and

(iii) redox processes involving the transfer of more than one electron.

Exchange reactions

The classical example of this type of process is the scrambling of the ^{55}Fe and ^{56}Fe isotopes between the (II) and (III) oxidation states:

$$^{55}Fe^{2+} + ^{56}Fe^{3+} \longrightarrow ^{55}Fe^{3+} + ^{56}Fe^{2+} \qquad (10.71)$$

which is followed by selectively precipitating the iron(III) from solution at various times after the initial mixing of the two isotopes. The reaction is **thermoneutral** in that $\Delta H^{\ominus}$ is zero. There is a slight entropy increase as the two isotopes become scrambled between the two oxidation states. The activation energy of the reaction is 41.4 kJ mol^{-1}, the bimolecular rate constant at 273 K being 0.87 l mol^{-1} sec^{-1}. The rates of ligand water replacement in the two iron species are 3×10^6 s^{-1}. and 3×10^3 s^{-1} respectively (Fe(II), Fe(III)), which become 5.4×10^4 l mol^{-1} s^{-1} and 54 l mol^{-1} s^{-1} when divided by the molarity of water to convert them into bimolecular rate constants. Both those rate constants are far greater than that observed for the exchange process, and this precludes the so-called **inner sphere** mechanism in which one ion loses a ligand by dissociation and a bridged complex is formed with the other ion which facilitates electron transfer. Reaction (10.71) takes place by the **outer sphere** mechanism in which both reactant complexes retain their inner coordination spheres.

Another factor which would rule out the formation of a bridged intermediate complex is that the water molecule has only one pair of non-bonding electrons of high enough energy to participate in bonding to a metal centre. There are two main factors which cause the reaction to have a reasonably high activation energy. These are (i) the overcoming of the electrostatic repulsion between the two ions, and (ii) the 'Franck–Condon' restrictions which permits electron transfer only when the two ions have identical Fe–O distances. The Franck-Condon Principle is important in electronic spectroscopy and indicates that 'the **vertical** transition between any two Morse curves is the most probable'. It arises from the recognition that electronic transitions occur in times which are far shorter than the time for a molecular vibration to occur. Applied to exchange processes it means that, in the transition state, the metal–ligand distances must be identical for the two participants, otherwise the first law of thermodynamics would be violated! The ground state ions would, if an electron were to be transferred, become vibrationally excited states which could relax to their respective ground states with the production of energy. The smaller Fe(III)–O bonds must therefore be stretched and the larger Fe(II)–O bonds must contract (both by vibrational excitation) before the electron may be transferred.

The mechanism whereby the two ions approach each other so that their primary solvation shells are within appropriate Van der Waals radii is known as the **outer sphere** mechanism. In this mechanism the inner spheres (the primary coordination spheres) of the two reactants remain intact throughout the process.

The most convincing evidence which supports the outer sphere mechanism is for the reaction:

$$Fe(CN)_6^{4-} + Fe(phen)_3^{3+} \longrightarrow Fe(CN)_6^{3-} + Fe(phen)_3^{2+} \qquad (10.72)$$

All four complexes undergo ligand hydrolysis at very low rates. Hydrolysis might be expected to be the first step in a process where one complex lost a ligand so that a bridged complex (by a cyanide ion) could be formed. The actual exchange rate for reaction (10.72) is extremely high.

The rate constants for isotopic scrambling in the couples $Fe(CN)_6^{3-/4-}$ and $Fe(phen)_3^{2+/3+}$ are 3×10^5 and $> 10^5 \, l \, mol^{-1} \, s^{-1}$ respectively, again indicating that the outer sphere mechanism must apply.

One-electron redox processes

When one electron is transferred between oxidation states of different elements there is no Franck–Condon restriction in operation. The mechanism of any one process may be either outer sphere or inner sphere. The classical work of H. Taube (Nobel Prize winner for Chemistry, 1983) established the criteria for the operation of the **inner sphere** mechanism.

The reaction:

$$Cr(H_2O)_6^{2+} + Co(NH_3)_5Cl^{2+} + 5H^+ \longrightarrow Cr(H_2O)_5Cl^{2+} + Co^{2+} + 5NH_4^+ \quad (10.73)$$

has a rate constant of $6 \times 10 \, l \, mol^{-1} s^{-1}$ at 298 K. The Cr(II) reactant has d^4 configuration and is expected to be substitutionally labile, the rate constant for water exchange being $10^9 \, s^{-1}$. The Co(III) complex has a d^6 configuration and is thus substitutionally inert and has a hydrolysis rate constant of $1.7 \times 10^{-7} \, s^{-1}$. The mechanism of the above reaction is thought to involve the replacement of a water ligand of the Cr(II) complex by the chloride ion ligand of the Co(III) complex to form a **bridged-activated complex** in which the electron transfer may take place:

$$(H_2O)_5Cr^{II}ClCo^{III}(NH_3)_5^{4+} \longrightarrow (H_2O)_5Cr^{III}ClCo^{II}(NH_3)_5^{4+} \quad (10.74)$$

Both formulations in equation (10.74) contribute to the transition state and when in the right-hand form there is a dissociation such that the chloride ion is retained by the Cr(III) (the d^3 configuration causing substitutional inertness), the other product is (initially) the substitutionally labile, d^7, Co(II) ion which loses all five of its ammonia ligands to the bulk solution. There is no other way in which the chloride ion could be transferred between the Cr(II) and Co(III) complexes. If the Co(III) complex released the chloride ion into the bulk solution (which is a very slow process) the labile Cr(II) complex would not be likely to react with it. If an electron were to be transferred by an outer sphere process then the chloride would appear in the bulk solution (released by the labile Co(II)) and would be unable to react with the Cr(III), which would appear as the hexaaquo-complex. Confirmatory evidence for the inner sphere mechanism is furnished by the observation that if the reaction is carried out in the presence of free $^{36}Cl^-$ there is no incorporation of the radioactive chloride ion by the Cr(III) product.

It should be noted that the chloride ion does possess a suitable pair of non-bonding electrons which allows it to act as a bridging ligand. The reaction of $[Co(NH_3)_6]^{3+}$ with $[Cr(H_2O)_6]^{2+}$ is a very slow process, with a rate constant of $10^{-3} \, l \, mol^{-1} \, s^{-1}$ — a factor of 6×10^8 times slower than the reaction involving a chloride ligand, the ammonia molecule being unable to participate in bridge formation. The mechanism must be of an outer sphere type.

Multiple-electron redox reactions

The basic rule for so-called non-complementary reactions in which the changes in oxidation state of the oxidizing and reducing agents are different is that the mechanism will consist of the smallest number of one-electron steps. In the reaction:

$$\text{Cr(VI)} + 3\text{Fe(II)} \longrightarrow \text{Cr(III)} + 3\text{Fe(III)} \tag{10.75}$$

the observed rate law is:

$$\text{Rate} = k[\text{Cr(VI)}][\text{Fe(II)}]^2(1 + k'[\text{Fe(II)}]/[\text{Fe(III)}]) \tag{10.76}$$

This may be explained by the three one-electron steps:

$$\text{Cr(VI)} + \text{Fe(II)} \xrightarrow{k_1} \text{Cr(V)} + \text{Fe(III)} \tag{10.77}$$

$$\text{Cr(V)} + \text{Fe(II)} \xrightarrow{k_2} \text{Cr(IV)} + \text{Fe(III)} \tag{10.78}$$

$$\text{Cr(IV)} + \text{Fe(II)} \xrightarrow{k_3} \text{Cr(III)} + \text{Fe(III)} \tag{10.79}$$

with the reverse of reaction (10.77) being important (rate constant given by k_{-1}). The mechanism involves two reactive intermediates — Cr(V) and Cr(IV) — whose respective concentrations would attain steady-state values consistent with the equations:

$$\begin{aligned} \text{d}[\text{Cr(V)}]/\text{d}t = k_1[\text{Cr(VI)}][\text{Fe(II)}] - k_{-1}[\text{Cr(V)}][\text{Fe(III)}] \\ - k_2[\text{Cr(V)}][\text{Fe(II)}] = 0 \end{aligned} \tag{10.80}$$

and

$$\begin{aligned} \text{d}[\text{Cr(IV)}]/\text{d}t = k_2[\text{Cr(V)}][\text{Fe(II)}] - k_3[\text{Cr(IV)}][\text{Fe(II)}] \\ = 0 \end{aligned} \tag{10.81}$$

The equation for the rate of production of Cr(III) — the final chromium product — is:

$$\begin{aligned} \text{d}[\text{Cr(III)}]\text{d}/t = k_3[\text{Cr(IV)}][\text{Fe(II)}] \\ = k_2[\text{Cr(V)}][\text{Fe(II)}] \end{aligned} \tag{10.82}$$

and the unknown [Cr(V)] may be derived from equation (10.80):

$$[\text{Cr(V)}] = k_1[\text{Cr(VI)}][\text{Fe(II)}]/(k_{-1}[\text{Fe(III)}] + k_2[\text{Fe(II)}]) \tag{10.83}$$

the final expression for the overall rate being:

$$\frac{d[Cr(III)]}{dt} = \frac{k_1k_2[Cr(VI)][Fe(II)]^2}{1 + k_2[Fe(II)]/k_{-1}[Fe(III)]} \tag{10.84}$$

which is consistent with the observed rate law with $k = k_1k_2$, and $k' = k_2/k_{-1}$.

10.6 HETEROGENEOUS REACTIONS

The only processes to be considered in this section are reactions taking place on a solid surface. Such reactions may be gas–solid or solution–solid phase combinations.

There are two general classes of adsorption process, depending upon the type of interaction between the adsorbed substance and the surface. If the interaction is due to the weak Van der Waals intermolecular forces the adsorption process is known as physical adsorption or **physisorption. Chemisorption** is reserved for the stronger interactions in which the adsorbed molecule enters into actual chemical bonding with the surface of the solid.

The process of adsorption may be regarded in kinetic terms as the establishment of an equilibrium between the solution (or gas) phase and the surface, with subsequent chemical reaction taking place on the surface of the solid.

The equilibrium of the reactant between the two phases may be written as:

$$M(aq) + S(s) \underset{k_d}{\overset{k_a}{\rightleftharpoons}} MS(s) \tag{10.85}$$

where k_a and k_d represent the rate constants for adsorption and desorption respectively. The rate of adsorption of M on the solid surface, S, will depend upon the concentration (or partial pressure) of M and the fraction of available surface, $1 - \vartheta$, where ϑ represents the fraction of the surface already occupied by reactant molecules:

$$\text{Rate of adsorption} = k_a[M](1 - \vartheta) \tag{10.86}$$

The rate of desorption is proportional to the quantity ϑ:

$$\text{Rate of desorption} = k_d\vartheta \tag{10.87}$$

At equilibrium the adsorption and desorption rates are equal so that:

$$k_a[M](1 - \vartheta) = k_d\vartheta \tag{10.88}$$

and solving for ϑ gives:

$$\vartheta = k_a[M]/(k_d + k_\vartheta[M]) \tag{10.89}$$

and if the ratio, k_a/k_d, is regarded as an equilibrium constant, K, then equation (10.89) may be written as:

$$\vartheta = K[\mathrm{M}]/1 + K[\mathrm{M}]) \tag{10.90}$$

The rate of the reaction (if any) occurring on the solid surface is proportional to ϑ and may be written as:

$$\begin{aligned} \text{Rate} &= k\vartheta \\ &= kK[\mathrm{M}]/(1 + K[\mathrm{M}]) \end{aligned} \tag{10.91}$$

The reciprocal of equation (10.91) is:

$$1/\text{Rate} = 1/k + 1/kK] \tag{10.92}$$

and a plot of 1/Rate versus 1/[M] should give a straight line, with a slope of $1/kK$ and an intercept of $1/k$, from which both k and K may be determined.

If the value of $K[\mathrm{M}] \gg 1$ then equation (10.91) reduces to:

$$\text{Rate} = k \tag{10.93}$$

and zero-order kinetics are observed.

The above scheme is that proposed by Langmuir, and assumes that only a single layer of reacting molecules is adsorbed onto the solid surface. There are instances where this is not the case, and multiple layers can build up so that more sophisticated treatments are necessary.

10.7 ENZYME KINETICS

Enzymes are proteins which have the capacity to catalyse reactions and which exhibit remarkable specificity. The protein molecules are characterized by their inclusion of metal ions which are important in determining their activity. They possess receptor sites which are specific to the type of molecule acceptable (substrate) to a particular enzyme and offer suitable molecules the opportunity to adhere in the reactive site long enough to undergo chemical change. After the reaction has occurred, the product molecule passes into the bulk solution easily since it no longer possesses the peculiar characteristics which the reactant had. There are complications to such processes, and the simple one outlined is the only one treated in this book.

The reactant molecule in enzyme kinetics is referred to as the **substrate**, S, and is considered to form an enzyme–substrate complex, ES, in a pseudo-equilibrium:

$$\mathrm{E} + \mathrm{S} \underset{k_{-1}}{\overset{k_1}{\rightleftharpoons}} \mathrm{ES} \tag{10.94}$$

which is followed by the release of the product(s) from the complex after reaction has taken place, suitably regenerating the enzyme:

$$\mathrm{ES} \xrightarrow{k_2} \mathrm{Products(s)} + \mathrm{E} \qquad (10.95)$$

If we regard the ES complex as a reactive intermediate this allows the application of steady-state theory so that the rate of change in the concentration of ES is equated to zero:

$$\mathrm{d[ES]/d}t = k_1[\mathrm{E}][\mathrm{S}] - k_{-1}[\mathrm{ES}] - k_2[\mathrm{ES}] = 0 \qquad (10.96)$$

giving a value for the steady-state concentration of ES as:

$$[\mathrm{ES}] = k_1[\mathrm{E}][\mathrm{S}]/(k_{-1} + k_2) \qquad (10.97)$$

The substrate concentration is usually easily measurable, but the free enzyme concentration has to be estimated as:

$$[\mathrm{E}] = [\mathrm{E}]_{\mathrm{total}} - [\mathrm{ES}] \qquad (10.98)$$

This allows equation (10.97) to be written as:

$$[\mathrm{ES}] = k_1([\mathrm{E}]_{\mathrm{total}} - [\mathrm{ES}])/(k_{-1} + k_2) \qquad (10.99)$$

and solving for [Es] gives:

$$[\mathrm{ES}] = [\mathrm{E}]_{\mathrm{total}}[\mathrm{S}]/((k_{-1} + k_2)/k_1 + [\mathrm{S}]) \qquad (10.100)$$

The group of rate constants in equation (10.100) may be given the symbol, K_{M}, as this is known as the Michaelis constant.

The overall rate of the reaction may be written as:

$$\begin{aligned}\mathrm{Rate} &= k_2[\mathrm{ES}] \\ &= k_2[\mathrm{E}]_{\mathrm{total}}[\mathrm{S}]/(K_{\mathrm{M}} + [\mathrm{S}])\end{aligned} \qquad (10.101)$$

When [S] is large compared to the value of K_{M}, equation (10.101) reduces to:

$$\mathrm{Rate}_{\mathrm{max}} = k_2[\mathrm{E}]_{\mathrm{total}} \qquad (10.12)$$

which expresses the maximum rate of the reaction. This corresponds to the situation where the receptor sites of the enzymes are filled.

A combination of equations (10.101) and (10.102) yields:

$$\mathrm{Rate} = \mathrm{Rate}_{\mathrm{max}}[\mathrm{S}]/(K_{\mathrm{M}} + [\mathrm{S}]) \qquad (10.103)$$

which is known as the Michaelis–Menten equation.

The value of K_M is equal to the substrate concentration which sustains the reaction at one half of its maximum rate, and is some measure of the sensitivity of the particular enzyme to that particular substrate.

In investigations of enzyme kinetics it is usual to rely upon the measurement of initial rates of reaction since many products of such reactions either inhibit the process or go on to further reaction with the same enzyme (either in the same type of receptor site or in a completely different one). It is not normal for rate constants to be estimated. The results generally quoted are the values of the Michaelis constants. The value of K_M is obtained by manipulating equation (10.103) in a suitable manner.

The usual method of obtaining experimental determinations of the values of K_M and the maximum rate is the Lineweaver–Burk plot.

The method involves the inversion of the Michaelis–Menten equation (10.103) to give:

$$1/\text{Rate} = 1/\text{Rate}_{max} + K_M/\text{Rate}_{max}[S] \qquad (10.104)$$

so that a graph of 1Rate versus 1/[S] gives a straight line, with a slope of K_M/Rate_{max} and an intercept of $1/\text{Rate}_{max}$. from which K_M and the maximum rate may be determined. The value of K_M is the ratio, slope/intercept.

10.8 THE KINETICS OF BACTERIAL GROWTH

Bacteria are important in two major modern areas of chemistry. These are (i) their interactions with metal ions leading to the subject of **bio-accumulation**, in which bacteria are used to extract polluting metal ions from aqueous effluent solutions, and (ii) their action as catalysts in promoting the **bio-oxidation** of sulphide and arsenosulphide minerals. In the latter subject there is particular interest in the bio-oxidation of those minerals which have a significant gold content. Such minerals are **refractory** or intractable, and are normally roasted with the production of the environmentally hostile materials SO_2 and As_2O_3. Bio-oxidation occurs in aqueous slurries of the minerals and produces sulphuric acid, with any arsenic content being disposable as insoluble iron(III) arsenate compounds. It is important in these respects to understand the kinetics of bacterial growth.

The population of a bacterial culture increases with time according to a very simple law. The growth period must be distinguished from the initial **lag phase** in which a newly inoculated medium contains bacterial cells which have to equilibrate with the constituents of that medium — their essential nutrients, which include phosphate and ammonium ions together with appropriate trace elements. The internal chemistry of the bacterial cell is complex, and until the reactions responsible for growth have their intermediates at their optimum steady-state concentrations, very little cell division can take place. Once the steady states are set up, the bacteria can then engage in cell division — their method of multiplying. The rate law for such a process is:

$$dN/dt = kN \qquad (10.105)$$

implying that the rate of increase in the number of bacteria in a system is proportional to that number at any time. Equation (10.105) is easily rearranged to:

$$dN/N = k\,dt \tag{10.106}$$

so that it can be integrated to give:

$$\ln N = kt + \text{constant} \tag{10.107}$$

The integration constant is evaluated by specifying that at time zero the number of bacteria in the system is N_0: so that

$$\ln N_0 = \text{constant} \tag{10.108}$$

and

$$\ln (N/N_0) = kt \tag{10.109}$$

The rate constant, k, may be obtained from a plot of $\ln N$ (or the logarithm of a quantity which is proportional to N such as the concentration of protein in the system) against time. It is normal for the **doubling time** to be quoted as the characteristic of a particular bacterium. If, in the doubling time, τ, the number of bacteria in the system increases from N_0 to $2N_0$, then equation (10.109) becomes:

$$\ln \quad 2 = k\tau \tag{10.110}$$

from which:

$$\tau = \ln 2/K \tag{10.111}$$

Equation (10.109) may be written in the exponential form:

$$N = N_0 e^{kt} \tag{10.112}$$

and the period of such growth is known as the **exponential** or **logarithmic phase**.

The exponential phase terminates when the bacterial medium becomes depleted of one or other of the essential nutrients. The rate of growth levels off to zero, and that state of the system is known as the **stationary phase**. If the medium is replenished then growth will start again after another lag phase.

The above scenario for bacterial growth describes the generally observed characteristics for many strains of bacteria.

10.9 THE PHOTOCHEMISTRY OF TRIOXYGEN (OZONE), O_3

The ozone molecule undergoes photochemical decomposition by absorbing a photon energetic enough to cause bond breakage. The mechanism depends upon the energy of the absorbed photon which determines the excited state of the ozone molecule.

Different dissociation products arise from different excited states, and two cases are described, together with a discussion of the effect of water vapour upon the reactions.

Ozone photolysis with orange light

With light of wavelength in the 500 nm region, the overall quantum yield (number of molecules destroyed/number of quanta absorbed) for the destruction of ozone is observed to be 2.0. This is consistent with the photolytic step being:

$$O_3(^1A_1) \xrightarrow{h\nu} O_2(^3\Sigma_g^-) + O(^3P) \qquad (10.113)$$

producing a dioxygen molecule and an oxygen atom which are both in their electronic ground states. The oxygen atom reacts with an ozone molecule to give two more dioxygen molecules:

$$O(^3P) + O_3 \longrightarrow 2O_2(^3\Sigma_g^-) \qquad (10.114)$$

The quantum yield of the photochemical stage (10.114) (in terms of ozone molecules destroyed per quantum absorbed) is 1.0, the destruction of a further ozone molecule in the thermal second step (10.114) causing the overall quantum yield to be 2.0.

Trioxygen (ozone) photolysis in the ultraviolet Region

The photolytic step in the ultra-violet irradiation of trioxygen is:

$$O_3 \xrightarrow{h\nu} O_2(^1\Delta_g) + O(^1D) \qquad (10.115)$$

with the production of electronically excited states of both the dioxygen molecule and the oxygen atom. The $^1\Delta_g$ state of dioxygen does not possess sufficient energy to participate in any further reaction and is collisionally deactivated to the $^3\Sigma_g^-$ ground state. The ^1D oxygen atom reacts with a trioxygen molecule to give one $^3\Sigma_g^-$ ground state dioxygen molecule and one $^3\Sigma_u^-$ dioxygen molecule which is both electronically and vibrationally (*) excited:

$$O(^1D) + O_3 \longrightarrow O_2^*(^3\Sigma_u^-) + O_2(^3\Sigma_g^-) \qquad (10.116)$$

The latter (providing that it is in at least the seventeenth level of vibrational excitation) can cause the destruction of a trioxygen molecule by the reaction:

$$O_2^*(^3\Sigma_u^-) + O_3 \longrightarrow 2O_2(^3\Sigma_g^-) + O(^1D) \qquad (10.117)$$

so allowing a chain reaction to occur, with ^1D oxygen atoms being the chain carriers. There are chain-breaking steps which involve the dimerization of the oxygen atoms in the system, but the chain causes the observed quantum yield to be about 6.0.

Another mechanistic feature presents itself in the case of trioxygen photolysis in the presence of water molecules when the quantum yield can be as high as 130. The reactions which cause this increase in trioxygen destruction involve the reaction of the. ^1D oxygen atoms with water to give two hydroxyl free radicals:

$$O(^1D) + H_2O \longrightarrow 2OH \tag{10.118}$$

The hydroxyl radicals can then react with trioxygen to give ground state dioxygen and the hydroperoxy free radical:

$$OH + O_3 \longrightarrow HO_2 + O_2 \tag{10.119}$$

which can destroy more trioxygen by the reaction:

$$HO_2 + O_3 \longrightarrow 2O_2 + OH \tag{10.120}$$

the OH proceeding to participate in reaction (10.119). The chain process is facilitated by the two chain carriers, OH and HO_2, and the reactions are relatively faster than the chain-terminating steps consistent with the larger observed quantum yields for trioxygen destruction.

10.10 THE RADIOLYSIS OF LIQUID WATER

The discussion of this topic is limited to the γ-irradiation of liquid water and dilute aqueous solutions of appropriate solutes, together with some reactions which are initiated by electron pulses.

Liquid water radiolysis

Gamma (γ)-ray quanta typically have energies between 0.5 and 2.0 MeV (50–200 GJ mol^{-1}). When a γ-quantum interacts with a molecule of water it causes ionization:

$$H_2O \xrightarrow{\gamma} H_2O^+ + e^- \tag{10.121}$$

producing a **photoelectron** which possesses the quantum energy minus the ionization energy of the water molecule. The photoelectron can cause a second ionization:

$$e^- + H_2O \longrightarrow H_2O^+ + 2e^- \tag{10.122}$$

and a large number of such processes take place until the electrons (which share the remaining energy) become **thermalized**—they then have average energies of kT and can no longer cause ionization. In such a condition the thermalized electrons become hydrated, $e^-(aq)$.

The result of the interaction of one γ-quantum with liquid water may be the production of a cascade of many H_2O^+/e^- pairs. Both species are very unstable in liquid water and undergo either recombination or reaction with surrounding water molecules:

$$H_2O^+ + H_2O \longrightarrow H_3O^+ + OH \qquad (10.123)$$

and

$$e^-(aq) + H_2O \longrightarrow H + OH^- \qquad (10.124)$$

In highly pure liquid water the resulting pairs of hydrogen atoms and hydroxyl radicals react with each other to give water molecules, and no overall chemical change is apparent. There are some H + H and OH + OH dimerizations to give molecular dihydrogen and hydrogen peroxide in small concentrations.

Radiolysis of dilute aqueous solutions

To demonstrate the presence of free radicals in a system, a molecule known as a **radical scavenger** is employed. For example, added acrylonitrile reacts with the hydroxyl radicals in producing a polyacrylonitrile polymer with OH end groups.

The hydrated electron may be demonstrated to be a distinct chemical entity by the use of chloroethanoic acid, $ClCH_2COOH$, as a scavenger. The scavenger also serves to distinguish hydrogen atoms from hydrated electrons. Free hydrogen atoms tend to react with available covalently bound hydrogen atoms to produce dihydrogen:

$$H + ClCH_2COOH \longrightarrow H_2 + ClCHCOOH \qquad (10.125)$$

the organic free radical undergoing dimerization to give the observed product, dichlorosuccinic acid, $(ClCHCOOH)_2$.

The hydrated electron is attracted towards the electronegative chlorine atom of the scavenger and causes the production of a chloride ion:

$$e^- + ClCH_2COOH \longrightarrow Cl^- + CH_2COOH \qquad (10.126)$$

with the organic fragment dimerizing to give the unchlorinated succinic acid, $(CH_2COOH)_2$. Which of the two reactions, (10.125) or (10.126), occurs depends upon the pH of the solution. In acid solutions (pH greater than 4.0) the reaction:

$$e^-(aq) + H^+ \longrightarrow H \qquad (10.127)$$

predominates and so reaction (10.125) follows to give H_2 and dichlorosuccinic acid as the products of scavenging. In neutral and alkaline solutions, reaction (10.126) predominates, and Cl^- and succinic acid are produced.

Pulse radiolysis

The ultimate confirmation of the existence of the hydrated electron came from a study of the effects of very short pulses of accelerated electrons upon liquid water, the immediate products of each pulse being observed by their effect upon a pulse of ultraviolet/visible radiation from a xenon discharge lamp which traversed the reaction cell at right angles to the direction of the electron pulse. The absorption spectrum of the hydrated electron was thus recorded and is shown in Fig. 10.2. The

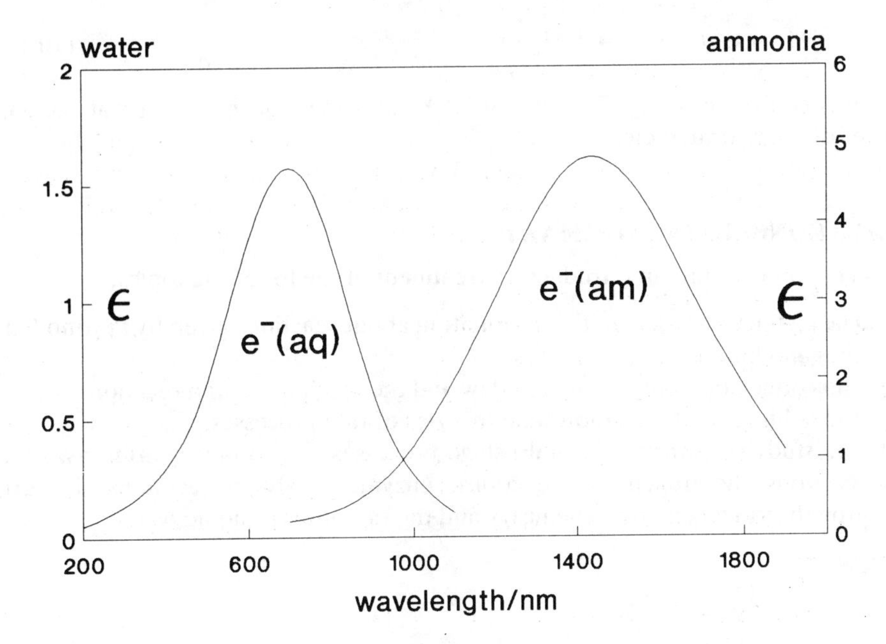

Fig. 10.2 — The absorption spectra of the electron solvated by water and ammonia.

observation of an absorption spectrum does not prove that the absorbing substance is a hydrated electron. Further work on the effects of solutes on the rates of disappearance of the electron absorption showed that it had the expected charge of -1. The spectrum of the hydrated electron is very similar to that of solutions of Group 1 metals in liquid ammonia — generally accepted to contain solvated electrons — the blue colour being independent of the Group 1 metal:

$$M(s) + nNH_3 \longrightarrow M^+(am) + e^-(am) \tag{10.128}$$

The spectrum of the electron, solvated by ammonia, is shown in Fig. 10.2.

Reducing power of the hydrated electron

The rate constant for reaction (10.124) has a value of $16\,s^{-1}$ at 298 K, the rate constant for the reverse reaction being $2 \times 10^7\,l\,mol^{-1}\,s^{-1}$. The equilibrium constant for the reaction is the ratio of the two rate constants and is thus 8×10^{-7} for the reaction written as (10.126). This corresponds to a $\Delta G^{\ominus}$ value of 34.8 kJ mol^{-1}. The equilibrium may be combined with that governing the self-ionization of water:

$$H^+ + OH^- \longrightarrow H_2O \tag{10.129}$$

which has the value 1×10^{14} at 298 K, and the change in standard Gibbs energy for the formation of dihydrogen from hydrogen atoms of $-203.3\,\text{kJ mol}^{-1}$ (of H atoms) to give a value of $\Delta G^{\ominus}$ of $-248.4\,\text{kJ mol}^{-1}$ for the reaction:

$$H^{+} + e^{-}(aq) \longrightarrow \tfrac{1}{2}\,H_2(g) \tag{10.130}$$

which is equivalent to an $E^{\ominus}$ value of 2.6 V, indicative of the very great reducing power of the hydrated electron.

10.11 CONCLUDING SUMMARY

This chapter contains an introductory treatment of the following topics.

(1) The distinction between the information about reactions given by thermodynamics and kinetics.
(2) The connexion between the rate law and the mechanism of a reaction.
(3) The collision and transition state theories of rate processes.
(4) The study of a variety of multi-stage processes — ligand substitutions, redox reactions, heterogeneous reactions, enzyme catalysed reactions, bacterial growth, trioxygen photochemistry and the radiolysis of liquid water.

Further reading guide

INTRODUCTION

This section contains lists of books from which the reader is encouraged to select to provide the next stages in the study of inorganic chemistry.

BOOK LIST

1. Data books

Although the large textbooks contain data and some of them have selective lists as appendices, it is advisable to have a data book. These are more comprehensive than any textbook and have the data in a concise form which is easy to use.

(A) *Chemistry Data Book* by J. G. Stark and H. G. Wallace, John Murray, 1978. This contains sixty-seven tables of data and is sufficient for most chemical requirements.
(B) *Nuffield Data Book* — a very useful set of data tables.
(C) *The Elements* by J. E. Emsley, Clarendon Press, Oxford, 1989. This is a novel data book, with two pages devoted to each of the elements. In addition there are thirty-five tables containing specific data in alphabetical order of the elements and in ranking order of the particular quantity.
(D) *Oxidation Potentials* by W. L. Latimer, Prentice-Hall, 1952. This magnificent book is an education in inorganic chemistry all by itself. It is a comprehensive collection of standard potentials, with an indication of their derivation and with discussion of the aqueous chemistry of the elements in their various oxidation states. The old convention is used (hence the title) and the present-day reader has to bear in mind that the quoted potentials are for the standard oxidation process and have opposite signs to the reduction potentials that are now adopted. The book is out of print but should be available in any library.
(E) *Tables for Group Theory* by P. W. Atkins, M. S. Child and C. S. G. Phillips, Oxford, 1975. After becoming conversant with group theory, all the reader will need

is a book of character tables for its application to any problem.
(F) *The Chemical Elements, Chemistry, Physical Properties and Uses in Science and Industry* by Liam P. Roche, Ellis Horwood, Chichester, 1990. This is a very up-to-date data book that includes sections on ores and their refinement, chemical engineering, and the current uses of the elements, in addition to the more conventional contents.
(G) *Atlas of Metal-Ligand Equilibria in Aqueous Solution* by J. Kragten, Ellis Horwood, Chichester, 1977. An invaluable source book dealing with the interactions between forty-five elements and twenty-nine ligands.

2. General textbooks

These are books that are suitable for wide, but not comprehensive, coverage of the subject of inorganic chemistry. They may cover the first two years of a course and may even be adequate for third-year courses. Almost certainly they would have to be supplemented by either specialist books or one or other of the major texts.

(A) *Modern Inorganic Chemistry* by W. L. Jolly, McGraw-Hill, 1984. This is an excellent general account of the subject and would be suitable for most first- and second-year university courses.
(B) *Inorganic Chemistry, a Modern Introduction* by T. Moeller, John Wiley, 1982. A lavish introduction to the subject, with enough material to cover some three-year courses.
(C) *Inorganic Chemistry, Principles of Structure and Reactivity* by J. E. Huheey, Harper and Row, 1983. More comprehensive treatment than (B) and nearly sufficient for most three-year courses.
(D) *Inorganic Chemistry* by D. F. Shriver, P. W. Atkins and C. H. Langford, Oxford, 1990. The very latest book in this category and very well worth reading. It contains an excellent blend of theoretical and descriptive chemistry. It would. need supplementation to cover a three-year course.
(E) *Main Group Chemistry* by Alan G. Massey, Ellis Horwood, Chichester, 1990. This is a very up-to-date treatment of the main group elements and is suitable for the main requirements of most degree courses.
(F) *Inorganic Reaction Chemistry, Volume I: Systematic Chemical Separation* by D.T. Burns, A. Townsend and A. G. Catchpole, *Volume II: Reactions of the Elements and Their Compounds* by D.T. Burns, A. Townsend and A. H. Carter, Ellis Horwood, Chichester, 1980/1981. This is a major work, detailing the descriptive chemistry of the elements and their compounds, and which contains a massive amount of analytical information.

3. Specialist books

These are sometimes helpful for particular courses, and only a small number are included in this list. The reader should await advice from the lecturer or tutor before deciding which to consult.

(A) *The Periodic Table of the Elements* by R. J. Puddephatt and P. K. Monaghan, Clarendon Press, Oxford, 1986. A very useful discussion of the modern periodic

table and a broader treatment than the one presented in Chapter 1 of this book.
(B) *Atomic Spectra and Atomic Structure* by G. Herzberg, Dover, 1944. A very clear and readable exposition of the subject matter. In spite of its age, the book is well worth reading, is full of information and is still available. The principles of the subject have not changed since 1944.
(C) *Chemical Applications of Group Theory* by F. A. Cotton, John Wiley, 1990. Although there are many books that deal with the subjects of symmetry and group theory, none of them is up to the standards of rigour and clarity set by Cotton's book.
(D) *The Chemical Bond* by J. N. Murrell, S. F. A. Kettle and J. M. Tedder, John Wiley, 1983. A very comprehensive treatment of the subject for most undergraduate courses.
(E) Ab Initio *Molecular Orbital Calculations for Chemists* by W. G. Richards and D. L. Cooper, Clarendon Press, Oxford, 1983. An introduction to quantitative and semi-quantitative molecular orbital calculations. A very clear exposition of the mathematical basis of the theory and of the approximations that may be made. This is an essential introduction for the student who wishes to carry out m.o. calculations.
(F) *Molecular Photoelectron Spectroscopy* by D. W. Turner, C. Baker, A. D. Baker and C. R. Brundle, John Wiley, 1969. A book full of useful spectra and their interpretation. There are a number of more modern books that deal with more specialized sections of the subject, and are available in libraries.
(G) *Introduction to Molecular Spectroscopy* by G. M. Barrow, McGraw-Hill, 1962. A very clear account of the principles of rotational, vibrational and electronic spectroscopy.
(H) *Molecular Spectra and Molecular Structure* by G. Herzberg, *I. Spectra of Diatomic Molecules*, *II. Infra-red and Raman Spectra of Polyatomic Molecules* and *III. Electronic Spectra and Electronic Structure of Polyatomic Molecules*, Van Nostrand Reinhold, 1950, 1945 and 1966. These three large volumes are a mine of information on their subjects. The principles are dealt with rigorously and the books contain a vast amount of data. All are out of print but are available in libraries.
(I) *Molecular Shapes — Theoretical Models of Inorganic Stereochemistry* by J. K. Burdett, John Wiley, 1980. A splendid discussion of all the theories that lead to the understanding of the factors influencing molecular shapes.
(J) *Structural Inorganic Chemistry* by A. F. Wells, Oxford, 1984. A book without equal, dealing with all aspects of the solid state. In addition to a comprehensive coverage of the principles of bonding in the solid state, the book contains a great deal of factual information about structures.
(K) *Ions in Solution* by J. Burgess, Ellis Horwood, Chichester, 1988. A very clear and wide-ranging discussion of all possible aspects of the understanding of the existence and behaviour of ions in solution.
(L) *Metal Ions in Solution* by J. Burgess, Ellis Horwood, Chichester, 1978. A valuable book for specialists in this area.
(M) *Kinetics and Mechanism* by A. A. Frost and R. G. Pearson, John Wiley, 1961. One of the best introductions to the subject.
(N) *Inorganic Electronic Spectroscopy* by A. B. P. Lever, Elsevier, 1984. A very comprehensive and satisfying treatment of all aspects of the subject.
(O) *NMR, NQR, EPR and Mössbauer Spectroscopy in Inorganic Chemistry* by

R.V. Parrish, Ellis Horwood, Chichester, 1990. A guide to these important topics and their use in inorganic chemistry.
(P) *Infra-red and Raman Spectra of Inorganic and Co-ordination Compounds* by K. Nakamoto, John Wiley, 1986. The only book of its type. A good treatment of the theoretical material and full of spectra and their interpretations.
(Q) *Bio-inorganic Chemistry* by R.W. Hay, Ellis Horwood, Chichester, 1984. A very good introduction to the very important subject matter.
(R) *Metals in Biological Systems* by M.J. Kendrick, M.T. May, M.J. Plishka and K.D. Robinson, Ellis Horwood, Chichester, 1990. A comprehensive overview of the subject.
(S) *Understanding Enzymes* by T. Palmer, Ellis Horwood, Chichester, 1985. The second edition of a successful work, dealing with all aspects of enzymes, their structures and reactions.
(T) *The Strange Story of the Quantum* by B. Hoffmann, Dover, 1959. An account of the history of the development of quantum ideas written for the general reader and well worth reading for the understanding it contains.
(U) *Popper* by Bryan Magee, Fontana Modern Masters, Ed. Frank Kermode, 1975. A superb account of Popper's philosophy and treatment of the scientific method; the book should be read by all aspiring scientists (and would do a power of good for some established ones!).

4. Major textbooks

(A) *Advanced Inorganic Chemistry* by F.A. Cotton and G. Wilkinson, John Wiley, 1988. This leading textbook has a comprehensive coverage of the chemistry of the elements and specialist sections of topical interest. The text is fully referenced, and this is the book of first preference to begin any project.
(B) *Chemistry of the Elements* by N.N. Greenwood and A. Earnshaw, Pergamon Press, 1984. A comprehensive coverage of the subject matter, with pleasing inclusions of interesting historical details, relevance to industrial processes and environmental impact.
(C) *Inorganic Chemistry* by K.F. Purcell and J.C. Kotz, Saunders, 1977. A major discursive work, leading to a very satisfying understanding of the subject. It should be regarded as complementary to either (A) or (B), and read in conjunction with your choice.
(D) *Inorganic Chemistry* by C.S.G. Phillips and R.J.P. Williams, Oxford, 1965. A two-volume exposition of theoretical and descriptive inorganic chemistry that is now (sadly) out of print, but readers are strongly advised to try to aquire a copy or to consult the book in the library.
(E) *Physical Chemistry* by P.W. Atkins, Oxford, 1989. An outstanding exposition of the principles of physical chemistry, with virtually no competition. The subject matter is dealt with rigorously and with clarity.
(E) *The Structure of Physical Chemistry* by C.N. Hinshelwood, Oxford, 1951. A magnificent discussion of the principles of physical chemistry and a book to read when a broad appreciation and a thorough understanding of the subject are required. Read Atkins first though! The book is out of print but should be in the library.

GENERAL APPROACH TO INORGANIC CHEMISTRY

There are two general strategies that may be adopted over a three-year course, with a view to gaining a comprehensive understanding of inorganic chemistry. They both require a good knowledge of the appropriate sections of physical chemistry.

I. (i) Begin with an introductory text such as this book.
   (ii) Move on to an intermediate text such as 2(A) (Jolly).
   (iii) Fill in any gaps by using appropriate specialist books plus major texts 4(A)+4(C) or 4(B)+4(C).

II. (i) As I.(i)
   (ii) Use appropriate specialist books together with either of the major text combinations 4(A)+4(C) or 4(B)+4(C).

Approach I is more gradual than II but is more expensive! With either approach, select the specialist texts only when the introductory text and/or the major text combinations fail to cover the particular course or your project requirements.

Appendix I

CHARACTER TABLES USED IN THE TEXT

The character tables used in the text are collected together in this section. The s orbitals of any central atom always transform as the completely symmetric representation of a point group, this never being indicated in a character table. The symmetry properties of x, y, z, xy, xz, yz, x^2-y^2 and z^2 are indicated in each table, together with those of the rotations about the three Cartesian axes, R_x, R_y and R_z.

C_{2v}	E	C_2	$\sigma_v(xz)$	$\sigma_v'(yz)$		
A_1	1	1	1	1	z	x^2, y^2, z^2
A_2	1	1	−1	−1	R_z	xy
B_1	1	−1	1	−1	x, R_y	xz
B_2	1	−1	−1	1	y, R_x	yz

C_{3v}	E	$2C_3$	$3\sigma_v$		
A_1	1	1	1	z	x^2+y^2, z^2
A_2	1	1	−1	R_z	
E	2	−1	0	$(x, y)(R_x, R_y)$	$(x^2-y^2, xy)(xz, yz)$

D_{2h}	E	$C_2(z)$	$C_2(y)$	$C_2(x)$	i	$\sigma(xy)$	$\sigma(xz)$	$\sigma(yz)$		
A_g	1	1	1	1	1	1	1	1		x^2, y^2, z^2
B_{1g}	1	1	−1	−1	1	1	−1	−1	R_z	xy
B_{2g}	1	−1	1	−1	1	−1	1	−1	R_y	xz
B_{3g}	1	−1	−1	1	1	−1	−1	1	R_x	yz
A_u	1	1	1	1	−1	−1	−1	−1		
B_{1u}	1	1	−1	−1	−1	−1	1	1	z	
B_{2u}	1	−1	1	−1	−1	1	−1	1	y	
B_{3u}	1	−1	−1	1	−1	1	1	−1	x	

D_{3h}	E	$2C_3$	$3C_2$	σ_h	$2S_3$	$3\sigma_v$		
A_1'	1	1	1	1	1	1		x^2+y^2, z^2
A_2'	1	1	−1	1	1	−1	R_z	
E'	2	−1	0	2	−1	0	(x, y)	(x^2-y^2, xy)
A_1''	1	1	1	−1	−1	−1		
A_2''	1	1	−1	−1	−1	1	z	
E''	2	−1	0	−2	1	0	(R_x, R_y)	(xz, yz)

D_{4h}	E	$2C_4$	C_2	$2C_2'$	$2C_2''$	i	$2S_4$	σ_h	$2\sigma_v$	$2\sigma_d$		
A_{1g}	1	1	1	1	1	1	1	1	1	1		x^2+y^2, z^2
A_{2g}	1	1	1	−1	−1	1	1	1	−1	−1	R_{zx}	
B_{1g}	1	−1	1	1	−1	1	−1	1	1	−1		x^2-y^2
B_{2g}	1	−1	1	−1	1	1	−1	1	−1	1		xy
E_g	2	0	−2	0	0	2	0	−2	0	0	(R_x, R_y)	(xz, yz)
A_{1u}	1	1	1	1	1	−1	−1	−1	−1	−1		
A_{2u}	1	1	1	−1	−1	−1	−1	−1	1	1	z	
B_{1u}	1	−1	1	1	−1	−1	1	−1	−1	1		
B_{2u}	1	−1	1	−1	1	−1	1	−1	1	−1		
E_u	2	0	−2	0	0	−2	0	2	0	0	(x, y)	

D_{6h}	E	$2C_6$	$2C_3$	C_2	$3C_2'$	$3C_2''$	i	$2S_3$	$2S_6$	σ_h	$3\sigma_d$	$3\sigma_v$		
A_{1g}	1	1	1	1	1	1	1	1	1	1	1	1		x^2+y^2, z^2
A_{2g}	1	1	1	1	−1	−1	1	1	1	1	−1	−1	R_z	
B_{1g}	1	−1	1	−1	1	−1	1	−1	1	−1	1	−1		
B_{2g}	1	−1	1	−1	−1	1	1	−1	1	−1	−1	1		
E_{1g}	2	1	−1	−2	0	0	2	1	−1	−2	0	0	(R_x, R_y)	(xz, yz)
E_{2g}	2	−1	−1	2	0	0	2	−1	−1	2	0	0		(x^2-y^2, xy)
A_{1u}	1	1	1	1	1	1	−1	−1	−1	−1	−1	−1		
A_{2u}	1	1	1	1	−1	−1	−1	−1	−1	−1	1	1	z	
B_{1u}	1	−1	1	−1	1	−1	−1	1	−1	1	−1	1		
B_{2u}	1	−1	1	−1	−1	1	−1	1	−1	1	1	−1		
E_{1u}	2	1	−1	−2	0	0	−2	−1	1	2	0	0	(x, y)	
E_{2u}	2	−1	−1	2	0	0	−2	1	1	−2	0	0		

T_d	E	$8C_3$	$3C_2$	$6S_4$	$6\sigma_d$		
A_1	1	1	1	1	1		$x^2+y^2+z^2$
A_2	1	1	1	−1	−1		
E	2	−1	2	0	0		$(2z^2-x^2-y^2, x^2-y^2)$
T_1	3	0	−1	1	−1	(R_x, R_y, R_z)	
T_2	3	0	−1	−1	1	(x, y, z)	(xy, xz, yz)

O_h	E	$8C_3$	$6C_2$	$6C_4$	$3C_2$ $(=C_4^2)$	i	$6S_4$	$8S_6$	$3\sigma_h$	$6\sigma_d$		
A_{1g}	1	1	1	1	1	1	1	1	1	1		$x^2+y^2+z^2$
A_{2g}	1	1	−1	−1	1	1	−1	1	1	−1		
T_g	2	−1	0	0	2	2	0	−1	2	0		$(2z^2-x^2-y^2,$ $x^2-y^2)$
T_{1g}	3	0	−1	1	−1	3	1	0	−1	−1	(R_x, R_y, R_z)	
T_{2g}	3	0	1	−1	−1	3	−1	0	−1	1		(xz, yz, xy)
A_{1u}	1	1	1	1	1	−1	−1	−1	−1	−1		
A_{2u}	1	1	−1	−1	1	−1	1	−1	−1	1		
E_u	2	−1	0	0	2	−2	0	1	−2	0		
T_{1u}	3	0	−1	1	−1	−3	−1	0	1	1	(x, y, z)	
T_{2u}	3	0	1	−1	−1	−3	1	0	1	−1		

$C_{\infty v}$	E	$2C_\infty^\phi$	…	$\infty\sigma_v$		
$A_1\equiv\Sigma^+$	1	1	…	1	z	x^2+y^2, z^2
$A_2\equiv\Sigma^-$	1	1	…	−1	R_z	
$E_1\equiv\Pi$	2	$2\cos\phi$	…	0	$(x,y)(R_x,R_y)$	(xz,yz)
$E_2\equiv\Delta$	2	$2\cos 2\phi$	…	0		(x^2-y^2,xy)
$E_3\equiv\Phi$	2	$2\cos 3\phi$	…	0		
…	…	…	…	…		

$D_{\infty h}$	E	$2C_\infty^\phi$	…	$\infty\sigma_i$	i	$2S_\infty^\phi$	…	∞C_2		
Σ_g^+	1	1	…	1	1	1	…	1		x^2+y^2, z^2
Σ_g^-	1	1	…	−1	1	1	…	−1	R_z	
Π_g	2	$2\cos\phi$	…	0	2	$-2\cos\phi$	…	0	(R_x,R_y)	(xz,yz)
Δ_g	2	$2\cos 2\phi$	…	0	2	$2\cos 2\phi$	…	0		(x^2-y^2,xy)
…	…	…	…	…	…	…	…	…		
Σ_u^+	1	1	…	1	−1	−1	…	−1	z	
Σ_u^-	1	1	…	−1	−1	−1	…	1		
Π_u	2	$2\cos\phi$	…	0	−2	$2\cos\phi$	…	0	(x,y)	
Δ_u	2	$2\cos 2\phi$	…	0	−2	$-2\cos 2\phi$	…	0		
…	…	…	…	……	…	…	…			

Appendix II

The Tanabe–Sugano diagrams for 3, 5, 6, 7 and 8 d electrons are shown below. The energy and Δ values are dimensionless, as explained in Chapter 8. (After Tanabe and Sugano, *J. Phys. Soc. Japan*, **9**, 753 (1954).)

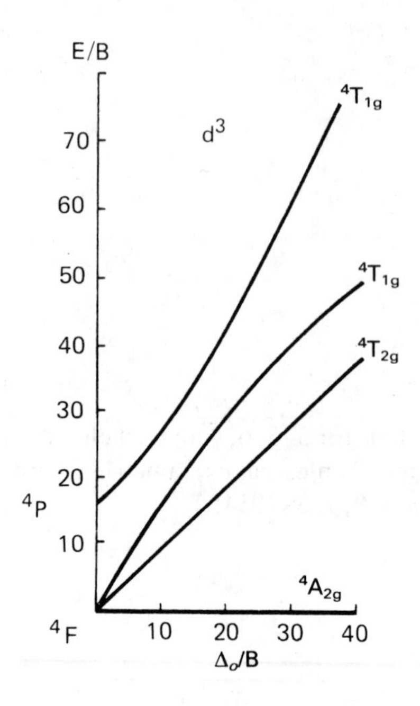
E/B
d3
70
60
50
40
30
20
10
4P
4F
10 20 30 40
Δo/B
4T1g
4T1g
4T2g
4A2g

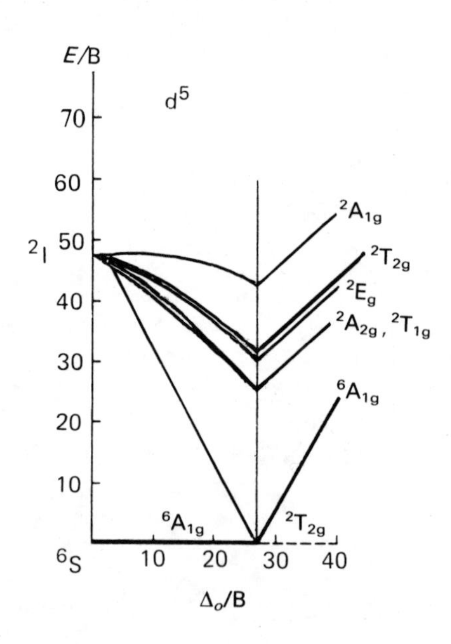
E/B
d5
70
60
50
40
30
20
10
2I
6S
10 20 30 40
Δo/B
2A1g
2T2g
2Eg
2A2g, 2T1g
6A1g
6A1g
2T2g

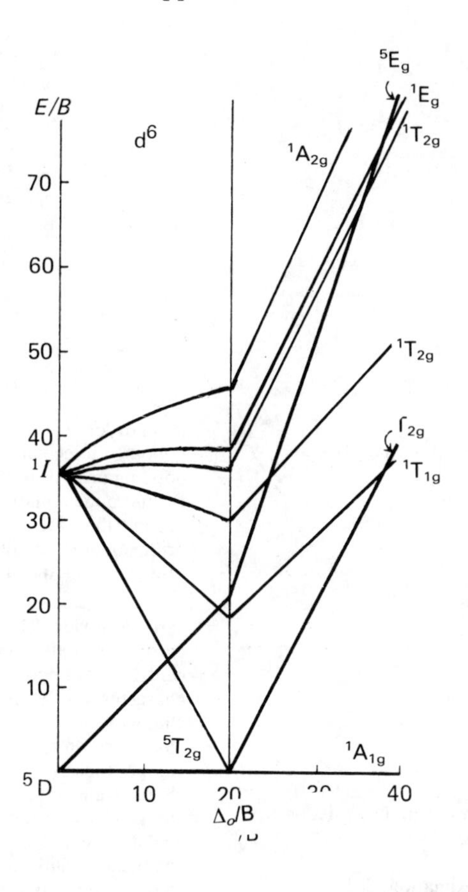
E/B
d6
70
60
50
40
30
20
10
1I
5D
10
20
40
Δo/B
5Eg
1Eg
1T2g
1A2g
1T2g
1T1g
5T2g
1A1g

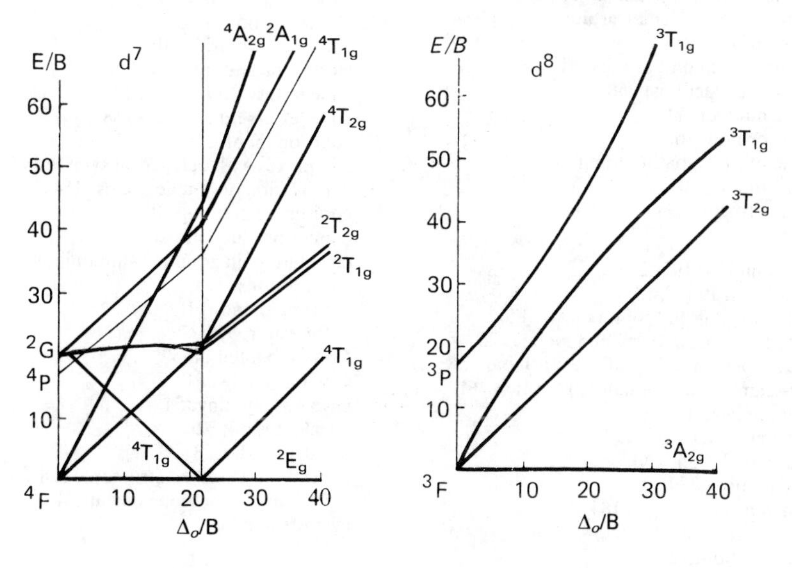
E/B
d7
60
50
40
30
10
2G
4P
4F
10
20
30
40
Δo/B
4A2g
2A1g
4T1g
4T2g
2T2g
2T1g
4T1g
4T1g
2Eg
E/B
d8
60
50
40
30
20
10
3P
3F
10
20
30
40
Δo/B
3T1g
3T1g
3T2g
3A2g

Index